Thermal Energy Management in Vehicles

Automotive Series

Series Editor: Thomas Kurfess

Thermal Energy Management in Vehicles

Vincent Lemort
University of Liège

Gérard Olivier
Renault

Georges de Pelsemaeker
Valeo Thermal Systems & University of Liège

The right of Vincent Lemort, Gérard Olivier, and Georges de Pelsemaeker to be identified as the authors of this work has been asserted in accordance with law.

Registered Offices
John Wiley & Sons, Inc., 111 River Street, Hoboken, NJ 07030, USA
John Wiley & Sons Ltd, The Atrium, Southern Gate, Chichester, West Sussex, PO19 8SQ, UK

Editorial Office
The Atrium, Southern Gate, Chichester, West Sussex, PO19 8SQ, UK

For details of our global editorial offices, customer services, and more information about Wiley products visit us at www.wiley.com.

Wiley also publishes its books in a variety of electronic formats and by print-on-demand. Some content that appears in standard print versions of this book may not be available in other formats.

Library of Congress Cataloging-in-Publication Data Applied for:
Hardback ISBN: 9781119251750

Cover Design: Wiley
Cover Images: © metamorworks/Shutterstock; Gorodenkoff/Shutterstock; Courtesy of Valeo

Set in 9.5/12.5pt STIXTwoText by Straive, Chennai, India
Printed and bound by CPI Group (UK) Ltd, Croydon, CR0 4YY

C9781119251750_050123

Contents

Acknowledgments

The authors would like to thank all the people who have contributed to the content of this book by sharing their knowledge. This content has relied heavily on the technical documentation prepared over the years by many Valeo and Renault experts. It is not possible to list all of them, but the authors hope that they will recognize each other.

In addition, many colleagues from the University of Liège and other universities as well as industrial partners took the time to answer the many technical questions they received. The authors would like to thank them for their time and consideration. Again, it is not easy to provide a complete list without forgetting anyone, and the authors hope that no one will take offense.

Finally, the authors would like to thank their families and their beloved for their support during this long project.

Above all, Vincent Lemort thanks his wife, children, family, and friends for their patience during these last two months of writing.

Nomenclature

List of Abbreviations

AC	accumulator
A/C	air-conditioning
ACAC	air-cooled charge air cooler
BDC	bottom dead center
BEV	battery electric vehicle
BMEP	brake mean effective pressure
BMS	battery management system
BPHEX	Brazed Plate Heat Exchanger
BTM	battery thermal management
BTMS	battery thermal management system
BOL	beginning of life
CAC	charge air cooler
CC	cooler core
CFC	chlorofluorocarbon
COP	coefficient of performance
CP	compressor
DN	direct normal
DOC	diesel oxidation catalyst
DP	damper
DPF	diesel particulate filter
ECV	externally controlled valve
EG	ethylene glycol
EGR	exhaust gas recirculation
EGRC	exhaust gas recirculation cooler
EHRS	exhaust heat recovery system
EM	electric motor
EOL	end of life
EREV	extended range electric vehicle
EV	electric vehicle
EXV	electronic expansion valve
HC	hydrocarbon
HEV	hybrid electric vehicle
HP	high pressure
FC	fuel cell

FCEV	fuel cell electric vehicle
FMEP	friction mean effective pressure
GWP	global warming potential
HC	heater core
HVAC	heating, ventilation, and air-conditioning
HFC	hydrofluorocarbon
HFO	hydrofluoroolefin
HHV	high heating value
ICD	internal condenser
ICE	internal combustion engine
ICT	information and communications technology
ICV	internally controlled valve
IEV	internal evaporator
IMEP	indicated mean effective pressure
IR	infrared
LHV	low heating value
LP	low pressure
LT	low temperature
MEP	mean effective pressure
NEDC	new European driving cycle
NTU	number of transfer units
ORC	organic Rankine cycle
OCR	oil circulation ratio
OCV	open circuit voltage
OHEX	outdoor heat exchanger
OT	orifice tube
PCM	phase change material
PE	power electronics
PHEV	plug-in hybrid electric vehicle
PMV	predicted mean vote
PPD	predicted percent dissatisfied
PTC	positive temperature coefficient
PVB	polyvinyl butyral
RC	Rankine cycle
RMS	root mean square
SCR	selective catalytic reduction
SHGC	solar heat gain coefficient
SHR	sensible heat ratio
SOC	state of charge
SOH	state of health
TDC	top dead center
TIM	thermal interface material
TXV	thermostatic expansion valve
WCAC	water-cooled charge air cooler
WCD	water-cooled condenser
WLTP	worldwide harmonized light vehicles test procedure
ZEV	zero emission vehicle

Nomenclature

a	specific Gibbs free energy [J kg^{-1}]
A	area [m^2]
AU	conductance [W K^{-1}]
B	bore [m]
c	specific heat [J kg^{-1} K^{-1}]
C	speed, velocity [m s^{-1}]
C	heat capacity [J K^{-1}]
C	clearance factor [−]
C	concentration [−]
e	specific total energy [J kg^{-1}]
e	thickness [m]
e	amount of excess air [−]
E	total energy [J]
E	emissive power [W m^{-2}]
f	fuel–air ratio [−]
F	force [N]
F	view factor [−]
g	gravitational acceleration [m s^{-2}]
g	specific Helmoltz free energy [J kg^{-1}]
G	irradiation [W m^{-2}]
h	specific enthalpy [J kg^{-1}]
h	convective heat transfer coefficient [W m^{-2}K^{-1}]
H	enthalpy [J]
H	height [m]
i	working cycle frequency [−]
I	irradiance [W m^{-2}]
I	electric current [A]
k	spring constant [N m^{-1}]
k	thermal conductivity [W m^{-1}K^{-1}]
L	length [m]
m	mass [kg]
$\dot{m}$	mass flow rate [kg s^{-1}]
MM	molar mass [kg kmol^{-1}]
n	number [−]
N	rotational speed [Hz]
P	pressure [Pa]
q	heat flux [W m^{-2}]
$\dot{Q}$	rate of heat transfer [W]
r	ratio [−]
R	heat transfer resistance [K W^{-1}]
RH	relative humidity [−]
rpm	rotational speed [rpm]
T	temperature [°C or K]
s	specific entropy [J kg^{-1} K^{-1}]
S	entropy [J K^{-1}]
S	stroke [m]

t	time [s]
T	torque [N m]
u	specific internal energy [J kg^{-1}]
U	internal energy [J]
U	overall heat transfer coefficient [W m^{-2} K^{-1}]
v	specific volume [m^3 kg^{-1}]
V	volume [m^3]
$\dot{V}$	volume flow rate [m^3 s^{-1}]
Vol	volume [m^3]
w	specific work [J kg^{-1}]
W	work [J]
$\dot{W}$	power [W]
x	displacement, distance [m]
x	quality [$-$]
X	ratio [$-$]
X	concentration [ppm]
z	elevation, altitude [m]

Subscripts

a	acceleration
a	air
$adiab$	adiabatic
amb	ambient
atm	atmospheric
avg	average
aux	auxiliaries
b	boundary
b	black body
bod	body
c	cold
c	cylinder
c	combustion
c	cutoff
c	convection
cab	cabin
cc	combustion chamber
cd	condenser
cl	cloth
$cond$	conduction
$cond$	condensate
$cool$	coolant
cp	compressor
cr	crank chamber
CV	control volume
d	displacement
d	diffuse

d	discharge
diff	diffusion
dh	diffuse horizontal
dp	dew point
el	electric, electrical
eng	engine
eq	equivalent
ex	exhaust
exf	exfiltration
exp	expander
ev	evaporator
f	saturated liquid
f	fluid
f	fuel
f	fin
f	free
f	final
form	formation
fric	friction
g	gravity
g	saturated vapor
g	gas
gc	gas cooler
gen	generated
gw	glycol water (coolant)
glaz	glazing
h	hydraulic
h	hot
ha	humid air
he	heat engine
i	initial
in	inside, indoor, internal
in	indicated
inf	infiltration
int	internally
k	kinetic
l	liquid
l	leakage
lat	latent
m	maximum
m	mechanical
m	metabolism
m	masses
mech	mechanical
mod	module
n	natural
o	operative
occ	occupant

out	outside, outdoor
p	constant pressure
p	potential
p	piston
plas	plastic
pp	pump
r	radiated
r	refrigerant
rad	radiator
rec	recirculated
ref	reference
rel	relative
rev	reversible
s	isentropic
s	surface
s	swept
s	solar
sa	sol-air
sat	saturated
sens	sensible
sf	secondary fluid
sh	shaft
sk	skin
st	stoichiometric
su	supply
surf	surface
th	thermal
th	theoretical
tot	total
tp	two-phase
turb	turbine
v	constant volume
v	vapor
vent	ventilation
w	water
w	wall
wb	wetbulb
wf	working fluid
wg	waste gate
0	at 0°C
0	clearance
II	second Law of Thermodynamics
∞	freestream

Exponents

°	ideal gas contribution
r	residual contribution

Greek Symbols

α	absorptivity [−]
β	solar altitude [rad]
γ	specific heat ratio [−]
Δ	difference [−]
ε	emissivity [−]
ε	effectiveness [−]
η	efficiency [−]
θ	specific total energy of flowing fluid [J kg^{-1}]
θ	crank angle [rad]
λ	wavelength [m]
μ	dynamic viscosity [kg m^{-1}s^{-1}]
ρ	density [kg m^{-3}]
ρ	reflectivity [−]
σ	Stefan–Boltzmann constant [5.67×10^{-8} W m^{-2}K^{-4}]
Σ	surface tilt angle [rad]
τ	transmissivity [−]
τ	time [s]
ϕ	solar azimuth [rad]
Φ	equivalence ratio [−]
ψ	surface azimuth [rad]
ω	specific humidity [kg kg^{-1}]

About the Companion Website

This book is accompanied by a companion website:

www.wiley.com/go/lemort/thermal

This website includes:

- EES files

Introduction

1 Genesis

The paternity of the automobile is still debated between several inventors among whom are Francesco di Giorgio Martin (1470), Roberto Valturio (1472), or Leonardo da Vinci whose sketches can be found in the Codex Atlantico (1478) and whose drawings are preserved in his engineering notebooks. A study of a self-propelled wagon probably for a theatrical machine, able to move for a short stretch on a stage, is known. For a long time, it was wrongly interpreted as a kind of ancestor of the automobile (Figure 1).

However, thanks to the first functional models of the Belgian Jesuit Ferdinant Verbiest (1623–1688), we can discover the description of a thermodynamic system that allows the movement of the vehicle. In 1672, to put into practice his studies on boilers, he installed one on a small cart. The jet of steam actuated a paddle wheel which drove the wheels through a set of gears.

The drawing in Figure 2 is by the hand of the inventor, as in his description, published in 1685, in Latin, in his treatise "Astronomia Europea."

The Frenchman Joseph Cugnot presented his "Fardier (or steamer)" developed during the period 1769–1771, a cart propelled by a steam boiler. As shown in Figure 3, it was difficult to brake the steamer, leading to probably the first car accident in history.

Other models followed, but steam propulsion was a stalemate in terms of the relationship between weight and performance. This is how the automobile evolved towards the electric car. The first electric car model was built by Sibrandus Stratingh (1835).

We could not resist quoting Camille Jenatzy's electric car, "La Jamais contente (or Never-Happy)" (Figure 4). This is the first motor vehicle to reach the $100\,km\,h^{-1}$ mark.

This electric car, in the shape of a torpedo on wheels, set this record on 29 April 1899 in Achères (France).

The first times of the electric car remained chaotic and inefficient. So, the German Carl Benz built the first automobile in history driven by a thermal engine (1886).

Several revolutions followed that led to changes to steam engines, electric, gasoline, diesel, fuel cell, and electric propulsion again.

Each time, the thermal systems have been adapted or reinvented themselves to meet the new challenges that the automotive industry has encountered. The necessary revolution towards carbon neutrality has accelerated those changes.

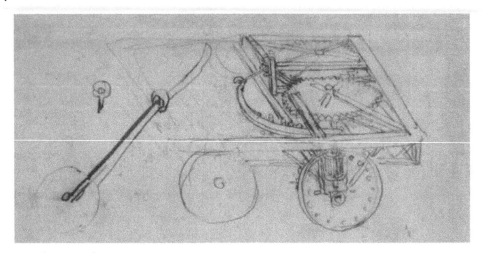

Figure 1 Self-propelled wagon as drawn by da Vinci. Source: Leonardo da Vinci – http://history-computer .com, public domain, https://commons.wikimedia.org/w/index.php?curid=14619567.

Figure 2 One of the first steam-driven cars by Belgian Ferdinant Verbiest. Source: Unknown author/Wikimedia/Public Domain.

Figure 3 Cugnot's Steamer ("Fardier de Cugnot"), tested in Paris in 1770.

Figure 4 "La Jamais contente (or Never-Happy)". Source: Unknown author/Wikimedia/Public Domain.

2 Vectors of Evolution of Thermal Systems

The vectors of the evolution of the automobile world and of its motorization were successively: a race for speed record, increase in the reliability of the engines, increase in the specific power of the engines, introduction of heating and then of air conditioning of the passenger compartment, reduction of vehicle consumption, regulatory constraints governing the environmental impact of engines, reduction in vehicle weight, conservation of the autonomy of electric vehicles, and finally, an improved comfort for passengers of electric and autonomous vehicles.

With each step, the thermal management of the vehicle has evolved toward more performance and functionality, less weight, and lower cost.

To cope with these new challenges, the number of independent thermal systems has increased initially, their interconnection has evolved, and today, many of these systems are fully connected to ensure optimal energy management.

3 The Regulatory Constraints of Change

Pollution regulations have been important vectors for the evolution of propulsion systems and they asked for the energy sobriety of the auxiliaries (all components and systems not directly contributing to propulsion, such as heating, air-conditioning, battery thermal management systems, etc.)

The evolution of the allowed emission limits, in CO_2 per kilometer, for the four main geographical areas, namely the USA, Europe, Japan, and China, is shown in Figure 5.

European CO_2 pollution standards imposed since 1992 refer to the New European Driving Cycle (NEDC). In addition to CO_2 reduction, the European regulations have imposed limitations on emissions of other pollutants, including NO_x, CO, particulate matter (PM), and $HC + NO_x$.

As an example, Figure 6 gives the allowed emission limits for diesel engines from July 1992 (Euro 1) to September 2015 (Euro 6).

To comply with these emission regulations, car manufacturers and tier one suppliers have developed major new systems such as turbocharger, fuel direct injection, high-pressure and low-pressure

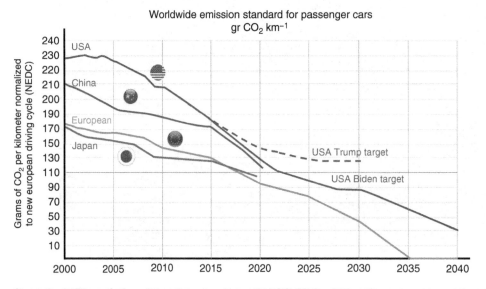

Figure 5 Yearly evolution of the allowed emission limits in CO_2 per kilometer.

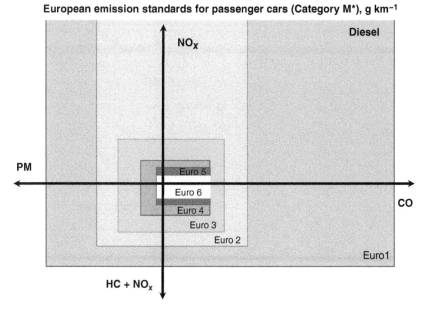

Figure 6 Allowed emission limits for diesel engines from Euro 1 (1992) to Euro 6 (2015) regulations.

exhaust gas recirculation systems (EGR), selective catalytic reduction (SCR), and diesel particulate filter (DFP).

Each of these systems requires optimal operating conditions and specific cooling or heating systems, which have complicated the thermal architecture of the vehicle.

The introduction of electrical motorization created new demands, which included cooling of the battery, fast cooling of the battery during charging, and compensation of the thermal deficit in winter for passenger comfort, and the problem is even more important for fuel cell systems.

The optimization of thermal energy for full electric vehicles is no more an option but a condition to secure vehicle range.

Despite the demands for reduction in the consumption of internal combustion engine vehicles following the oil crises (1973 and 1979) and finally since 1992, the increasingly stringent depollution regulations enacted, the GHG (greenhouse gas) emissions of the transport sector are the only one increasing compared to other sectors responsible of GHG emissions (power generation, industry, buildings, etc.). The index shown in Figure 7 is a relative measurement of the emissions of gases responsible for the greenhouse effect.

In addition, the share of road transport represents 11.9% of GHG emissions. Figure 8 shows the distribution of the GHG emission per sector. The energy sector represents 73.2% of the global emissions.

For this reason and following the Diesel Gate (2008–2015), state and city standards have been tightened, and the NEDC standard has been replaced by the worldwide harmonized light vehicles test procedure (WLTP) standard, which represents more real-time driving of the vehicle by integrating the consumption of accessories.

Furthermore, real driving emissions (RDE) pollution standards were introduced. These standards refer to a fleet of vehicles in real use during their lifetime and not only for a new vehicle.

Figure 9 shows that the reduction of the pollution has accelerated mainly after the Diesel Gate.

Figure 10 shows a schematic illustration of average CO_2 emission levels in the EU between 2014 and 2030, assuming a 3.9% per year and 6.8% per year CO_2 reduction scenario.

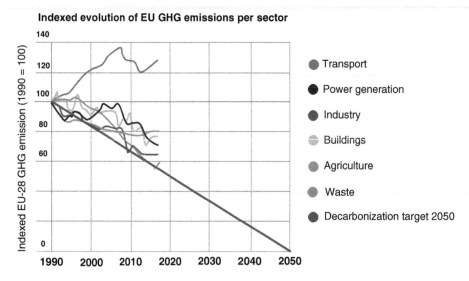

Figure 7 Evolution of the European GHG emissions relative to 1990 per sector. Source: Data from Transport & Environment (1998), UNFCC (1990-2016 data) and EEA's approximated EU greenhouse inventory (2017 data).

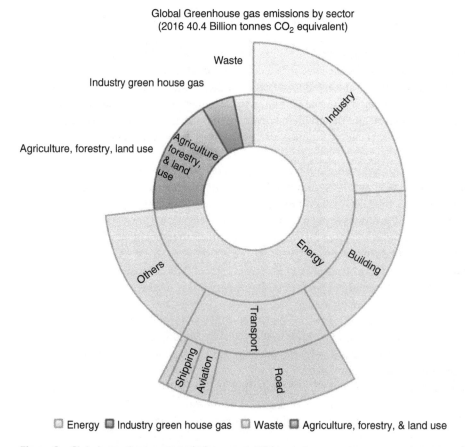

Figure 8 Global greenhouse gas emissions per sector.

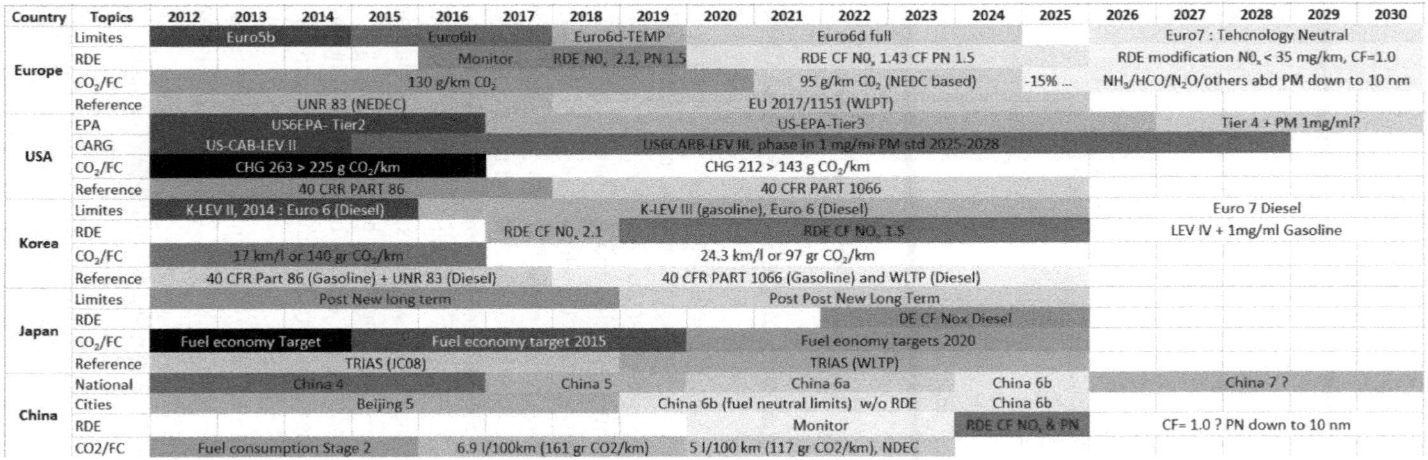

Country	Topics	2012	2013	2014	2015	2016	2017	2018	2019	2020	2021	2022	2023	2024	2025	2026	2027	2028	2029	2030
Europe	Limites		Euro5b			Euro6		Euro6d-TEMP				Euro6d full					Euro7 : Tehnology Neutral			
	RDE					Monitor	RDE NO$_x$ 2.1, PN 1.5				RDE CF NO$_x$ 1.43 CF PN 1.5						RDE modification NO$_x$ < 35 mg/km, CF=1.0 NH$_3$/HCO/N$_2$O/others abd PM down to 10 nm			
	CO$_2$/FC				130 g/km CO$_2$					95 g/km CO$_2$ (NEDC based)					-15% ...					
	Reference		UNR 83 (NEDEC)						EU 2017/1151 (WLTP)											
USA	EPA		US6EPA: Tier2			US6EPA-Tier3					US-EPA-Tier3							Tier 4 + PM 1mg/ml?		
	CARG	US-CAB+LEV II							US6CAB-LEV III, phase in 1 mg/mi PM std 2025-2028											
	CO$_2$/FC		CHG 263 > 225 g CO$_2$/km							CHG 212 > 143 g CO$_2$/km										
	Reference			40 CRR PART 86							40 CFR PART 1066									
Korea	Limites		K-LEV II, 2014 : Euro 6 (Diesel)							K-LEV III (gasoline), Euro 6 (Diesel)								Euro 7 Diesel		
	RDE						RDE CF NO$_x$ 2.1					RDE CF NO$_x$ 1.5						LEV IV + 1mg/ml Gasoline		
	CO$_2$/FC		17 km/l or 140 gr CO$_2$/km								24.3 km/l or 97 gr CO$_2$/km									
	Reference	40 CFR Part 86 (Gasoline) + UNR 83 (Diesel)							40 CFR PART 1066 (Gasoline) and WLTP (Diesel)											
Japan	Limites		Post New long term								Post Post New Long Term									
	RDE												DE CF Nox Diesel							
	CO$_2$/FC	Fuel economy Target				Fuel economy target 2015					Fuel economy targets 2020									
	Reference				TRIAS (JC08)						TRIAS (WLTP)									
China	National			China 4				China 5		China 6a				China 6b				China 7 ?		
	Cities				Beijing 5				China 6b (fuel neutral limits) w/o RDE					China 6b						
	RDE										Monitor			RDE CF NO$_x$ & PN						
	CO2/FC	Fuel consumption Stage 2				6.9 l/100km (161 gr CO2/km)				5 l/100 km (117 gr CO2/km), NDEC							CF= 1.0 ? PN down to 10 nm			

Figure 9 Evolution of regulations regarding pollutant emissions.

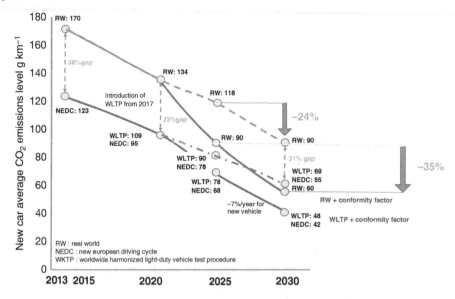

Figure 10 CO_2 emission level for RW, NEDC, and WLTP regulation evolutions.

4 The First Three Revolutions of the Twenty-First Century

Three major revolutions are shaking the automotive industry, namely decarbonization, driving automation, and connectivity, which makes it possible to create more and more embedded digital services and a better transition between modes of transport, home, and road.

Because of their emissions of nitrogen oxide and fine particles, real or perceived values, diesel engines have been discarded to the benefit of the hybridization of gasoline engine (hybrid electric vehicle [HEV]) and electric motors (battery electric vehicle [BEV]).

The ratio of sales of diesel to gasoline vehicles is reversed in less than a year and several developments of heat engines have been discontinued and replaced by developments of electric motors.

Fuel cell motorization is seen as a potential alternative to electric vehicles.

This electric revolution is causing new constraints such as cooling the batteries to prevent battery cell degradation during use or rapid charging under hot conditions.

With the arrival of electric motors, the energy consumed by the auxiliaries and, first and foremost, the comfort in winter leads to a strong reduction in the driving range of vehicles when the temperature falls.

In addition to decarbonization (electricity) constraints, new needs are emerging for autonomous, connected, and shared vehicles, such as cooling of on-board computers or adaptation and personalization of passenger comfort.

Since 1769, thermal systems have evolved as multiple, connected, complex, efficient, and adaptive.

Challenges linked to the revolutions of the twenty-first century, electrification of the powertrain, automation of connected vehicles, and digital mobility will allow once again the thermal management of the vehicle to evolve with elegance toward global management.

5 Ambition of the Authors

The target audience of this book are not engineering students but also more experienced engineers who wish to perfect their knowledge in thermal sciences (heat transfer, thermodynamics, and fluid mechanics) applied to vehicle thermal management systems.

The authors were inspired by several reference works published; some of them were published several decades ago and have been the reference of several generations of students. Among these reference works, we would like to mention the following books:

- "Automotive Climatization (La climatisation Automobile)" from André Colinet (1993)
- "Technical refrigeration manual – Le Pohlman" translated in French by Maake et al. (1993). This reference is the evolution of a pocketbook published in 1908 (Taschenbuch der Kältetechnik) for refrigeration technicians. In 1908, the 1st International Cold Congress was held in Paris, which brought together all the minds interested in low temperatures.
- "Automotive Air conditioning and Climate control systems" by Steven Daly (2006)
- The classroom Manual For Automotive Heating & Air Conditioning written by Schnubel (2016)

The present book goes one step further by presenting the impact of electrification on the overall thermal management of vehicles. The present book also introduces some simple modeling techniques of vehicle thermal management components and systems.

6   Organization of the Book

The book is organized as follows. The first chapter covers the fundamentals of thermal systems and highlights applications to vehicle thermal management components. The second chapter presents the thermal management systems associated with internal combustion engines. The third chapter is dedicated to passenger thermal comfort and cabin air conditioning. Finally, the last chapter explains the specific thermal management solutions for hybrid and electric vehicles.

In Chapters 2–4, several examples of numerical problems are proposed. Equations implemented in Engineering Equation Solver (EES) software are presented. The EES files are available for download on the website associated with the book. EES is a software developed by Prof. Sanford Klein from the University of Wisconsin–Madison. This software combines an acausal equation solver and an accurate database of thermodynamic and transport properties of working fluids typically used in vehicle thermal management systems (humid air, refrigerants, aqueous solution of glycol, etc.). EES allows for solving large systems of coupled nonlinear algebraic equations and differential equations. The acausal feature of the solver allows to focus on the physical modeling of thermal systems (in both steady-state and transient regimes) rather than on the numerical solving of the models. One very interesting feature of EES is its tool for conducting parametric analyses. This is very convenient to investigate the sensitivity of a thermal component or system to a change in operating conditions or design parameters. More information is available on the F-Chart website (https://fchartsoftware.com/ees/).

References

Colinet, A. (1993). *La climatisation automobile*. Editions techniques pour l'automobile et l'industrie.

Daly, S. (2006). *Automotive Air Conditioning and Climate Control Systems*. Butterworth-Heinemann.

Maake, W., Eckert, H.-J., and Cauchepin, J.L. (1993). *Le Pohlmann : manuel technique du froid*, vol. 2, 1, 1204. PYC Livres.

Schnubel, M. (2016). *Today's Technicians™: Classroom Manual for Automotive Heating & Air Conditioning*, 6the. Boston, USA: Cengage Learning.

Transport & Environment (1998). EU publishes climate strategy to exit oil. Available: https://www .transportenvironment.org/discover/eu-publishes-climate-strategy-exit-oil/ (accessed 9 May 2022).

1

Fundamentals

1.1   Introduction

This textbook deals with the study of different vehicle thermal systems and components from an energy engineering point of view. It is therefore necessary to recall the fundamentals of heat transfer as well as thermodynamics and some elements of fluid mechanics for a good understanding of the content of the next Chapters 2, 3, and 4. This is the objective of the present chapter, the content of which has been largely summarized from major reference textbooks, especially those of Incropera and DeWitt (2002), Çengel and Boles (2006), Braun and Mitchell (2012), and Klein and Nellis (2016).

1.2   Fundamental Definitions in Thermodynamics

Thermodynamics is the branch of physics that studies conversions between heat and work in one or the other direction. Thermodynamics is particularly useful for the analysis of components and systems presented in this book.

Thermodynamics makes use of some important notions to which the reader should become familiar.

1.2.1   System, Surroundings, and Universe

In thermodynamics, a *system* is defined as a delimited region of space or a quantity of matter that is investigated. The concept of "investigation" may still be a little bit fuzzy and will progressively develop. Let's say that investigating a system means quantifying its energy performance and the relation between this performance and operating conditions. The system is delimited by a *boundary* (Figure 1.1). A boundary has neither mass nor thickness. The *surroundings* of the system are the region of space or the quantity of matter that is outside the system. Hence, the boundary is the surface that separates the system from its surroundings. The system and its surroundings constitute the *universe*.

Among the systems, one can distinguish the *closed systems* and the *open systems*. A closed system does not exchange any mass with its surroundings. Consequently, its mass is constant. An open system, also called *control volume* (CV), exchanges mass with its surroundings. Such a system is represented in Figure 1.2. The system consists of the region in space delimited partially by plain lines and partially by two dashed lines. Some fluid can enter or leave the system through physical connections to the surroundings. Consequently, the mass of the system may vary.

Thermal Energy Management in Vehicles, First Edition. Vincent Lemort, Gérard Olivier, and Georges de Pelsemaeker.
© 2023 John Wiley & Sons Ltd. Published 2023 by John Wiley & Sons Ltd.
Companion website: www.wiley.com/go/lemort/thermal

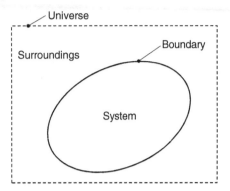

Figure 1.1 System, boundary, surroundings, and universe.

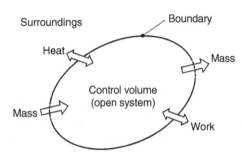

Figure 1.2 Mechanisms of energy transfer between an open system and its surroundings.

In Figure 1.2, part of the boundary is real. It is represented by solid lines and can correspond to the physical envelope of the system. Contrarily, dashed lines represent imaginary boundaries. They are the openings of the system that allow for mass exchange with the surroundings.

As depicted in Figure 1.2, an open system can exchange energy with its surroundings through three mechanisms: heat transfer, work transfer, and mass transfer. It will be shown later that the energy transfer associated with the mass transfer is computed based on the *enthalpy* of the flow.

In the specific case where the open system does not exchange mass with its surroundings, it becomes a closed system. If the system does not exchange heat with its surroundings, it is said to be *adiabatic*. If the system exchanges neither heat nor work nor mass, it is said to be *isolated*. It will be shown later that engineering applications of thermodynamics are particularly interested by (useful) energy transfers between a system and its surroundings.

A very common example of an open system, largely described in textbooks, is the cylinder-piston assembly equipped with valves represented in Figure 1.3. Since one desires to describe the state of the fluid inside the cylinder, the content of the cylinder is defined as the thermodynamic system. The dashed line represents the boundary of the system. It consists of the cylinder wall, cylinder

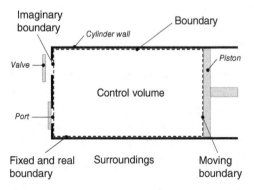

Figure 1.3 Example of an open system with a moving boundary.

head (comprising the ports), and the inner surface of the piston. The latter portion of the boundary is moving. Some fluid can enter or leave the control volume through the ports, provided they are not covered by the valves.

In this book, many open and closed systems will be investigated. Among the major open systems, one can mention heat exchangers, compressors, turbines, and pumps. All these components have inlet and outlet ports that allow for mass exchange with their surroundings. The vehicle cabin can also be considered as an open system, since it exchanges air with the vehicle outdoor through the ventilation system. When the ventilation is in the recirculation mode (which will be described in Chapter 3) and if infiltrations and exfiltrations of air are neglected, the vehicle cabin can be seen as a closed system. Another example of a closed system could be the whole engine coolant loop. During regular operation, the mass of coolant in the loop is constant.

An appropriate choice of the boundary will simplify the thermodynamic description of the system under investigation.

1.2.2 Properties

When investigating vehicle thermal management, one has to describe numerous thermodynamic systems.

A system can be described by its characteristics, which are named as thermodynamic *properties*. Describing a system means describing its thermodynamic *state*. The state of a system is defined when the latter is in equilibrium. The most known properties are pressure P [Pa], temperature T [K], mass m [kg], and volume V [m³]. These properties are *internal* properties. Speed C [m s⁻¹] and elevation z [m] are *external* properties and do not depend on the molecular structure of the matter (Klein and Nellis, 2016). Note that these properties are measurable properties. It will be shown later that other properties that cannot be directly measured are also very useful for the description of a system, such as internal energy U [J], enthalpy H [J], or entropy S [J K⁻¹]. Such nonmeasurable properties can be calculated based on measurable ones and thermodynamic relations.

A *specific property* is a property expressed per unit of mass of the system. Specific properties are usually denoted with lowercase letters. For instance, specific volume v [m³kg⁻¹] is volume V [m³] divided by mass m [kg]. Other properties that will be used in this book are specific internal energy u [J kg⁻¹], specific entropy s [J kg⁻¹K⁻¹], specific enthalpy h [J kg⁻¹], and specific heat at constant pressure c_p [J kg⁻¹K⁻¹].

Among the properties, the distinction can also be done between the *extensive* and *intensive* properties. The extensive properties depend on the size of the system and vary linearly with its mass. Examples are mass m [kg], volume V [m³], and internal energy U [J]. On the contrary, the intensive properties do not depend on the system size and mass. Examples are pressure P [Pa] and temperature T [°C] and also any specific properties.

As it will be explained later, to specify the state of a system in internal equilibrium (and in the absence of electrical, magnetic, or other effects), two independent intensive properties are needed.

1.2.3 Process

A *process* is a transformation that brings the system from a given state A to another state B. The initial and final states are described by two independent intensive properties. Therefore, the state of a system is a "picture" of the system and does not depend on its "history" (the way this state was obtained). Indeed, different paths, involving different heat and work transfers, can bring the system from state A to state B. Hence, the properties of the system are called *point functions*.

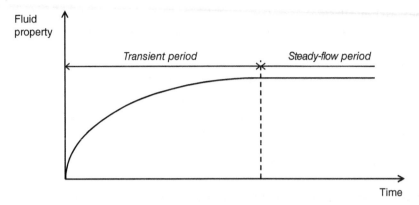

Figure 1.4 Steady-flow period following a transient period.

In contrary, heat and work are called *path functions*. Examples of processes involved in vehicle thermal management are fluid compression, expansion, cooling, and heating.

A process that brings the system back to its initial state is a *cycle*. This type of process is encountered in thermal machines where a fluid is circulating in a cyclic manner. One major example of the thermodynamic cycle is the refrigeration cycle that is exploited by the vehicle air-conditioning loop. It will be described in detail in Chapter 3.

There is one particular operation of open systems that largely simplifies their analysis. This is the steady-state regime. A fluid flows steadily through a control volume if the properties inside the control volume do not vary with time. The fluid properties can vary from one point to another inside the control volume, but at a given location inside the control volume, the fluid properties do not vary with time (Çengel and Boles, 2006). In such a situation, the control volume is said to undergo a *steady-flow* (or steady-state) process. During such a process, the energy and the mass contained in the system are constant in time. Steady-flow processes are approximated in practice when a system is operating for a long period of time with no variation in operating conditions.

For instance, when an internal combustion engine is switched on, it will first undergo a transient period during which the temperature of the metal and internal fluids will increase. After this period, the engine temperature and consequently the temperature of the fluids leaving the engine will stabilize. This is represented in Figure 1.4. When acceptable, the assumption of the steady-flow process simplifies the analysis of the performance of the systems.

1.2.4 Energy

This book will extensively make use of the concept of energy. Actually, it will describe different components and systems of components used in vehicles that transfer energy or convert it from one form to another. Defining the energy is not an easy task. Our everyday experience teaches us that energy can appear under different forms, among which thermal energy, mechanical energy, kinetic energy, potential energy, electric energy, or nuclear energy. By summing all the quantities of energy contained in a system under its different forms, one obtains the total energy E [J] of the system. The specific total energy is defined as $e = E/m$ [J kg^{-1}].

It should also be mentioned that the energy contained in a system depends on the reference state for which this energy is null. This is not of primary importance, since the description of thermal systems mainly implies the quantification of *energy variations*.

Among the different forms of energy, one can also distinguish the *microscopic* forms and the *macroscopic* forms of energy. The microscopic forms of energy are sensible energy (energy associated with the movement of molecules, atoms, and nucleons), latent energy (energy associated with the binding forces between the molecules; these forces decrease from the solid phase to the liquid phase and to the gaseous phase), chemical energy (energy associated with the atomic bonds in a molecule), and nuclear energy (energy associated with the bonds between the nucleons inside the nucleus of the atom). The macroscopic forms of energy of a system are associated with its velocity C [m s^{-1}] and altitude z [m], i.e. its kinetic and potential energies.

The internal energy U [J] of a system is the sum of all microscopic forms of energy. The specific internal energy is defined as $u = U/m$ [J kg^{-1}] . The total and internal energies are related by

$$E = U + m\frac{C^2}{2} + mgz \tag{1.1}$$

where

C is the system velocity, [m s^{-1}]
g is the gravitational acceleration, [m s^{-2}]
z is the elevation of the system from a reference altitude, [m].

In the right-hand side of the previous equation, the second and third terms are the kinetic and potential energies, respectively.

1.2.5 Heat

Heat Q [J] is the form of energy that is exchanged between a system and its surroundings because of their difference in temperatures. A system does not contain heat, but thermal energy. Heat is the visualization of thermal energy transfer through a system boundary under the action of a temperature gradient. To be rigorous, one should not talk about "heat transfer," but "thermal energy transfer." However, the latter expression is commonly accepted.

The heat transfer rate $\dot{Q}$ [W] is defined as the heat exchanged between a system and its surroundings per unit of time. In the case of an adiabatic system, $\dot{Q} = 0$ [W].

1.2.6 Work

Work is the form of energy that is transferred when a force acts on the system over a distance (Çengel and Boles, 2006). The different forms of work can be categorized as mechanical forms and nonmechanical forms. Different forms of mechanical work can also be distinguished.

1.2.6.1 Mechanical Forms of Work

A mechanical work is exchanged between a system and its surroundings when a force is acting on the system boundary and when this system or its boundary is moving. If these two conditions are met, a work interaction exists between the system and its surroundings. The work is either done *by* the system or done *on* the system. In the former case, the external force acting on the system and its motion have opposite directions. In the latter case, the external force acting on the system and its motion has the same direction.

The mechanical forms of work that will be met in the rest of this book are the moving boundary work, the shaft work, the spring work, and the work necessary to raise or to accelerate a system.

1.2.6.1.1 Moving Boundary Work When a fluid is compressed or expanded, the boundary separating the fluid from its surroundings is moving. As a consequence of the displacement of the boundary, moving boundary work W_b [J] is exchanged between the fluid and its surroundings. Such a work interaction is represented in Figure 1.5.

Piston cross-sectional area = A
Moving boundary

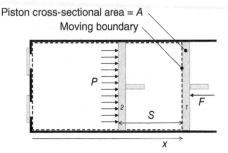

During the compression (related to distance $S = x_1 - x_2$ travelled by the piston), moving boundary work W_b can be computed by integrating the product of the force acting on the piston and incremental displacement dx. That is,

$$W_b = - \int_1^2 F\,dx \tag{1.2}$$

In Eq. (1.2), the force can be related to the pressure acting on the inner surface of the piston, leading to Eq. (1.3). In the case of a *reversible* evolution (1-2), since the pressure is uniform within the cylinder, the pressure in Eq. (1.3) is the system pressure. The notion of reversible evolution will be further developed later.

$$W_b = - \int_1^2 P A\,dx = - \int_1^2 P\,dV \tag{1.3}$$

1.2.6.1.2 Shaft Work Shaft work is the work associated with a rotating shaft. This is, for instance, the work transmitted at the shaft of the engine of a car or the work absorbed at the shaft of a compressor (Figure 1.6).

Shaft work W_{sh} [J] exchanged after X revolutions of a shaft on which a torque T [N·m] of moment arm r [m] is acting is given by

$$W_{sh} = \frac{T}{r} 2\pi r X = 2\pi X T \tag{1.4}$$

Shaft power $\dot{W}_{sh}$ [W] (shaft work per unit of time) can be calculated based on the rotational speed N [Hz] of the shaft, which is the number of revolutions done per second.

$$\dot{W}_{sh} = 2\pi N T \tag{1.5}$$

1.2.6.1.3 Spring Work According to Hooke's law, the force F [N] needed to compress a linear-elastic spring of spring constant k [N m^{-1}] by a distance X [m] varies linearly with this distance (Figure 1.7).

$$F = kx \tag{1.6}$$

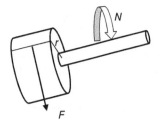

Figure 1.6 Work transmitted at a shaft. Source: Reproduced from Çengel and Boles (2006).

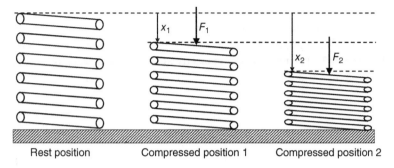

Rest position   Compressed position 1   Compressed position 2

Figure 1.7 Spring work.

Hence, the work needed to compress (or extend) a spring from a position where its displacement is x_1 to a position where its displacement is x_2 is

$$W_{spring} = \frac{1}{2}k\left(x_2^2 - x_1^2\right) \tag{1.7}$$

Springs are commonly used in thermostatic actuators met in thermal management systems. Two examples are the engine coolant thermostat (described in Chapter 2) and the thermostatic valve of the air-conditioning loop (studied in Chapter 3).

1.2.6.1.4 Work Necessary to Raise or to Accelerate a System
Work must be transferred to a system to accelerate it or to increase its altitude. Once transferred to the system, the kinetic and potential energies are increased, respectively. Similarly, a work is done by a system when it is decelerated or when its altitude is decreased. In that case, the kinetic and potential energies of the system are decreased, respectively.

Hence, work W_a transferred to a system of mass m to increase its velocity from C_1 to C_2 is equal to

$$W_a = \frac{1}{2}m\left(C_2^2 - C_1^2\right) \tag{1.8}$$

Similarly, work W_g transferred to a system of mass m to increase its altitude from z_1 to z_2 is equal to

$$W_g = m\,g\,(z_2 - z_1) \tag{1.9}$$

Such works are illustrated in Figure 1.8, depicting a car of mass m [kg] accelerating from velocities C_1 to C_2 and climbing a hill from altitudes z_1 to z_2.

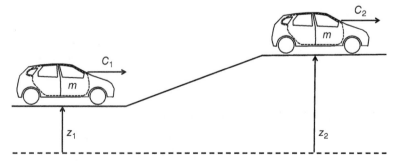

Figure 1.8 Work necessary to raise and accelerate a system.

1.2.6.2 Nonmechanical Forms of Work

Two usual nonmechanical forms of work are magnetic work and electrical work. Transfers of electrical work are common in automotive applications (for instance, positive temperature coefficient [PTC] heaters introduced in Chapter 3). The rate transfer of electrical work is electrical power. It is associated with current I [A] flowing through resistance R [Ω] under the action of a difference of potential V [V].

$$\dot{W}_{el} = V I = R I^2 \tag{1.10}$$

1.2.7 Enthalpy

When a fluid flows into the control volume, it does not bring only the total energy e [J kg^{-1}] but also an additional energy named *flow energy*. Flow energy, or flow work, is the work necessary to push a fluid element into the control volume. This work is done by the fluid upstream the flow element. Per unit of mass of fluid element, flow energy is equal to Pv [J kg^{-1}]. Hence, the total energy of a flowing fluid is equal to

$$\theta = e + Pv = u + Pv + ke + pe = h + ke + pe \tag{1.11}$$

As indicated in Eq. (1.11), internal energy u and flow energy Pv have been merged into a variable h [J kg^{-1}], which is named *specific enthalpy*. This variable is very convenient when investigating open systems, since it includes both the internal energy and energy involved when pushing the fluid into or out of the control volume.

1.3 Fluids

Vehicle thermal management systems involve different fluids, the properties of which make them suitable for specific functions.

This book focuses on thermodynamic properties, which are properties that allow describing the thermodynamic state of the fluid. Among fluids, the distinction can be done between pure fluids and mixtures. Some mixtures can be treated as pseudo-pure fluids. We will see hereunder that if the fluid is a pure fluid or pseudo-pure fluid, its thermodynamic state can be easily described.

1.3.1 Pure and Pseudo-Pure Fluids

A pure fluid (also named pure substance or chemical substance) is a substance that has a homogeneous and stable chemical composition. Homogeneous mixtures can be considered as pure fluids. This is, for instance, the case of air, which contains nitrogen and oxygen. Heterogeneous mixtures cannot be considered as pure fluids.

A pure fluid can exist under different phases: solid, liquid, and gas. Each phase has a different molecular structure. The strength of the intermolecular bonds decreases from solid to liquid and from liquid to gas. The molecular spacing increases from solid to liquid and from liquid to gas (phase changes). A mixture of several phases of a pure fluid is still a pure fluid as long as all phases have the same chemical composition (Çengel and Boles, 2006).

1.3.2 Liquid–Vapor Phase Change for a Pure or Pseudo-Pure Fluid

Some phase changes are largely exploited in vehicle thermal management systems. This is the case of liquid–vapor phase changes (exploited in air conditioning (A/C) loops or to explain the phenomenon of mist formation on the windshield) and to a lesser extent solid–liquid phase

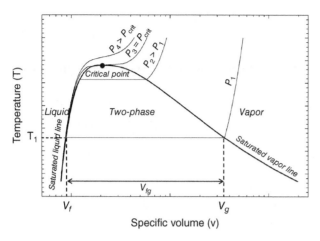

Figure 1.9 Temperature–volume diagram of a pure fluid (different lines of constant pressure are represented).

changes (exploited in wax-type thermostats or in phase change material [PCM] storage). This section emphasizes the liquid–vapor phase change.

A pure fluid can be in different equilibrium states such as liquid, two-phase, and vapor states. The temperature–volume and pressure–enthalpy diagrams of Figures 1.9 and 1.10 are commonly used to represent the state of a fluid that can possibly undergo a phase change. At a given pressure P_1, if the temperature of the fluid is low enough (lower than temperature T_1 in Figures 1.9 and 1.10), the fluid is in *subcooled liquid state*. If the fluid is heated at a constant pressure P_1, its temperature and specific volume increase. When this temperature reaches T_1, the liquid is about to vaporize. In this particular state, the fluid is in *saturated liquid state*. Any slight heat addition to the fluid would result in the vaporization of some liquid. Temperature T_1 is the saturation temperature at pressure P_1. From the saturated liquid state, as heat is provided to the fluid at constant pressure P_1, its temperature remains constant (and equal to T_1), and its specific volume increases. The quantity of liquid decreases, and the quantity of vapor increases. This is the liquid–vapor phase change process. The phase change ends when the last particle of liquid turns into vapor. In this particular state, the fluid is in *saturated vapor state*. Actually, any slight cooling of the fluid would result into

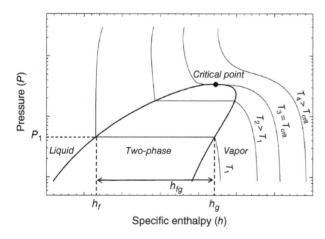

Figure 1.10 Pressure–enthalpy diagram of a pure fluid (different lines of constant temperature are represented).

the condensation of some vapor. During the phase change, the fluid is in *two-phase state*, which corresponds to an equilibrium state between a saturated liquid and a saturated vapor. During the phase change, the mass fractions of saturated liquid and saturated vapor change, but the intensive properties of each phase remain constant. If the saturated vapor is heated at constant pressure P_1, its temperature and specific volume increase. The fluid is in *superheated vapor state*.

The transition from subcooled liquid to superheated vapor at a constant pressure P_1 can be visualized in the temperature–volume diagram given in Figure 1.9. The constant pressure curve associated with pressure P_1 is represented. The specific volume of the saturated liquid is v_f, and the specific volume of the saturated vapor is v_g. The variation of the specific volume during the phase change is $v_{fg} = v_g - v_f$.

The transitions from the subcooled liquid state to the two-phase state and from the two-phase state to the superheat vapor state can be visualized for pressure P_2 higher than P_1 in Figure 1.9. It can be observed that the variation of the specific volume v_{fg} during the phase change is smaller than that for pressure P_1. The same transitions can be drawn for a large range of pressures. The larger the pressure, the shorter is the saturation line. By connecting the saturated liquid states corresponding to the different pressures, the *saturated liquid line* is obtained. Similarly, the *saturated vapor line* is obtained by connecting the saturated vapor states corresponding to the different pressures. The saturated liquid and vapor lines can be visualized in Figure 1.9. The two lines intersect at the *critical point* and form a dome.

The fluid inside the dome is in the two-phase state, which is a mixture of saturated liquid and saturated vapor. The zones on the left-hand and right-hand sides of the dome correspond to the subcooled liquid state and superheated vapor state, respectively.

At the critical point, the variation v_{fg} of specific volume is null. The saturated liquid state and the saturated vapor state are identical. At the critical point, the pressure is equal to the critical pressure and the temperature is equal to the critical temperature. For pressures equal to or larger than the critical pressure, there is no clear transition between the subcooled liquid state and the superheated vapor state. For those pressures, as the temperature increases, at constant pressure, from temperatures lower than the critical temperature to temperatures higher than the critical temperature, the specific volume increases, but there is one single phase all along the process. It is not possible to distinguish a phase change, even if the fluid finally looks like vapor.

Figure 1.10 shows the pressure–enthalpy diagram of a pure fluid. Similar to the temperature–volume diagram, this diagram makes appear a dome that delimits liquid, two-phase, and vapor regions. In this diagram, some constant-temperature curves are overlaid. It can be observed that the curve associated with the critical temperature goes through the critical point. As explained in Chapter 3, this diagram will be used to describe the vapor–compression refrigeration cycles.

To undergo a liquid–vapor phase change, heat must be provided to the fluid. The amount of heat necessary to vaporize 1 kg of a substance at a given pressure is *the latent heat of vaporization* (or *condensation*). Figure 1.10 indicates the latent heat of vaporization h_{fg} [J/kg] at pressure P_1. It corresponds to

$$h_{fg}(P_1) = h_g(P = P_1) - h_f(P = P_1) \tag{1.12}$$

The saturated vapor enthalpy and the saturated liquid enthalpy are given by

$$h_g = h(P = P_1, x = 1) \tag{1.13}$$

$$h_f = h(P = P_1, x = 0) \tag{1.14}$$

The latent heat of vaporization at a temperature of 0°C for 3 fluids typically used in vehicles is provided in Table 1.1.

Table 1.1 Heat of vaporization of three different fluids at 0°C.

	Water	R1234yf	CO_2
$h_{fg}(T = 0°C)$ [J/kg]	2,501E+06	163283	230884

The latent heat of vaporization can also be defined for another temperature T_1. The enthalpy of vaporization at 0°C is generally written as $h_{fg,0}$.

Similarly, the latent heat of fusion (or solidification) is the heat necessary to melt 1 kg of a substance at the normal melting temperature or at the triple-point temperature.

1.3.3 Computing the Properties of Pure and Pseudo-Pure Fluids

1.3.3.1 Phase Rule

To specify the state of system in internal equilibrium, a given number of internal intensive properties must be known.

The Phase Rule states that for a nonreacting thermodynamic system, the number F of internal intensive properties required to fix the state of the system is given by

$$F = C - \Pi + 2 \tag{1.15}$$

where

C is the number of distinguishable chemical species, [−]
Π is the number of phases, [−]

For a pure fluid or pseudo-pure fluid, $C = 1$. A pseudo-pure fluid is a mixture characterized by a constant composition, which allows to treat it as a pure fluid (Klein and Nellis, 2016).

1.3.3.1.1 Single-Phase State For a pure or pseudo-pure fluid in the single-phase state, $F = 2$. This means that two internal intensive properties must be specified to fix the thermodynamic state of the system. It is not possible to specify the value of a third variable.

For a pure or pseudo-pure fluid in two-phase state, $F = 1$. This means that only 1 internal intensive property must be specified to fix the intensive state of each phase. For instance, if the pressure P of the system is specified, the temperature of both phases is fixed.

Coming back to the case of pure or pseudo-pure fluid in single-phase state, the number of internal intensive properties needed to fix the system is 2. Hence, any intensive property can be expressed as a function of two other independent properties.

These two properties could be temperature T and volume v or pressure P and volume v. For instance, the specific enthalpy can be computed as

$$h = f_1(T, v) \text{ or } h = f_2(P, v) \tag{1.16}$$

For single-phase fluids (f.i. subcooled liquids and superheated vapors), the pair P–T could also be used, since these two properties are independent. However, the pressure and temperature are not independent for saturated liquids, saturated vapors, and mixtures of both. Figure 1.11 shows the evolution of the saturation pressure with the saturation temperature for different working fluids that are typically met in automotive applications. Temperature ranges extend from the triple point to the critical temperature (the latter being 30.98°C, 94.7°C, and 373.9°C for CO_2, R1234yf, and water, respectively).

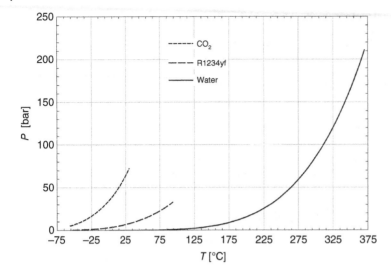

Figure 1.11 Liquid–vapor saturation curves for different fluids commonly used in automotive applications.

1.3.3.1.2 Two-Phase State In the two-phase state or mixed phase, (i.e. a mixture of saturated liquid and saturated vapor), temperature T and volume v can be, for instance, used to describe other properties. Another property that is typically used to describe mixtures of saturated liquid and vapor is the quality. The latter is defined as the ratio of the mass of vapor to the total mass of the mixture.

$$x = \frac{m_g}{m_f + m_g} \tag{1.17}$$

where

m_g is the mass of saturated vapor, [kg]
m_f is the mass of saturated liquid, [kg].

In practice, it is not usual to distinguish the mass of liquid and the mass of vapor inside the mixture. The mixture is considered as a fictitious homogeneous mixture showing average properties. For instance, the average specific volume of the mixture can be defined as

$$v_{avg} = (1-x)v_f + x\,v_g \tag{1.18}$$

The average specific internal energy and enthalpy of the mixture can be computed in a similar way.

$$u_{avg} = (1-x)u_f + x\,u_g \tag{1.19}$$

$$h_{avg} = (1-x)h_f + x\,h_g \tag{1.20}$$

1.3.3.2 The Equations of State Relating *P*, *T*, and *v* (Relation Between Measurable Properties)

Relations between intensive properties have been established for fluids used in engineering applications and are available in the forms of tables, equations, or graphs (Klein and Nellis, 2016).

Among these relations, the equations of state for pressure, temperature, and volume are largely used. They are expressed in the form $P = f(T,v)$. The interest in such relations is that they involve measurable internal intensive variables.

The equation of state of the ideal gas is well known and very convenient to use. When a gas is at low density, it can be accurately described by the ideal gas law given by Eq. (1.21).

$$Pv = rT = \frac{R_u}{MM}T \tag{1.21}$$

where

R_u is the gas constant, [8314 J kmol^{-1} K^{-1}]
MM is the molar mass of the gas, [].

Respecting this equation is actually the condition for a gas to be considered as ideal. Air in cabin and also its water vapor content can be considered as ideal gases.

At high pressures or low temperatures, more accurate equations of state can be used. Among the most famous equations of state, one can mention the Peng–Robinson Equation, which has the following form

$$P = \frac{rT}{v - b} - \frac{a}{v(v + b) + b(v - b)} \tag{1.22}$$

Coefficients a and b can be expressed as function of the critical temperature and pressure, reduced temperature, and acentric factor of the fluid. The interested reader is invited to refer to the book of Klein and Nellis (2016) for more information.

Equations of state such as Peng–Robinson can also be used to describe the liquid phase. However, the accuracy of the prediction of the saturated liquid specific volume is limited. Specific correlations are proposed in the literature for the prediction of the saturated liquid specific volume v_f (see Klein and Nellis, 2016).

1.3.3.3 Computing Non-Measurable Properties (*u*, *h*, and *s*) in General Case of Real Pure Fluids

Pressure, temperature, and specific volume are the properties that can be measured. However, first and second laws of thermodynamics involve other properties that cannot be directly measured, such as specific internal energy, specific enthalpy, and specific entropy. The meaning and use of entropy will be presented later.

1.3.3.3.1 Computing Thermodynamic Properties Based on an Equation for P, T, and v These nonmeasurable properties can be calculated provided that an equation of state of the working fluid is known as well as the specific heat capacity at a pressure low enough for the working fluid to behave as an ideal gas (Klein and Nellis, 2016). The latter condition means that the specific heat capacity is a function of only temperature and not of pressure.

The previous equations of state relate the pressure to the specific volume and temperature. If we want to determine other properties of pure fluids, such as specific internal energy u, specific enthalpy h, or specific entropy s, we need complete equations of state. The latter equations are, for instance,

$$u = u(s, v)$$

$$h = h(s, P)$$

$$a = a(T, v)$$

$$g = g(T, v) \tag{1.23}$$

In the latter equations, a is the specific Helmoltz free energy and g is the specific Gibbs free energy. The meaning of the entropy will be explained in Section 1.6.5.

If one of the complete equations of state is known, all thermodynamic properties can be determined using the fundamental properties relations for pure fluids (Klein and Nellis, 2016). These four fundamental properties equations (Eq. (1.24)) are also called Gibbs equations.

$$du = Tds - Pdv$$

$$dh = Tds + vdP$$

$$da = -sdT - Pdv$$

$$dg = -sdT + vdP \tag{1.24}$$

The Maxwell relations express that the properties u, h, a, and g are state properties, and therefore, the fundamental properties relations du, dh, da, and dg are exact differentials of the complete equations of state. Therefore,

$$\left(\frac{\partial T}{\partial v}\right)_s = -\left(\frac{\partial P}{\partial s}\right)_v$$

$$\left(\frac{\partial T}{\partial P}\right)_s = \left(\frac{\partial v}{\partial s}\right)_P$$

$$\left(\frac{\partial s}{\partial v}\right)_T = \left(\frac{\partial P}{\partial T}\right)_v$$

$$\left(\frac{\partial s}{\partial P}\right)_T = -\left(\frac{\partial v}{\partial T}\right)_P \tag{1.25}$$

Actually, one can note that the fundamental properties relations are the total differentials of the complete equations of state. For instance, the temperature can be obtained by the following partial derivative

$$T = \left(\frac{\partial h}{\partial s}\right)_P \tag{1.26}$$

Using the fundamental property relations and Maxwell's relations, one can express specific internal energy u, specific enthalpy h, and specific entropy s as functions of measurable variables: $h(T,P)$, $u(T,v)$, and $s(T,v)$, respectively. For instance, the differential of $h(T,P)$ is given by

$$dh = c_p\, dT + \left[v - T\left(\frac{\partial v}{\partial T}\right)_P\right] dP \tag{1.27}$$

In this equation, all quantities can be measured. The derivative of the specific volume with respect to the temperature at constant pressure can also be assessed by means of an equation of state, such as Peng–Robinson (Klein and Nellis, 2016).

Enthalpy $h(T,P)$ at a given temperature and pressure can be determined by integrating the previous differential. The methodology to conduct this integration is detailed in Klein and Nellis (2016).

With a similar procedure as that for the specific enthalpy, differentials of u and s involving only measurable quantities can be developed.

$$du = c_v dT + \left[T\left(\frac{\partial P}{\partial T}\right)_v - P\right] dv \tag{1.28}$$

$$ds = \frac{c_v}{T} dT + \left(\frac{\partial P}{\partial T}\right)_v dv = \frac{c_p}{T}\, dT - \left(\frac{\partial v}{\partial T}\right)_P dP \tag{1.29}$$

1.3.3.3.2 Computing Thermodynamic Properties Based on Helmoltz-Energy-Explicit EOS Another way to derive thermodynamic properties is to use the EOS based on the Helmoltz energy a and consider temperature T and density ρ as independent properties. The nondimensionalized Helmoltz energy is obtained by dividing Helmoltz energy a by rT. The nondimensionalized Helmoltz energy can be represented as the sum of an ideal gas contribution (denoted by superscript 0 in Eq. (1.30)) and a residual contribution (denoted by superscript r), leading to the following nondimensionalized Helmoltz energy equation of state:

$$\alpha(\tau,\delta) = \frac{a}{r\,T} = \frac{u - Ts}{r\,T} = \alpha^0(\tau,\delta) + \alpha^r(\tau,\delta) \tag{1.30}$$

with

$$\delta = \frac{\rho}{\rho_{crit}}$$
$$\tau = \frac{T_{crit}}{T} \tag{1.31}$$

The ideal gas reduced Helmoltz energy $\alpha^0(\tau,\delta)$ and the residual reduced Helmoltz energy $\alpha^r(\tau,\delta)$ can be expressed by correlations built from experimental data.

High-accuracy equations of states are based on this formalism of nondimensionalized Helmoltz energy, since all other thermodynamic properties can be obtained by analytical derivatives of the terms $\alpha^0(\tau,\delta)$ and $\alpha^r(\tau,\delta)$. The analytical expressions are derived using the fundamental property relations. More details can be found in Bell et al. (2014), and a detailed example is provided in the thermodynamics textbook of Klein and Nellis (2016). Also, Bell et al. (2014) explain how to compute saturated liquid and vapor densities.

$$\frac{P}{\rho\,r\,T} = 1 + \delta\left(\frac{\partial\alpha^r}{\partial\delta}\right)_\tau \tag{1.32}$$

$$\frac{h}{r\,T} = \tau\left[\left(\frac{\partial\alpha^0}{\partial\tau}\right)_\delta + \left(\frac{\partial\alpha^r}{\partial\tau}\right)_\delta\right] + \delta\left(\frac{\partial\alpha^r}{\partial\delta}\right)_\tau + 1 \tag{1.33}$$

$$\frac{s}{r} = \tau\left[\left(\frac{\partial\alpha^0}{\partial\tau}\right)_\delta + \left(\frac{\partial\alpha^r}{\partial\tau}\right)_\delta\right] - \alpha^0 - \alpha^r \tag{1.34}$$

1.3.3.4 Computing Non-measurable Properties (*u*, *h*, and *s*) in the Specific Case of Ideal Fluids
The equations developed in Section 1.3.3.3.1 simplify in the case of ideal behavior.

1.3.3.4.1 Ideal Gas An ideal gas is a gas that obeys the equation of state given in Eq. (1.21). The differential of $h(T,P)$, $u(T,v)$, and $s(T,P)$ reduces to

$$dh = c_p dT \tag{1.35}$$

$$du = c_v dT \tag{1.36}$$

$$ds = \frac{c_p}{T}dT - \frac{r}{P}dP \tag{1.37}$$

The two first equations indicate that the enthalpy and internal energy depend only on the temperature. This is true since, for ideal gases, c_p and c_v are function of only T. This can be demonstrated by expressing the exactness of both expressions of the total differential ds (Eq. (1.29)) developed previously (Chapter 12 of Çengel and Boles (2006)).

Therefore, the variations of specific enthalpy and internal energy between two temperatures T_1 and T_2 are expressed by the relations:

$$\Delta h = h_2 - h_1 = \int_{T_1}^{T_2} c_p(T)\,dT \tag{1.38}$$

$$\Delta u = u_2 - u_1 = \int_{T_1}^{T_2} c_v(T)\,dT \tag{1.39}$$

If the specific heat at constant pressure and the specific heat at constant volume are assumed constant (possibly as a first approximation, assuming a linear dependency with respect to temperature, the values calculated at the average temperature between T_1 and T_2 can be considered), the previous relations reduce to

$$\Delta h = h_2 - h_1 = c_p \, (T_2 - T_1)$$ (1.40)

$$\Delta u = u_2 - u_1 = c_v \, (T_2 - T_1)$$ (1.41)

Regarding the variation of the specific entropic, still considering a constant value of c_p, the integration of the differential ds yields

$$\Delta s = c_p \ln \frac{T_2}{T_1} - r \ln \frac{P_2}{P_1}$$ (1.42)

Previous equation is particularly useful to derive relations between pressure and temperature in the case of an isentropic process ($\Delta s = 0$). Assuming that the specific heats are constant (the average values can be considered) and recognizing that $r = c_p - c_v$, it yields

$$\frac{T_{2,s} \, [\mathrm{K}]}{T_1 \, [\mathrm{K}]} = \left(\frac{P_2}{P_1} \right)^{(\gamma-1)/\gamma}$$ (1.43)

where $\gamma = c_p/c_v$ is the specific heat ratio.

1.3.3.4.2 *Incompressible Fluid*
Liquids (but also solids) often behave as incompressible substances, meaning that a very large pressure difference is required to observe a variation of the specific volume. In that case, the specific volume is a function of only temperature ($v = v(T)$). For incompressible substances, the specific heats at constant pressure and volume are equal ($c_v(T) = c_p(T) = c(T)$). This can be demonstrated using the Mayer relationship (Çengel and Boles, 2006). Similar to ideal gases, the dependence of c only on T can be demonstrated by expressing the exactness of the differential ds.

In the ideal case, the specific volume is constant. An average specific volume can also be considered in the range of temperature variations considered. In both cases, the specific volume is constant and assuming a constant specific heat, the differentials of specific enthalpy, specific internal energy, and specific entropy reduce to

$$dh = c \, dT + v \, dP$$ (1.44)

$$du = c \, dT$$ (1.45)

$$ds = \frac{c}{T} dT$$ (1.46)

As for ideal gases, the specific internal energy is a function of only temperature ($u = u(T)$). That is,

$$\Delta u = c \, (T_2 - T_1)$$ (1.47)

However, the variation of the enthalpy depends also on the variation of pressure, even though the second term can be often neglected

$$\Delta h = c \, (T_2 - T_1) + v \, (P_2 - P_1)$$ (1.48)

Also,

$$\Delta s = c \ln \frac{T_2 [K]}{T_1 [K]}$$ (1.49)

Therefore, in the case of an incompressible liquid undergoing an isentropic evolution, the temperature remains constant. For instance, assuming an isentropic compression from P_1 to P_2 of an ideal liquid in a pump, the temperature would not increase, and the specific work would be equal to

$$w_s = \Delta h_s = v(P_2 - P_1) \tag{1.50}$$

1.3.4 Fluids Commonly Used in Automotive Applications

1.3.4.1 Oil

Lubricating oil is mainly composed of hydrocarbons plus some additives. It can be considered as an incompressible liquid with density and specific heat, which are functions of only temperature. The following correlations can be used in the range of temperatures $[-20; 200\,°C]$

$$\rho\,[\text{kg m}^{-3}] = 895.75 - 0.65 \cdot T\,[°C] \tag{1.51}$$

$$c[\text{J kg}^{-3}\text{K}^{-1}] = 1787.33 + 3.60125 \cdot T[°C] \tag{1.52}$$

1.3.4.2 Coolant

As described later in this book, many components inside the vehicle must be cooled down, and liquid water is a convenient fluid. However, it has the drawback to freeze if its temperature is decreased below 0°C. This is likely to happen if the vehicle is at a standstill under negative outdoor temperatures. To decrease the freezing point of the coolant, an antifreeze agent is mixed with water. Antifreeze agents are typically ethylene glycol or propylene glycol.

Figure 1.12 shows the evolution of the freezing point of an aqueous solution of ethylene glycol with the mass concentration of glycol in the solution. For instance, the freezing point of a 50% in mass aqueous solution of ethylene glycol is −36°C.

Adding ethylene glycol in water also increases the boiling point. For instance, the boiling point of a 50% in mass aqueous solution of ethylene glycol is close to 107°C.

However, adding glycol in water decreases the thermal capacity and the thermal conductivity of the coolant and increases its viscosity. This increases pump consumption and decreases heat exchanger performance (lower convective heat transfer coefficient than water).

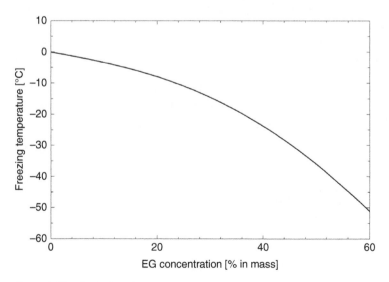

Figure 1.12 Evolution of the freezing temperature of an aqueous solution of ethylene glycol with the mass concentration of glycol.

The following correlations can be used to assess the density and specific heat of an aqueous solution of ethylene glycol 50% in mass as a function of temperature in the range $[-20; 100\,°C]$

$$\rho\,[\text{kg m}^{-3}] = 1074.91 - 0.597356 \cdot T\,[°C] \tag{1.53}$$

$$c[\text{J kg}^{-3}\text{K}^{-1}] = 3204.02 + 5.72974 \cdot T[°C] - 0.012805 \cdot T^2[°C] \tag{1.54}$$

1.3.4.3 Refrigerant

In 1989, the Montreal Protocol banned fluids using chlorine, such as CFC (chlorofluorocarbon) R12 used in automotive applications. R12 was replaced by HFC (hydrofluorocarbon) R134a. Until recently, refrigerant R134a has been largely used in automotive air conditioners. However, this refrigerant shows a large global warming potential (GWP) of 1430. The GWP is a measure of heat trapped in the atmosphere by a given mass of greenhouse gas over a given time horizon relative to the heat trapped by the same mass of carbon dioxide. Recent European regulations have forbidden the use of refrigerants with GWP larger than 150 in all vehicles. Consequently, HFC R134a has been phased out and is replaced by HFO (hydrofluoroolefin) R1234yf in most of the vehicles. This fluid shows a very low GWP of 4. It has been classified as mildly flammable.

Carbondioxide (R744) is another replacement fluid for R134a. This natural fluid has a GWP of 1. Its major thermophysical properties are given in Table 1.2. R744 shows the major advantages of being nontoxic and nonflammable. The use of CO_2 in refrigerators is not new: it has already been used in the early times of artificial refrigeration. However, there is renewed interest for R744 in automotive air-conditioners since the 1990s. R744 is currently used in domestic and industrial vapor compression machines. Because of its low critical temperature, R744 cycles used in A/C loops operate in transcritical regime. In such a regime, the compressor discharge pressure is larger than the critical pressure. This will be discussed more in detail in Chapter 3.

1.3.4.4 Humid Air

Air is involved in many components encountered in automotive thermal management (air-cooled condenser, air-conditioning evaporator, radiator, air circulation inside the cabin, etc.). This air is atmospheric air and contains not only oxygen and nitrogen but also water in vapor state. It is called humid air. The mixture of air and water vapor is a gas–vapor mixture, meaning that the water vapor is close to saturation and may condense during processes. For instance, water condensation may occur in an air-conditioning evaporator when the air temperature is decreased. It may also occur on the glazing surface when the glazing surface temperature is lower than that of the air. The condition for condensation is that the air temperature is decreased down to the dew point temperature, defined hereunder.

To describe the properties of humid air, it is assumed that the air and water behave as two ideal gases. As a consequence, the total pressure is the sum of the partial pressure of air and the partial pressure of water vapor. The latter is also called the *vapor pressure*. That is,

$$P = P_a + P_w \tag{1.55}$$

Table 1.2 Major characteristics of mobile A/C refrigerants.

	HFC-134a	HFO-1234yf	R744
Molecule	$C_2F_4H_2$	$C_3F_4H_2$	CO_2
Critical temperature [°C]	101	94.7	30.98
Critical pressure [bar]	40.59	33.82	73.77
Molar mass [g mol^{-1}]	102	114	44.01
100-year GWP	1430	4	1

Note that the total pressure is the atmospheric pressure if the humid air is at the atmospheric pressure. It could be higher if humid air is compressed.

Being ideal gases, both the air and water vapor can be described by the equation of state (1.21), paying attention that quantity r is not the same for air and water.

In engineering problems involving humid air, it is useful to quantify the content of water in the mixture as well as to express energy and mass balances across components. The quantities introduced in the following sections will be of primary importance.

1.3.4.4.1 Specific Humidity *Specific humidity* ω (also called *absolute humidity* or *humidity ratio*) is the mass of water vapor m_w [kg] per kilogram of dry air m_a [kg]. It can be expressed as a function of the vapor pressure P_w and total pressure P by

$$\omega = \frac{m_w}{m_a} = \frac{0.622 \, P_w}{P - P_w} \tag{1.56}$$

Humidifying the air increases its specific humidity, while drying the air decreases its specific humidity.

1.3.4.4.2 Relative Humidity Dry air contains no moisture. Hence, its specific humidity is equal to 0. As long as the air absorbs water vapor, its specific humidity increases. Simultaneously, its vapor pressure P_w increases until it reaches the saturation pressure of water $P_g(T)$ corresponding to atmospheric air temperature T. When this pressure is reached, the air cannot absorb any additional moisture, and it is said to be saturated. Any additional water vapor injection in the air would result in condensation. If m_g is the maximum mass of water vapor that the atmospheric air at temperature T can hold, the relative humidity is defined as the ratio of the actual mass of water vapor m_w to m_g.

$$RH = \frac{m_w}{m_g} = \frac{P_w}{P_g(T)} \tag{1.57}$$

Hence, the relative humidity of dry air is equal to 0, and the relative humidity of saturated air is equal to 1. It is important to observe that the relative humidity depends on the air temperature. Increasing the air temperature decreases its relative humidity, while its specific humidity is constant. Decreasing the air temperature increases its relative humidity, while its specific humidity is constant (as long as no water condensation occurs). This observation is of primary importance when investigating windshield misting and demisting mechanisms.

It should also be noted that the thermal comfort is affected by the relative humidity. The latter can be controlled by adjusting the air temperature or by adding or removing moisture in the air.

By combining previous equations, the specific humidity can be related to the relative humidity by

$$\omega = \frac{0.622 \, RH \, P_g(T)}{P - RH \, P_g(T)} \tag{1.58}$$

The relative humidity can be directly measured by relative humidity sensors. This measurement must be combined with the measurement of the temperature and total pressure to determine the specific humidity.

1.3.4.4.3 Dew-Point Temperature As explained previously, when atmospheric air is progressively cooled down at a constant pressure, its relative humidity increases. The temperature at which air becomes saturated is the dew-point temperature. The dew-point can be directly measured by means of a dew-point meter. This delicate measurement consists of observing the apparition of condensates on the surface of a mirror that is progressively cooled down.

At the dew-point temperature, observing that RH = 1, Eq. (1.58) becomes

$$\omega = \frac{0.622\ P_g(T_{dp})}{P - P_g(T_{dp})} \tag{1.59}$$

Equation (1.59) indicates that only the measurement of the total pressure P must be associated with the measurement of the dew-point temperature to determine the specific humidity of the air. The dew-point measurement is hence a highly accurate measurement technique of the moisture content of the air.

1.3.4.4.4 Wet-Bulb Temperature

The wet-bulb temperature is another indicator of the moisture content of the air. The wet-bulb temperature is the temperature read by a thermometer the bulb of which is covered by a wetted wick. The wet-bulb thermometer is often part of the psychrometer, which consists of two thermometers whose bulbs are in contact with the same air stream. Such an apparatus is illustrated in Figure 1.13. The first thermometer measures the dry-bulb temperature, which is the atmospheric air temperature. The second thermometer measures the wet-bulb temperature. The bulb of such a thermometer is covered by a wick wetted with water. As the air flows through the wick, water from the wick vaporizes, leading to saturated air leaving the wick (if original air is unsaturated). The heat of vaporization of water is taken from the water, yielding a decrease in water temperature inside the wick. Ultimately, because of the temperature gradient, the air temperature decreases down to the wet-bulb temperature, which is read by the thermometer. The larger the difference between the dry-bulb and the wet-bulb temperature, the drier the air is. If the air were saturated, both thermometers would read the same temperature. In that specific case, the dry-bulb temperature, the wet-bulb temperature, and the dew-point temperature are equal.

This measurement is close to the adiabatic saturation temperature (or thermodynamic wet-bulb). The latter is the temperature of air that would be obtained if the air was brought to saturation by vaporizing water at a temperature equal to that of the air after saturation.

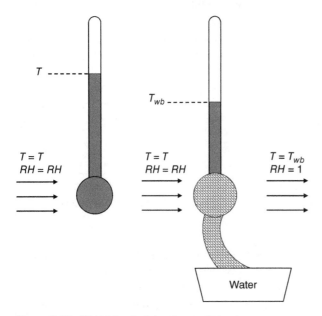

Figure 1.13 Working principle of a psychrometer.

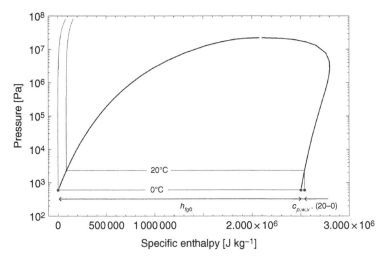

Figure 1.14 Visualization of the specific enthalpy of water vapor in the pressure–enthalpy diagram.

1.3.4.4.5 Specific Enthalpy Having an expression of enthalpy of humid air is very convenient for expressing energy balances across open components involving humid air. The specific enthalpy h [J kg^{-1}] of humid air is given by

$$h = h_a + \omega h_w \tag{1.60}$$

Taking 0°C as the reference temperature, the enthalpy of dry air is given by

$$h_a = c_{p,a}T = 1005\,T \tag{1.61}$$

Taking saturated liquid at 0°C as the reference state, the specific enthalpy of water vapor is given by

$$h_w = h_{fg,0} + c_{p,w,v}\,T = 2500561 + 1820\,T \tag{1.62}$$

This equation assumes that water vapor behaves as an ideal gas, and hence, its enthalpy is function of only temperature. Figure 1.14 illustrates the computation of the water vapor enthalpy for a temperature of 20°C. It can be observed that the iso-thermal line associated with 20°C is quasi vertical in the vapor phase, indicating that the enthalpy is not a function of pressure.

1.3.4.4.6 Psychrometric Diagram The thermodynamic state of humid air can be conveniently visualized in the psychrometric diagram. Such a diagram is represented in Figure 1.15 for a given total pressure. The x- and y-axes indicate the dry-bulb and the specific humidity, respectively. Lines of constant relative humidity and lines of constant wet-bulb temperature are drawn. The diagram is delimited on the top by the thicker line associated with $RH = 1$ (saturation state).

1.3.4.4.7 Typical Processes Encountered by Moist Air Among others, the psychometric chart is commonly used to represent air-conditioning processes and any other processes characterized by moisture content variation. As shown in Figure 1.16, these processes include heating, cooling, humidification, and dehumidification processes or combinations of some of them. Examples of applications to vehicles are as follows: a is the humidity increase by water vapor produced by occupants (if their sensible heat is neglected), c represents air heating at windshield position, f represents the air evolution in the A/C loop evaporator (if water condensation occurs), and h is the adiabatic cooling of the air produced by water pulverization in the air.

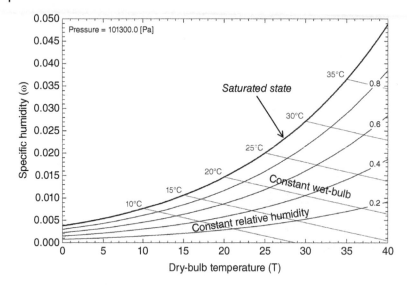

Figure 1.15 Example of a psychrometric diagram.

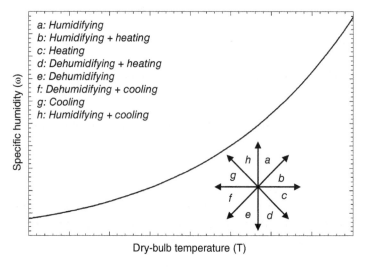

Figure 1.16 Basic processes in the psychrometric diagram. Source: Reproduced and adapted from Çengel and Boles (2006).

The energy analysis of such processes involves expressing the energy balance as well as mass balance (in the case of an open system). The mass balance is split into both water mass balance and dry air mass balance.

1.4 Heat Transfers

This textbook will largely illustrate heat transfers between bodies. The purpose of this section is to better understand the physical mechanisms associated with heat transfers.

The thermal energy of a body seen as a thermodynamic system corresponds to the kinetic energy at the microscopic scale (molecules, atoms, free electrons, and nucleons) plus the latent energy. These microscopic particles can exchange part of their kinetic energy with the system surroundings, leading to an increase or a decrease in the thermal energy of the body.

When two bodies are at temperatures that are different, the hottest body exchanges thermal energy with the coldest one. A heat transfer process occurs between the two bodies. This heat transfer process can be characterized by its intensity and direction.

Three mechanisms, corresponding to different physical phenomena, can explain the heat transfer processes described in this book:

- Conduction
- Convection
- Radiation

In this section, the main mechanisms of heat transfer are described. Heat exchangers, which ensure the function of heat transfer, will be described in a future Section 1.8 of this chapter.

1.4.1 Conduction

Heat transfer by conduction occurs within the body and is characterized by its thermal conductivity k, [W/m · K]. The thermal conductivity is the capacity of the body material to conduct heat.

Conduction of heat results from the *"transfer of energy from the more energetic to the less energetic particles of a substance due to interactions between the particles"* (Incropera and DeWitt, 2002), without any bulk motion of the body. At the microscopic scale, moving or vibrating particles transfer their kinetic and potential energies to the neighboring particles. In gases and liquids, the transfer of energy results from the collision between the molecules. In solids, this transfer results from the oscillation of the lattice due to the movement of atoms and, for solid conductors, from the translational motion of free electrons (Incropera and DeWitt, 2002).

Conduction is the main mechanism of heat transfer within solid bodies. In solid conduction, as mentioned above, thermal energy is transferred from one molecule to another by vibration. A similar mechanism occurs in liquids and gases. However, in the latter phases, heat transfer by conduction is often negligible with respect to convection.

Conduction within a body only occurs when a temperature gradient is imposed within the body under an external action.

The inverse of thermal conductivity is thermal resistivity. For the same temperature gradient, materials characterized by higher thermal conductivity (i.e. low thermal resistivity) conduct more heat than materials characterized by lower thermal conductivity (i.e. high thermal resistivity). Materials characterized by large thermal resistivity are insulating materials.

In the steady-state regime, the heat conduction can be described by Fourier's law, which relates the heat flux through a surface to the temperature gradient. The differential and one-dimensional form of Fourier's law can be expressed as

$$q_x = -k\frac{dT}{dx} \tag{1.63}$$

where

q_x is the heat flux in the x-direction (normal to the cross-sectional surface), [W m^{-2}]

k is the material thermal conductivity, [W m^{-1} K^{-1}]

dT/dx is the temperature gradient, [K m^{-1}].

The minus sign indicates that the heat flux and temperature gradient have opposite directions. Assuming that the temperature is constant over the surface (as explained in Incropera and DeWitt

(2002), the heat flow direction is perpendicular to an isothermal surface), and integrating the previous equation over the surface gives

$$\dot{Q} = -k A \frac{dT}{dx} \tag{1.64}$$

where

$\dot{Q}$ is the heat flow rate across surface A, [W]
A is the cross-sectional area normal to the heat flow, [m^2]

Applying Fourier's law of conduction to a homogeneous 1-D wall, whose surface area is A [m^2] and thickness is e [m], gives the following relation to express the heat flow rate that is transferred through the wall

$$\dot{Q} = A \cdot \frac{k}{e} \cdot (T_1 - T_2) \tag{1.65}$$

where

$\dot{Q}$ is the heat transfer rate, [W]
A is the wall surface area, [m^2]
e is the wall thickness, [m]
k is the thermal conductivity, [W m^{-1} K^{-1}]
$\frac{e}{k}$ is the thermal resistivity of the material for a thickness e, [m^2 K W^{-1}]
T_1 is the temperature of the hottest side of the wall, [°C]
T_2 is the temperature of the coldest side of the wall, [°C]

Expression (1.65) will be used, for instance, in Chapter 4, when describing the heat transfer through the casing of a battery module.

1.4.2 Convection

This mechanism of heat transfer is associated with mass transfer. More exactly, there are two super-posed mechanisms that explain the energy transfer: the random molecular motion (also called diffusion) and the bulk motion of the fluid (also called advection). Convection heat transfer between a fluid in motion and a surface occurs when the fluid is at temperature T_f different from surface temperature T_s.

1.4.2.1 Forced Convection
In the forced convection regime, the fluid motion results from a pressure difference created by an external mechanism. This mechanism can be a fan or a pump, which typically forces fluid (such as air, coolant, or oil) to flow through a pipe or a heat exchanger. The motion of a body in a fluid also creates a fluid flow around the body surfaces. This is, for instance, the case of a moving vehicle whose surfaces are in contact with airflows.

1.4.2.2 Natural Convection
In the natural convection regime (also called free convection), the fluid motion is induced by buoyancy forces caused by density differences resulting from temperature differences within the fluid. Actually, as a liquid or gas is heated, its density decreases. Due to buoyancy forces, the heated fluid moves up, transferring its heat to colder regions. As the fluid is cooled down, its density increases, and the fluid moves down. A convection cell is created. The fluid velocity within the cell is quite low.

1.4.2.3 Mixed Forced and Natural Convection

In mixed forced and natural convection regime, the flow induced by pressure difference and the flow induced by temperature difference are of similar magnitudes.

1.4.2.4 Sensible and Latent Heat Transfer by Convection

A phase change can create fluid motion. For instance, if a fluid is heated until it vaporizes, bubbles are formed and move upwards due to buoyancy forces. In that case, latent heat is transferred by convection. Condensation of a vapor also involves latent heat transfer. Without phase change, only the sensible heat is transferred.

1.4.2.5 Convection Heat Transfer Rates

Regardless of whether sensible or latent heat is transferred, the general equation to compute the heat transfer by convection has the following form:

$$q = h(T_s - T_f) \tag{1.66}$$

where

q is the heat flux, [W m^{-2}]
h is the convection heat transfer coefficient, [W m^{-2} K^{-1}]
T_s is the surface temperature, [°C]
T_f is the fluid temperature, [°C].

The heat transfer rate by convection $\dot{Q}[W]$ can be obtained by multiplying the heat flux by surface area A. Therefore,

$$\dot{Q} = A\,h\,(T_s - T_f) \tag{1.67}$$

1.4.2.6 Laminar and Turbulent Regimes

Irrespective of the type of convection, the distinction can be done between:

- The laminar regime in which the fluid motion is ordered. The fluid follows in a streamline manner.
- The turbulent regime where the fluid motion is irregular. Any direction can be taken by the fluid.

The Reynolds number, which is a dimensionless number, is used to determine whether the flow is laminar or turbulent. At low Re ($Re < 2300$ for fully developed flows in pipes), the flow is laminar. At high Re ($Re > 2900$ for fully developed flows in pipes), the flow is turbulent. The Reynolds number is defined as

$$Re = \frac{\rho\,C\,L}{\mu} \tag{1.68}$$

where

ρ is the density of the fluid, [kg m^{-3}]
C is the speed of the fluid, [m s^{-1}]
L is a characteristic dimension, [m]
μ is the dynamic viscosity of the fluid, [kg m^{-1} s^{-1}].

In the particular case of a flow in a pipe, the characteristic dimension is the hydraulic diameter of the pipe. Often, the natural convection is laminar and the forced convection is turbulent.

1.4.2.7 Convection Heat Transfer Coefficients

Convection heat transfer coefficient h depends on many parameters: the characteristics of the fluid, the nature and velocity of flow, the temperature of the fluid and surface, and the heat transfer surface geometry. There is an exhaustive literature providing correlations for assessing convective heat transfer coefficients involving dimensionless numbers (Reynolds, Prandtl, Grashof, etc.).

Order of magnitude of convection heat transfer coefficients is given in Table 1.3.

1.4.3 Radiation

Radiation is the third mechanism of heat transfer. Atoms and molecules of matter emit energy as electromagnetic waves transmitted at the speed of around $300\,000\,\text{km s}^{-1}$. Unlike conduction and convection, energy transfer by radiation does not require any medium. Hence, a body in vacuum emits energy by radiation, while it can transfer energy neither by conduction nor convection.

The distinction must be done between the radiation that is emitted by the surface of a body and the radiation that is incident to a body.

1.4.3.1 Emitted Radiation

The rate of energy emitted by a surface per unit area of surface is called the *emissive power* E [W m^{-2}]. For a blackbody, which is not only a perfect emitter but also absorber (it absorbs any incident radiation), the emissive power is given by the Stefan–Boltzmann law

$$E_b = \sigma\, T_s^4 \tag{1.69}$$

where

σ is the Stefan–Boltzmann constant, $(\sigma = 5.67 \times 10^{-8}\,\text{W m}^{-2}\,\text{K}^{-4})$
T_s is the blackbody absolute temperature, [K]

A real surface emits less energy than a blackbody at the same temperature. The ratio of the emissive power of a real surface to that of a blackbody at the same temperature is the surface *emissivity* ε. Therefore,

$$E = \varepsilon\, \sigma\, T_s^4 \tag{1.70}$$

Table 1.3 Typical values of convection heat transfer coefficients.

Process	Convection heat transfer coefficient h [W m^{-2} K^{-1}]
Free convection	
Gases	2–25
Liquids	50–1 000
Forced convection	
Gases	25–250
Liquids	100–20 000
Convection with phase change	
Boiling or condensation	2500–100 000

Source: Incropera and DeWitt, 2002.

The emissivity is a function of the material and of the finish of the surface (Incropera and DeWitt, 2002). In the previous equation, ε stands for the *total hemispherical emissivity*. Actually, a surface emits energy by radiation through electromagnetic waves of different wavelengths and in all possible directions. The emissivity varies with both the wavelength λ and the direction.

The *spectral hemispherical emissive power* E_λ [W m^{-2}μm^{-1}] is defined as *"the rate at which radiation of wavelength λ is emitted in all directions from a surface per unit wavelength interval $d\lambda$ about λ and per unit surface area"* (Incropera and DeWitt, 2002). Therefore,

$$E = \int_0^\infty E_\lambda \, d\lambda \tag{1.71}$$

The same equation applies for a blackbody, whose radiant energy also comprises electromagnetic waves at different wavelengths. Note that a blackbody is a diffusive emitter, which means that the radiation it emits is independent of the direction. For a blackbody, the spectral distribution of the emissive power is given by Planck's law

$$E_{\lambda,b} = \frac{3.742 \times 10^8}{\lambda^5 (\exp(1.439 \times 10^4 / \lambda T_s) - 1)} \; [\text{W m}^{-2}\text{μm}^{-1}] \tag{1.72}$$

where

λ is the wavelength, [μm]

Wien's displacement law states that, for a blackbody at temperature T_s, there exists a wavelength λ_m for which the spectral emissive power is maximum. The relation between temperature T_s [K] and wavelength λ_m is given by

$$\lambda_m T_s = 2897.8 \, \text{μm K} \tag{1.73}$$

Figure 1.17 shows the spectral distribution of the emissive power divided by the maximum emissive power for two blackbodies: the sun (which can be seen as a blackbody at 5760 K) and a blackbody at room temperature. It can be seen that the two spectra are clearly distinct: they go through maxima for two distinct wavelengths. The range of wavelengths associated with the room temperature blackbody corresponds to "longwave infrared" radiation. The solar radiation spectrum comprises ultraviolet, visible, and near-infrared regions. This will be discussed in Chapter 3.

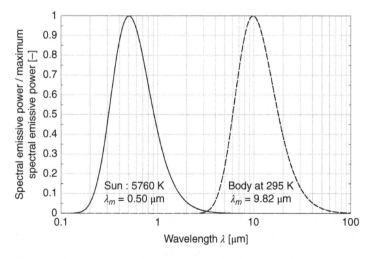

Figure 1.17 Spectral distribution of the ratio of the emissive power to the maximum emissive power for the sun (which can be viewed as a blackbody at 5760 K) and a blackbody at room temperature.

Figure 1.17 explains the greenhouse effect, which contributes to the heating of car radiated by the sun: glazing is transparent to solar radiation but opaque to the longwave infrared radiation emitted by the seats, the dashboard, and other surfaces inside the car cabin, which tends to "trap" the solar energy inside the cabin.

The emissive power of a real surface can be related to the spectral emissive power of a black body by

$$E = \int_0^\infty \varepsilon_\lambda E_{\lambda,b} \, d\lambda \tag{1.74}$$

This gives a relation between the total hemispherical emissivity ε [−] and the spectral hemispherical emissivity ε_λ [−]. These two quantities are related by

$$\varepsilon = \frac{1}{\sigma \, T_s^4} \int_0^\infty \varepsilon_\lambda E_{\lambda,b} \, d\lambda \tag{1.75}$$

1.4.3.2 Incident Radiation

The rate at which radiation is incident on a surface *"per unit area from all directions and at all wavelengths"* (Incropera and DeWitt, 2002) is called the *total irradiation G* [W m^{-2}]. The irradiation is thus the intensity of the incident radiation. *Spectral irradiation G_λ* [W m^{-2}μm^{-1}] is *"the rate at which radiation of wavelength λ is incident on a surface per unit area of the surface and per unit wavelength interval $d\lambda$ about λ"* (Incropera and DeWitt, 2002). Total and spectral irradiations are related to each other by

$$G = \int_0^\infty G_\lambda(\lambda) \, d\lambda \tag{1.76}$$

When radiant energy is incident to a surface, it can be absorbed, transmitted (if the surface is semitransparent), or reflected. Hence, the spectral irradiation can be split as

$$G_\lambda = G_{\lambda,\alpha} + G_{\lambda,\tau} + G_{\lambda,\rho} \tag{1.77}$$

The spectral hemispherical absorptivity α_λ is defined as the fraction of the spectral irradiation that is absorbed by the surface

$$\alpha_\lambda = \frac{G_{\lambda,\alpha}}{G_\lambda} \tag{1.78}$$

The total hemispherical absorptivity α is defined as the fraction of the total irradiation that is absorbed by the surface

$$\alpha = \frac{G_\alpha}{G} \tag{1.79}$$

Spectral and total transmissivity and reflectivity can be defined in the same way. The spectral absorptivity, transmissivity, and reflectivity are related by

$$1 = \alpha_\lambda + \tau_\lambda + \rho_\lambda \tag{1.80}$$

Averaging these properties over the full spectrum of wavelengths, it follows

$$1 = \alpha + \tau + \rho \tag{1.81}$$

Note that the latter two equations are valid for semitransparent surfaces. In the case of an opaque surface, $\tau = 0$. These thermal radiative properties will be extensively used in Chapter 3 when describing radiative exchanges with vehicle glazing and opaque body.

1.4.3.3 Kirchhoff's Law and the Gray Surfaces

The most general Kirchhoff's law states that

$$\varepsilon_{\lambda,\theta} = \alpha_{\lambda,\theta} \tag{1.82}$$

with no restriction on spectral and directional properties.

A surface is said to be *diffuse* if $\varepsilon_{\lambda,\theta}$ and $\alpha_{\lambda,\theta}$ are independent of direction θ. In that case,

$$\varepsilon_{\lambda} = \alpha_{\lambda} \tag{1.83}$$

A surface is said to be gray if ε_{λ} and α_{λ} are independent of λ. Hence (Incropera and DeWitt, 2002), for a gray and diffuse surface, Kirchhoff's law reduces to

$$\varepsilon = \alpha \tag{1.84}$$

It should be stressed that a surface could be gray in a limited spectral region, comprising irradiation and emission, where surface spectral properties are constant.

1.4.3.4 Radiation Exchange Between Surfaces

The description of vehicle thermal management systems involves radiative heat exchanges between surfaces at different temperatures, with different radiative properties, and at different orientations. A typical example is the radiative exchanges between surfaces inside the vehicle cabin, which largely impacts the air and surface temperatures and therefore the occupants' comfort. This example will be treated in Chapter 3.

This section will only treat the radiative exchange between two surfaces. To learn a more exhaustive approach to treat exchange between more than two surfaces, the reader is invited to refer to Incropera and DeWitt (2002) textbook.

Some specific cases are described in the following paragraphs and should allow to treat most of radiative heat transfer problems introduced in future chapters.

1.4.3.4.1 *General Case of Radiation Exchange Between two Diffuse and Gray Surfaces* Let's imagine an enclosure made of two diffuse and gray surfaces. The net rate of radiation leaving surface *1* is $\dot{Q}_1$. The latter can be seen as the heat transfer rate to provide to surface at T_1 to maintain its temperature constant (Incropera and DeWitt, 2002). Since there are only two surfaces, $\dot{Q}_1$ must be equal to $\dot{Q}_2$. The latter heat flow rate is the net radiative rate arriving on surface 2. It can also be written $\dot{Q}_{12}$ to express that the net rate of radiation leaving surface 1 is transferred to surface 2.

It can be demonstrated (Incropera and DeWitt, 2002) that the net radiative exchange between the two surfaces can be computed by

$$\dot{Q}_{12} = \frac{\sigma \left(T_1^4 - T_2^4 \right)}{\frac{1-\varepsilon_1}{\varepsilon_1 A_1} + \frac{1}{A_1 F_{12}} + \frac{1-\varepsilon_2}{\varepsilon_2 A_2}} \tag{1.85}$$

Equation (1.85) introduces the view factor F_{12}. The view factor is "*the ratio of radiant energy leaving a surface i and directly incident on a surface k to the total radiant energy leaving i in all directions*" (Incropera and DeWitt, 2002). A very convenient relation, named the reciprocity relation, states that

$$F_{ik} A_i = F_{ki} A_k \tag{1.86}$$

From Eq. (1.85), simplified expressions can be derived for specific cases listed hereunder.

1.4.3.4.2 Radiation Exchange Between two Parallel Diffuse and Gray Surfaces Since $F_{12} = 1$ and $A_1 = A_2$, Eq. (1.85) reduces to

$$\dot{Q}_{12} = \frac{\varepsilon_1 \varepsilon_2 A_1 \sigma \left(T_1^4 - T_2^4\right)}{\varepsilon_1 - \varepsilon_1 \varepsilon_2 + \varepsilon_2} \tag{1.87}$$

Such heat transfer occurs typically between the two layers of a multilayer material if they are separated by a nonparticipating gas, such as the surfaces of a wall cavity (Braun and Mitchell, 2012).

1.4.3.4.3 Radiation Exchange Between a Small and Convex Diffuse and Gray Surface 1 Surrounded by a Large Diffuse and Gray Surface 2 In this case, $F_{12} = 1$ and $A_1/A_2 \ll$. Therefore, Eq. (1.85) reduces to

$$\dot{Q}_{12} = \varepsilon_1 A_1 \sigma \left(T_1^4 - T_2^4\right) \tag{1.88}$$

The previous equation to compute the net rate of radiation heat transfer between surface 1 surrounded by surface 2 seen as its surroundings can be linearized in the following way

$$\dot{Q}_{12} = A_1 h_r \left(T_1 - T_2\right) \tag{1.89}$$

with

$$h_r[\text{Wm}^{-2}] = \varepsilon_1 \sigma \left(T_1 + T_2\right) \left(T_1^2 + T_2^2\right) \tag{1.90}$$

1.4.3.4.4 Radiation Exchange Between a Diffuse and Gray Surface 1 and a Blackbody 2 Assuming surface 2 to be a blackbody, $\varepsilon_2 = 1$. That is

$$\dot{Q}_{12} = \varepsilon_1 A_1 F_{12} \sigma \left(T_1^4 - T_2^4\right) \tag{1.91}$$

If surface 2 surrounds surface 1, then $F_{12} = 1$ and it follows that

$$\dot{Q}_{12} = \varepsilon_1 A_1 \sigma \left(T_1^4 - T_2^4\right) \tag{1.92}$$

1.4.3.4.5 Radiation Exchange Between Two Blackbodies In this case, ε_1 and ε_2 are equal to unity. It follows that

$$\dot{Q}_{12} = A_1 F_{12} \sigma \left(T_1^4 - T_2^4\right) \tag{1.93}$$

1.5 First Law of Thermodynamics

The First Law of Thermodynamics states that energy can be neither created nor destroyed, but only converted from one form to another. Therefore, the First Law of Thermodynamics is the principle of conservation of energy. In its more general form, it states that the variation of the total energy of a system ΔE_{system} over a process is equal to the total energy E_{su} entering the system minus the total energy E_{ex} leaving the system during the process.

$$E_{su} - E_{ex} = \Delta E_{system} = \Delta U_{system} + \Delta KE_{system} + \Delta PE_{system} \tag{1.94}$$

Often, the variation of the kinetic energy ΔKE_{system} and the variation of potential energy ΔPE_{system} of the system can be neglected. The former assumption is valid if the system is at rest or if it evolves at a constant speed. The latter assumption is valid if the altitude of the system is constant. The variation of the total energy reduces to the variation of internal energy ΔU_{system}.

A system can exchange energy with its surroundings under different forms:

Heat Q

- Work W: it could be, among others, a mechanical work or an electrical work.
- Mass m: any mass stream entering or leaving the system contains energy equal to the enthalpy of the flow plus its kinetic energy and potential energy.

Therefore, the previous equation can be written as

$$(Q_{su} - Q_{ex}) + (W_{su} - W_{ex}) + \sum_{su} m\left(h + \frac{C^2}{2} + gz\right) - \sum_{ex} m\left(h + \frac{C^2}{2} + gz\right) = \Delta U_{system} \quad (1.95)$$

where

Q_{su} is the heat transferred to the system from its surroundings, [J]

Q_{ex} is the heat transferred from the system to its surroundings, [J]

W_{su} is the work transferred to the system from its surroundings, [J]

W_{ex} is the work transferred from the system to its surroundings, [J]

h is specific enthalpy of each mass stream entering/leaving the system, [J kg^{-1}]

$C^2/2$ is the kinetic energy of each mass stream entering/leaving the system, [J kg^{-1}]

gz is the potential energy of each mass stream entering/leaving the system, [J kg^{-1}]

$h + C^2/2 + gz$ is the total energy of each mass stream entering/leaving the system, [J kg^{-1}]

ΔU_{system} is the variation of internal energy of the system, [J]

The previous equation can be written in its rate form:

$$(\dot{Q}_{su} - \dot{Q}_{ex}) + (\dot{W}_{su} - \dot{W}_{ex}) + \sum_{su} \dot{m}\left(h + \frac{C^2}{2} + gz\right) - \sum_{ex} \dot{m}\left(h + \frac{C^2}{2} + gz\right) = \frac{dU_{system}}{dt}$$

$$(1.96)$$

The rate of work can be, among others, a mechanical power (for instance, a shaft power) or an electrical power.

Equations (1.95) and (1.96) are of paramount importance for evaluating the performance of any component or system in vehicle thermal management.

1.5.1 Closed System

As explained previously, a closed system (or "control mass") is a system that does not exchange mass with its surrounding. As a consequence, the mass of such a system is constant. In closed systems, the only forms of energy transfer are heat and work. The First Law of Thermodynamics for closed systems is therefore:

$$(\dot{Q}_{su} - \dot{Q}_{ex}) + (\dot{W}_{su} - \dot{W}_{ex}) = \frac{dU_{system}}{dt} \quad (1.97)$$

In the particular case of a closed system undergoing a cycle, the initial and final states of the system are equal. Hence,

$$(\dot{Q}_{su} - \dot{Q}_{ex}) + (\dot{W}_{su} - \dot{W}_{ex}) = 0 \quad (1.98)$$

In the particular case of a closed system that exchanges neither heat nor work with its surroundings, the system is "isolated." Then,

$$\frac{dU_{system}}{dt} = 0 \quad (1.99)$$

1.5.2 Open System

An open system (or "control volume," indicated by "CV" hereafter) can exchange mass with its surroundings. Therefore, besides the conservation of energy, the conservation of mass must also be expressed. Note that for closed systems, the conservation of mass is trivial and states that the mass of the system is constant.

1.5.2.1 Mass Balance

Let's assume an open system with multiple supply ports and exhaust ports. The rate of variation of the mass of the control volume is equal to the total mass flow rate entering the system minus the total mass flow rate leaving the system.

$$\frac{dm_{CV}}{dt} = \sum_{su} \dot{m}_{su} - \sum_{ex} \dot{m}_{ex} \tag{1.100}$$

An open system can undergo a steady-flow process when the fluid flows steadily through the control volume. As shown previously, the opposite process is the transient process. The steady-flow process is approached when a system operates for a long time under the same conditions. We will see that this assumption is very useful to describe components such as compressors and heat exchangers. In the particular case of a steady-flow process, the properties at a given point of the control volume remain constant during the process, even if these properties can vary from point to point inside the control volume (for instance, the temperature of one stream at the inlet of a heat exchanger may be different to that at the outlet). Since the properties at each point of the control volume do not vary with time, the mass contained in the control volume is constant. Therefore, in the steady-state regime, the total mass flow rate entering the system is equal to the total mass flow rate leaving the system.

$$\sum_{su} \dot{m}_{su} = \sum_{ex} \dot{m}_{ex} \tag{1.101}$$

In the specific case of a system with one single inlet and one single outlet, the previous equation reduces to

$$\dot{m}_{su} = \dot{m}_{ex} \tag{1.102}$$

If the working fluid is incompressible (and hence its specific volume is constant), the previous equation can be expressed in terms of volume flow rate and speed, if the cross-sectional area A of the pipe is known.

$$\dot{V}_{su} = \dot{V}_{ex} \Longleftrightarrow A_{su} C_{su} = A_{ex} C_{ex} \tag{1.103}$$

where

$\dot{V}$ is a volume flow rate, $[\text{m}^3\text{s}^{-1}]$

A is a cross sectional area, $[\text{m}^2]$

C is a velocity, $[\text{m s}^{-1}]$

1.5.2.2 Energy Balance

The most general form of the energy balance applied to an open system is Eq. (1.96).

As explained for the mass balance, an open system can undergo either a steady flow process or a transient process. An example of a system that undergoes a transient process is a thermal storage system during charging or discharging processes. This will be addressed in Chapter 4 when modeling PCM energy storage. Another example of transient process is the heating-up or cooling-down phases of a car cabin. During these processes, the internal energy of the system is increasing or decreasing.

In the case of a steady-flow process, if the properties at each point of the control volume are constant in time, it means that the internal energy content is constant.

Consequently, in the steady-flow regime, the energy balance for a control volume can be expressed as:

$$(\dot{Q}_{su} - \dot{Q}_{ex}) + (\dot{W}_{su} - \dot{W}_{ex}) + \sum_{su} \dot{m} \left(h + \frac{C^2}{2} + gz \right) - \sum_{ex} \dot{m} \left(h + \frac{C^2}{2} + gz \right) = 0 \qquad (1.104)$$

As it will be shown in the next chapters, for many thermodynamic systems, the kinetic and potential energies, or their variations, can be neglected. Also, many systems are characterized by one single flow supply and exhaust. For such systems, the previous equation reduces to

$$\dot{m}(h_{ex} - h_{su}) = (\dot{Q}_{su} - \dot{Q}_{ex}) + (\dot{W}_{su} - \dot{W}_{ex}) \qquad (1.105)$$

The latter equation is of particular use to describe heat exchangers, compressors, pumps, and turbines.

1.6 Second Law of Thermodynamics

The First Law of Thermodynamics alone cannot explain all evolutions. Let's imagine, for instance, a car with a cabin indoor temperature of 15°C in surroundings with a temperature of 5°C (instead of the example of the hot coffee cup described by Çengel and Boles (2006)). Let's also assume the solar radiation to be null and there is no heat source inside the cabin (no occupant and heating system switched off). Our everyday experience teaches us that the cabin indoor temperature cannot spontaneously rise, since the cabin outdoor temperature is lower than the cabin indoor temperature. Actually, the cabin indoor temperature is expected to decrease because of heat losses from the cabin to the surroundings through the cabin envelope. Heat cannot flow spontaneously from the surroundings to a cabin. To achieve such a heat transfer, a *heat pump* is needed. However, the First Law does not prevent heat from flowing from the surroundings to the cabin, resulting in an increase of its temperature. The First Law applied to the cabin, as a thermodynamic system, will be satisfied: when the quantity of thermal energy lost by the surroundings is equal to the one received by the cabin and energy is conserved.

Another example of impossible thermodynamic evolution from our everyday life is the spontaneous transformation of heat into work. If a body is sliding along a surface, there is a friction force that is opposed to the force applied to the body to move. This friction force leads to an additional friction work to be provided to move the body. Ultimately, this friction work is converted into heat released to the body and to the surroundings. This heat cannot be spontaneously transformed into work that would contribute to move the body. To transform heat into work, a *heat engine* is required.

Hence, another law should explain why some evolutions are possible (heat flowing spontaneously from hot bodies to cold bodies, work transformed directly and entirely into heat) and not possible (heat flowing spontaneously from cold bodies to hot bodies, heat transformed directly and entirely into work). This other law is named the Second Law of Thermodynamics.

While the First Law of Thermodynamics can be "simply" summarized as the conservation of energy, the Second Law can be expressed through different statements that are equivalent. These statements will be progressively introduced in this section.

1.6.1 Concepts and Definitions

To understand and express the Second Law of Thermodynamics, it is necessary to introduce some concepts and definitions.

1.6.1.1 Heat Reservoir, Source, and Sink

A *heat reservoir* is a body that can exchange heat with its surroundings without any variation of its temperature. An example of a heat reservoir is the ocean. Our everyday experience teaches us that swimming in the ocean does not yield any noticeable variation of the ocean temperature, while the human body transfers thermal power to the water. Another example of thermal reservoir is a mixture of ice and liquid water or any two-phase mixture of a pure fluid.

A *heat source* is a thermal reservoir that gives heat. A *heat sink* is a thermal reservoir that receives heat.

1.6.1.2 Heat Engines

A heat engine is an apparatus that receives heat from a heat source at a high temperature T_{high}, converts part of it into work, and releases the remaining part to a heat sink at a low temperature T_{low}.

As depicted in Figure 1.18a, a heat engine comprises a working fluid that undergoes a cycle. During this evolution, the fluid receives and gives heat and work from/to its surroundings. The First Law expressed over one entire cyclic process can be written as

$$Q_{su} - Q_{ex} = W_{ex} - W_{su} = W_{net,ex} = -W_{net,su} \tag{1.106}$$

The thermal efficiency of a heat engine is defined as the ratio of the net work produced by the system to the heat it receives from the heat source.

$$\eta = \frac{W_{net,ex}}{Q_{su}} \tag{1.107}$$

The major example of heat engines covered in this book is the internal combustion engine. In the case of an internal combustion engine, the heat received by the engine corresponds to its fuel consumption m_{fuel} multiplied by the low heating value LHV_{fuel} of the fuel. In terms of rate of heat transfer and power, the thermal efficiency can be written as

$$\eta = \frac{\dot{W}_{net,ex}}{\dot{m}_{fuel}LHV_{fuel}} \tag{1.108}$$

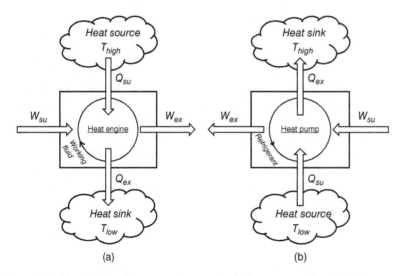

Figure 1.18 Coupling between a heat engine (a)/heat pump (b) and the heat source/heat sink.

The low heating value *LHV* [J kg^{-1}] is defined as the thermal energy released by the combustion of 1 kg of fuel if the water in the combustion gases stays in vapor phase. The high heating value *HHV* [J kg^{-1}] is defined as the heat that is released by the combustion of 1 kg of fuel if the water in the combustion gases is condensed to liquid state. Hence, the high heating value of a fuel is higher than its low heating value. For instance, for gasoline, *LHV* = 44 [MJ kg^{-1}].

1.6.1.3 Refrigerators and Heat Pumps

As explained previously, heat cannot flow spontaneously from a cold body to a hot body. To achieve such energy transfer, a refrigerator or heat pump is needed. A refrigerator or heat pump is an apparatus that absorbs a quantity of heat Q_{su} from a heat source at low temperature T_{low}, absorbs a net quantity of work $W_{net,su}$, and rejects a quantity of heat Q_{ex} into a heat sink at a high temperature T_{high}. Such a system is represented in Figure 1.18b. In a similar way to a heat engine, a heat pump or refrigerator comprises a working fluid that undergoes a cycle. This working fluid is called refrigerant. If the useful effect of the system is to produce a cooling effect, it is called a refrigerator. If the useful effect of the system is to produce a heating effect, it is called a heat pump. Because the system undergoes a cycle, Eq. (1.106) also applies.

The coefficient of performance (COP) of the refrigerator or heat pump is defined as the thermal energy exchanged (useful effect) divided by the work consumed (required input). In the refrigerator mode, this yields

$$COP_{cool} = \frac{Q_{su}}{W_{net,su}} \tag{1.109}$$

In the heat pump mode, the COP is defined as

$$COP_{heat} = \frac{Q_{ex}}{W_{net,su}} \tag{1.110}$$

Considering Eq. (1.106), the COPs in refrigerator and heat pump modes can be related by

$$COP_{heat} = COP_{cool} + 1 \tag{1.111}$$

The latter equation means that, in the absence of heat transfer with other thermal reservoirs than the heat source/sink (for instance, heat losses to the ambient), the heating COP is at least equal to unity. This equation is of primary importance to understand why it is more relevant to heat an electric vehicle with a heat pump rather than a resistive heater (*COP* = 1). This will be discussed in Chapter 4. Note that there exist many technologies of heat pump/refrigerator. However, the most used is the vapor compression one. It will be described later.

1.6.2 Kelvin Planck and Clausius Statements of the Second Law

The Second Law of Thermodynamics can be expressed in the form of two statements that are equivalent.

(1) The Kelvin–Planck Statement of the Second Law of Thermodynamics can be written as (Çengel and Boles, 2006)

> *It is impossible for any device that operates on a cycle to receive heat from a single reservoir and produce a net amount of work*

As a consequence to this statement, it is not possible to build a heat engine that would have an efficiency of 100%, since part of the heat absorbed from the heat source must be rejected to the heat sink (see Figure 1.18a).

(2) The Clausius Statement of the Second Law of Thermodynamics states that (Çengel and Boles, 2006)

> *It is impossible to construct a device that operates in a cycle and produces no effect other than the transfer of heat from a lower-temperature body to a higher-temperature body*

As a consequence of this statement, it is not possible to build a refrigerator or heat pump that would allow for the thermal energy transfer from a low temperature heat source to a high temperature heat sink without any work consumption (which would correspond to a COP of infinity).

1.6.3 Reversible Processes

Summarizing the two previous statements, the Second Law of Thermodynamics imposes that the thermal efficiency of a heat engine cannot be 100%, and the coefficient of performance of a refrigerator/heat pump cannot be infinity. Hence, we would like to know what the performance of an ideal heat engine/refrigerator/heat pump is. To address this question, the definition of an ideal machine and of a reversible process should be expressed.

A *reversible process* is a process that can be reversed by bringing back the system and its surroundings to their initial states. In contrary, a process that is not reversible is said to be *irreversible*. Among reversible processes, the distinction can be done between the internally and externally reversible processes.

Internally reversible processes are processes occurring without any irreversibilities within the system boundaries. Examples of internal irreversibilities are friction between moving elements in contact, sudden compressions and expansions, and pressure drops. A process is internally reversible if it can be reversed and undergo the same intermediate equilibrium states.

Externally reversible processes are processes without any irreversibilities outside the system boundaries. In such systems, heat transfer between the system and its surroundings is achieved without any temperature difference between the heat source/heat sink and the working fluid.

Both internally and externally reversible processes are theoretical processes that cannot be achieved in practice. A process that is both internally and externally reversible is said to be *totally reversible*.

1.6.4 Ideal Heat Engines, Refrigerators, and Heat Pumps

An ideal machine is a machine whose working fluid follows internally and externally reversible processes. Consequently, the cycle described by the working fluid is totally reversible. A machine describing a totally reversible cycle is said to be reversible. In such machines, work consumption W_{su} is minimal, work production W_{ex} is maximal, and the heat exchanges with the heat source and heat sink are achieved with no temperature difference between the heat sink/source and the working fluid. A reversible heat engine produces a maximum of work and a reversible heat pump consumes a minimum of work.

An example of a totally reversible cycle is the Carnot cycle (Figure 1.19). A Carnot heat engine, which is a reversible heat engine, describes such a cycle. In this engine, a gas undergoes successively four reversible processes:

Processes 1-2: Isothermal expansion. To maintain constant the temperature of the gas during the expansion, a quantity of heat Q_{su} is transferred from the heat source to the gas. During this heat transfer, both the heat source and the gas are at temperature T_{high}.

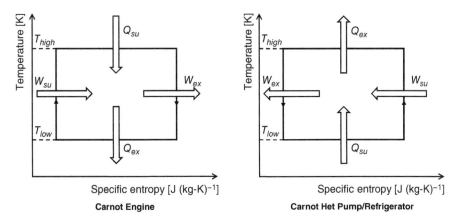

Figure 1.19 Carnot heat engine and refrigerator.

Processes 2-3: Adiabatic expansion. The expansion process continues, but without heat exchange between the gas and its surroundings. Since the expansion is adiabatic, the gas temperature decreases during the expansion from T_{high} to T_{low}.

Processes 3-4: Isothermal compression. To maintain constant the temperature of the gas during the compression, a quantity of heat Q_{ex} is transferred from the gas to the heat sink. During this heat transfer, both the heat sink and the gas are at temperature T_{low}.

Processes 4-1: Adiabatic compression. The compression process continues, but adiabatically. As a consequence, the gas temperature increases from T_{low} to T_{high}.

Other totally reversible heat engines are the Stirling and the Ericsson heat engines, which describe the Stirling and Ericsson cycles, respectively. Those cycles won't be described here, because of their limited use in vehicle applications at the present time.

Knowing what a reversible heat engine is, the two Carnot statements can be expressed as follows (Çengel and Boles 2006)

> *"The efficiency of an irreversible heat engine is always less than the efficiency of a reversible one operating between the same two reservoirs"*

> *"The efficiencies of all reversible heat engines operating between the same two reservoirs are the same"*

It could be demonstrated that violating these principles would violate the Kelvin–Planck Statement of the Second Law of Thermodynamics, which is the meaning of "equivalence." Therefore, up to now, we have expressed the Second Law of Thermodynamics under the form of 4 equivalent statements.

The Second Carnot principle states that the efficiency of a reversible heat engine is independent of the working fluid and of the type of heat engine and of the cycle described by the fluid (Çengel and Boles 2006). Hence, the efficiency of a reversible cycle is only function of the temperatures of the heat source and the heat sink. It could be demonstrated that the thermal efficiency of a reversible heat engine is equal to the *Carnot efficiency* expressed as

$$\eta_{rev} = 1 - \frac{T_{low} \, [K]}{T_{high} \, [K]} \tag{1.112}$$

Being a reversible engine, a Carnot heat engine, a Stirling heat engine, and an Ericsson heat engine operating between a heat source at temperature T_{high} and a heat sink at temperature T_{low} show the same efficiency that is equal to the Carnot efficiency (Eq. (1.112)).

Because it is reversible, a Carnot heat engine can be inversed and operated as a Carnot refrigerator or a heat pump between a heat source at temperature T_{low} and a heat sink at temperature T_{high}. The coefficients of performance of a Carnot refrigerator and of a Carnot heat pump are given by

$$COP_{cool,rev} = \frac{T_{low}\,[K]}{T_{high}\,[K] - T_{low}[K]} \tag{1.113}$$

$$COP_{heat,rev} = \frac{T_{high}\,[K]}{T_{high}\,[K] - T_{low}\,[K]} \tag{1.114}$$

An irreversible heat engine and an irreversible refrigerator/heat pump will show a thermal efficiency and a COP that are fractions of the Carnot efficiency and Carnot COP. These fractions are called the second law efficiencies of heat engines, refrigerators, and heat pumps.

For a heat engine, we have

$$\eta = \varepsilon_{II} \cdot \eta_{rev} \tag{1.115}$$

For a refrigerator, we have

$$COP_{cool} = \varepsilon_{II} \cdot COP_{cool,rev} \tag{1.116}$$

Typical values of second law efficiencies of around 50% can be met in practice. However, less and more efficient machines could show values lower and higher than 50%. To some extent, the value of the second law efficiency depends on a trade-off between cost and performance, which is the object of thermoeconomics.

The previous equations indicate that the thermal efficiency of a heat engine and the COP of a refrigerator/heat pump could be increased by

- decreasing irreversibilities associated with the machine, thus increasing the second law efficiency
- increasing the heat source temperature and/or decreasing the heat sink temperature, thus increasing the Carnot efficiency and Carnot COP.

For a refrigerator, this means increasing the temperature at which the cooling effect is produced and rejecting the heat into a heat sink at a lower temperature. For a heat pump, this means reducing the temperature at which the heating effect is produced and absorbing the heat from a heat source at a higher temperature.

1.6.5 Entropy

Another inequality associated with the Second Law of Thermodynamics is the Clausius inequality. It can be demonstrated based on the Kelvin–Planck Statement. The Clausius inequality states (and this is our fifth statement) that

$$\oint \frac{\delta Q}{T} \leq 0 \tag{1.117}$$

The symbol on the integral sign means that the integration is conducted over one entire cycle. In this expression, $T\,[K]$ is the system boundary temperature, and δQ is the differential amount of heat exchanged between the system and its surroundings.

The equality occurs in the specific case of an internally reversible process:

$$\oint \left(\frac{\delta Q}{T} \right)_{int,rev} = 0 \tag{1.118}$$

By definition, the expression $(\delta Q/T)_{int,rev}$ must be the variation of a state variable, since this variation on a cycle is equal to zero. This state variable is called *entropy* and is denoted by the symbol $S\,[J\,K^{-1}]$.

Since the entropy is a state variable, its variation can be calculated by considering a hypothetical process that is internally reversible. This gives

$$\Delta S = S_2 - S_1 = \int_1^2 \left(\frac{\delta Q}{T}\right)_{int,rev} \tag{1.119}$$

According to the Clausius inequality, we have

$$\Delta S = S_2 - S_1 \geq \int_1^2 \frac{\delta Q}{T} \tag{1.120}$$

The equality is valid for an internally reversible process. In the case of an irreversible process, the variation of entropy has two contributions: the entropy transferred associated with the heat transfer and the entropy generation due to irreversibilities. The term S_{gen} can only be positive or equal to zero, but never negative.

$$\Delta S = S_2 - S_1 = \int_1^2 \frac{\delta Q}{T} + S_{gen} \tag{1.121}$$

If the system is isolated, the variation of its entropy corresponds to the internal production of entropy. In this case, the previous equation simplifies to

$$\Delta S = S_2 - S_1 = S_{gen} \geq 0 \tag{1.122}$$

This means that the entropy of an isolated system can only be constant or increase, but never decrease.

In the case of an adiabatic and reversible evolution, the variation of entropy is null, and the process is said to be isentropic. Indeed, the two terms of the right-hand side of Eq. (1.121) are null. The isentropic process is often considered as an ideal reference process to evaluate the performance of pumps, fans, compressors, and turbines. Actually, in the absence of cooling or heating mechanisms, these components ideally operate adiabatically.

The isentropic effectiveness of a compressor is defined as the ratio of the power that would be consumed if the fluid were compressed isentropically to the actual power that is consumed by the compressor.

$$\varepsilon_{s,cp} = \frac{\dot{W}_{cp,s}}{\dot{W}_{cp}} \tag{1.123}$$

If the variations of kinetic and potential energies can be neglected, the previous equation becomes

$$\varepsilon_{s,cp} = \frac{\dot{m}\,(h_{ex,s} - h_{su})}{\dot{W}_{cp}} \tag{1.124}$$

where

$\dot{m}$ is the fluid mass flow rate flowing through the compressor, [kg s^{-1}]
$h_{ex,s}$ is the enthalpy of the fluid at the exhaust of the compressor if the compression was isentropic, [J kg^{-1}]
h_{su} is the enthalpy of the fluid at the supply of the compressor, [J kg^{-1}]
$\dot{W}_{cp}$ is the actual power consumed by the compressor, [W].

The same definition applies for fans and pumps. For a pump, using Eq. (1.50), the isentropic effectiveness can be expressed as

$$\varepsilon_{s,pp} = \frac{\dot{m}\,v\,(P_{ex} - P_{su})}{\dot{W}_{cp}} = \frac{\dot{V}\,(P_{ex} - P_{su})}{\dot{W}_{cp}} \tag{1.125}$$

For a turbine, the isentropic effectiveness is defined as the ratio of the actual power produced by the turbine to the power that would be produced if the fluid were expanded isentropically. If the variations of kinetic and potential energies can be neglected, the turbine isentropic effectiveness is given by

$$\varepsilon_{s,turb} = \frac{\dot{W}_{turb}}{\dot{m}(h_{su} - h_{ex,s})} \qquad (1.126)$$

1.7 Flows in Hydraulic Circuits

Automotive thermal management systems comprise many different hydraulic circuits, with air, coolant (glycol water), or oil as major fluids. Circulation of the fluids through these circuits is generally made possible by pumps and fans. It will be shown in Chapter 2 that the vehicle ram effect also contributes to create the motion of air through the front-end module. The energy consumption of fans and pumps depends not only on their efficiencies, but also on the mechanical energy losses to overcome. The following section introduces basic relationships to describe flows in hydraulic and aeraulic circuits and assess the consumption of fans and pumps.

The flow in a hydraulic circuit can be described using the Bernoulli equation and the First Law of Thermodynamics. Let's assume a portion of hydraulic circuit represented in Figure 1.20. The fluid enters the portion of circuit through section 1 at pressure P_1 [Pa], velocity C_1 [m s^{-1}], and elevation z_1 [m] and leaves it through section 2 at pressure P_2 [Pa], velocity C_2 [m s^{-1}], and elevation z_2 [m]. A pump is placed on the circuit to increase the mechanical energy of the flow. Mechanical energy losses occur on the circuit due to two types of pressure losses: a friction pressure drop and a local pressure drop (the latter can be due to a section reduction, a valve, an elbow, etc.).

The fluid is assumed to be incompressible and characterized by a specific volume v [m^{3}kg^{-1}]. Furthermore, the portion of the circuit is assumed to be in the steady-state regime.

The conservation of mass between sections 1 and 2 expresses that the mass flow rate entering the portion is equal to that leaving the portion and is simply written as $\dot{m}$ [kg s^{-1}].

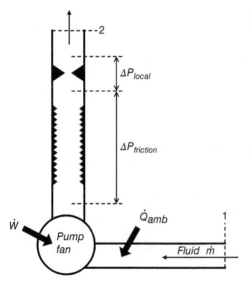

Figure 1.20 Description of the flow in a hydraulic circuit.

To determine the power of fan/pump necessary to move the fluid in this portion of hydraulic circuit, the generalized Bernoulli equation is used. It relates the rate of mechanical energy at the inlet of the portion of circuit to that at the outlet. That is,

$$\dot{m}\left(v P_1 + \frac{C_1^2}{2} + g z_1\right) + \eta \, \dot{W} = \dot{m}\left(v P_2 + \frac{C_2^2}{2} + g z_2\right) + \dot{m} \, e_{mech,loss} \tag{1.127}$$

In the latter equation, $C^2/2$ [J kg^{-1}] represents the kinetic energy per kilogram of fluid and $g z$ [J kg^{-1}] the potential energy per kilogram of fluid. $\dot{W}$ [W] is the power consumed by the fan or fan/pump. It is mechanical power if the fan/pump is mechanically driven or electrical power if it is driven by an electrical motor. The efficiency of the fan/pump is written as η (for electrically driven fans and pumps, it includes the electric motor efficiency). Therefore, $\eta \, \dot{W}$ represents the minimum power required to move the fluid. Pressure losses yield a rate of mechanical energy loss $\dot{E}_{mech,loss} = \dot{m} \, e_{mech,loss}$ [W].

The conservation of energy for the control volume represented in Figure 1.20 can be expressed to compute the heating of the fluid between sections 1 and 2. It follows that

$$\dot{m}\left(h_1 + \frac{C_1^2}{2} + g z_1\right) + \dot{W} + \dot{Q}_{amb} = \dot{m}\left(h_2 + \frac{C_2^2}{2} + g z_2\right) \tag{1.128}$$

Substituting h by $u + pv$ and expressing the variation of internal energy as function of the variation of temperature for an ideal liquid, it follows that

$$\dot{m} \, c \, (T_2 - T_1) = \dot{m}\left(v P_1 + \frac{C_1^2}{2} + g z_1\right) - \dot{m}\left(v P_2 + \frac{C_2^2}{2} + g z_2\right) + \eta \dot{W} + (1 - \eta)\dot{W} + \dot{Q}_{amb}$$

$$\tag{1.129}$$

Combining this equation with Bernoulli's equation yields

$$\dot{m} \, c \, (T_2 - T_1) = \dot{E}_{mech,loss} + (1 - \eta)\dot{W} + \dot{Q}_{amb} \tag{1.130}$$

In the particular case of a closed circuit (such as the coolant loop of an engine), having section 1 corresponding to section 2, the generalized Bernoulli's equation reduces to Eq. (1.131), meaning that the minimum power of the fan/pump overcomes the rate of mechanical energy loss through the circuit.

$$\eta \, \dot{W} = \dot{E}_{mech,loss} \tag{1.131}$$

Also, if the closed circuit is in the steady-state regime, the ambient losses must balance the fluid heating-up due to the rate of mechanical energy loss and fan/pump inefficiency.

$$\dot{E}_{mech,loss} + (1 - \eta)\dot{W} = -\dot{Q}_{amb} \tag{1.132}$$

As mentioned earlier, mechanical energy can be lost through two types of irreversibilities: frictional losses along the walls of the pipe and losses in valves, elbows, etc. The former losses correspond to friction pressure losses ($\Delta P_{friction}$) and the latter ones to local pressure losses (ΔP_{local}).

The friction pressure losses can be calculated by

$$\Delta P_{friction} = f \frac{L}{D_h} \frac{C^2}{2 v} \tag{1.133}$$

In the latter equation, f is the friction factor. For laminar flows, $f = Constant/Re$. For turbulent flows, the friction factor also depends on the Reynolds number and whether the walls are smooth or rough. In the latter case, the friction factor is function of the relative roughness: $f = f(\varepsilon/D_h)$. Such relations can be easily found in fluid mechanics textbooks.

The local pressure losses can be assessed by the following relationship involving a local loss coefficient K_{local} that is function of the type of piping singularities:

$$\Delta P_{local} = K_{local} \frac{C^2}{2\,v} \tag{1.134}$$

All the previous developments will be useful when investigating the aeraulics associated with a vehicle front-end module (Chapter 2).

1.8 Heat Exchangers

A heat exchanger is a device that transfers thermal energy from a hot fluid to a cold fluid. There could be some variants to this general definition:

- The heat transfer can occur between a solid surface and a fluid or between solid particulates and a fluid (Shah and Sekulić, 2003).
- Mass transfer can occur simultaneously with heat transfer.
- The heat exchanger can involve more than two fluids.
- The heat exchanger can involve internal heat production or combustion or chemical reactions (Shah and Sekulić, 2003).

Heat exchangers are key components in the vehicle overall thermal management system. Actually, the latter comprises several fluid loops that exchange heat among themselves (e.g. oil to coolant, coolant to oil, coolant to air, refrigerant to air, and air to refrigerant). Different technologies of heat exchanger are used, depending on the nature of the working fluids, operating conditions, and weight/packaging constraints. The performance of the heat exchangers is of paramount importance to optimize the performance of the whole thermal management system. It will be shown later that this performance is expressed in terms of both thermal and hydraulic performance.

1.8.1 Classification of Heat Exchangers

There exist many technologies of heat exchangers. Several classifications can be used to emphasize the major characteristics of each of them.

1.8.1.1 Classification According to the Mechanism of Energy Transfer

1.8.1.1.1 Heat Exchangers Without Mass Transfer Most of the heat exchangers only exchange thermal energy ("heat"). In such heat exchangers, both fluids are generally separated by a wall, which is called "primary surface." This wall prevents any mixing of both fluids.

1.8.1.1.2 Heat Exchangers with Mass Transfer Heat exchangers that also allow for mass transfer between both fluids do not comprise any separating wall (or the wall is porous). This is the case of direct contact cooling towers, which find a lot of applications in stationary HVAC and industry. In such cooling towers, water to be cooled down is in direct contact with air. Heat and mass are exchanged between both fluids. Some liquid water is actually vaporized to increase the vapor content of humid air. In a cooling tower, as the temperature of water decreases, the wet-bulb temperature of the air increases. Example of porous heat exchangers are "enthalpy" heat exchangers that allow for the heat and water vapor transfer between a stream of hot and humid air and a stream of cold and dry air.

1.8.1.1.3 Heat Exchangers with Intermediate Energy Storage (Regenerators) Another distinction can be done within all mechanisms of energy transfer. In most heat exchangers, heat (and possibly mass) is directly transferred from one fluid to the other. There exist heat exchangers, named "regenerators," where heat (and possibly mass) is stored in an intermediate way in a medium during its transfer from one fluid to the other. This medium is usually a solid matrix. In this type of heat exchanger, the hot fluid flows through the matrix that stores the heat. Later in time, the cold fluid flows through the same matrix that releases its heat. Hence, both fluids flow in the same space, but at different positions in time. Regenerators combine the function of heat transfer and heat storage. Such heat exchangers are widely used for ventilation heat recovery in buildings.

1.8.1.2 Classification According to the Phases of Both Fluids

In conventional heat exchangers, where fluids are separated by a wall, only thermal energy is transferred from a fluid to another. The fluids can be in liquid or gas phase, or they can experience a phase change when flowing through the heat exchanger. Examples of heat exchangers with different fluid phases applied to automotive thermal management are provided in Table 1.4.

In the specific case of air-heated evaporators, the heat exchanger is fed by humid air on one side and refrigerant on the other side. If the temperature of the separating wall on the air-side (called "contact temperature") is decreased underneath the air dew-point, water vapor in the air condenses. If the contact temperature is negative, liquid water solidifies. This yields to the formation of frost, a mixture of ice and air.

1.8.1.3 Classification According to the Flow Arrangement

Heat exchangers can show counter-flow, parallel-flow, and cross-flow configurations.

In *counter-flow* heat exchangers, illustrated in Figure 1.21, the hot and cold fluids enter the heat exchanger through opposite ends. The fluids flow parallel to each other and in opposite directions. The exhaust temperature of the cold fluid can be larger than that of the hot fluid (Incropera and DeWitt, 2002).

Table 1.4 Classification of the heat exchangers according to the phases of the fluids – Example of applications in the automotive domain.

Phase of fluid 1–phase of fluid 2	Example of applications
Gas–gas	Air-cooled charge air cooler (Chapter 2)
Liquid–gas	Air-cooled oil cooler (Chapter 2)
	Water-cooled charge air cooler (Chapter 2)
	Water-cooled exhaust gas recirculation cooler (Chapter 2)
Liquid–liquid	Water-cooled oil cooler (Chapter 2)
Two-phase–liquid	Water-cooled condenser (Chapter 3)
	Water-heated evaporator ("chiller") (Chapter 4)
Two-phase–gas	Air-heated evaporator (Chapter 3)
	Air-cooled condenser (Chapter 3)
Two-phase–two-phase	Air-heated evaporator with water condensation on the air side (Chapter 3)
	Internal heat exchanger if condensation and evaporation occur on both sides (Chapter 3)

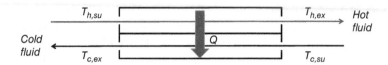

Figure 1.21 Flow arrangement in a counter-flow heat exchanger.

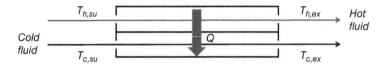

Figure 1.22 Flow arrangements in a parallel-flow heat exchanger.

In *parallel-flow* heat exchangers, illustrated in Figure 1.22, the hot and cold fluids enter and leave the heat exchanger at the same ends. They also flow parallel to each other, but in the same direction. Such heat exchanger allows for "quickly" decreasing the temperature of the hot fluid in the vicinity of the heat exchanger inlet, but also allows to bring the exhaust temperatures of both fluids close to each other (Incropera and DeWitt, 2002).

In *cross-flow* heat exchangers, both fluids flow perpendicular to each other. Within cross-flow configurations, one can also distinguish the cases where both fluids are mixed (a), un-mixed (c), or only one of them is mixed (b). Un-mixed flow means that the overall flow is divided into a number of separate and parallel flows. This can, for instance, represent the air flow on the fin-side of a tube-fin heat exchanger. Mixed and un-mixed flows are two ideal cases. In practice, no flow fits exactly in one or another category (Figure 1.23).

The performance of counter-flow heat exchangers is usually better than that of parallel-flow. Indeed, for the same supply and exhaust temperatures, the mean temperature difference between both fluids is higher. Therefore, for the same heat transfer rate $\dot{Q}$, a lower heat transfer area A is needed assuming the same overall heat transfer coefficient U (Incropera and DeWitt, 2002). The performance of a cross-flow heat exchanger lies between that of parallel-flow and counter-flow heat exchangers.

1.8.1.4 Classification According to the Pass Arrangement
A single-pass flow indicates that the fluid flows only once along the length of the heat exchanger. The fluid can flow several times along the length of the heat exchanger and show a multi-pass arrangement. With a large number of passes, the cross-flow configuration tends to be a counter-flow configuration.

1.8.1.5 Classification According to the Type of Construction
Another classification is based on the type of construction of heat exchangers. One can distinguish prime-surface heat exchangers and extended-surface heat exchangers.

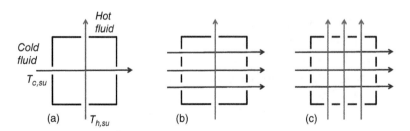

Figure 1.23 Flow arrangements in a cross-flow heat exchanger.

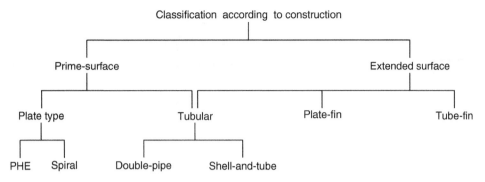

Figure 1.24 Classification of heat exchangers according to the type of construction. Source: Adapted from Shah and Sekulić (2003).

Prime-surface heat exchangers only use the primary surface, i.e. the surface of the dividing wall that separates the fluids passages, as the heat transfer surface.

Extended-surface heat exchangers use fins, also called secondary surface, to increase the heat transfer area. The increase of the heat transfer area can compensate for the decrease of the heat transfer coefficient on one side or both sides of the heat exchanger. Hence, the use of fins reduces the thermal resistance on the side of the heat exchanger where they are employed. It should be mentioned that the presence of fins can either increase or decrease the heat transfer coefficient. Louvered fins, which will be introduced later, create flow interruption and increase the heat transfer coefficient by 2–4 in comparison to plain fins (Shah and Sekulić, 2003).

Figure 1.24 proposes a classification of the heat exchangers most commonly used in vehicle thermal management. It is inspired from Shah and Sekulić (2003) who proposed a much wider classification of heat exchangers according to their construction. Prime-surface heat exchangers include plate type and tubular heat exchangers. Extended surface heat exchangers include plate-fin and tube-fin heat exchangers. Tubular heat exchangers having fins on the inner or outer surfaces of the tubes are classified as extended surface heat exchangers.

The heat exchangers mentioned in the tree structure of Figure 1.24 are described in detail hereunder, with examples of realizations dedicated to vehicle thermal management

1.8.1.5.1 *Tubular Heat Exchangers* Among the most commonly used tubular heat exchangers, one can mention:

- The double pipe (or "concentric tubes") heat exchanger. In such configuration, two tubes are nested into each other concentrically. One fluid flows through the inner tube, while the other fluid flows in the annulus section between both tubes.
- The shell-and-tube heat exchanger. It is composed of a bank of tubes bundled in a shell. One of the fluids flows through the tubes, while the other one flows outside the tubes. Baffles can be used to enhance the convective coefficient of the fluid flowing outside the tubes and to support the bank of tubes. Examples of different architectures of shell-and-tube heat exchangers used as exhaust gas recirculation (EGR) coolers (the role of which is explained in Chapter 2) are given in Figure 1.25.

1.8.1.5.2 *Plate Heat Exchangers* Plate heat exchangers are made of a stack of thin plates. The assembly of plates creates parallel thin channels for the working fluids. Usually, each plate has four ports at the corners. All these ports form distribution headers for the two fluids (Shah and Sekulić, 2003). These headers allow for two adjacent channels to be fed by different fluids.

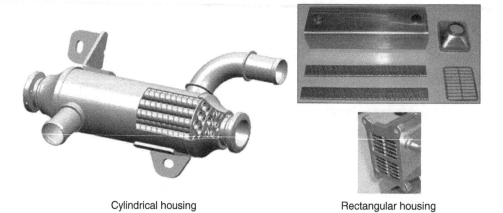

Cylindrical housing   Rectangular housing

Figure 1.25 Shell-and-tube heat exchangers used as EGR coolers. (a) Cutaway photograph of a heat exchanger with a cylindrical housing and round tubes. (b) Components of a heat exchanger with rectangular housing and flat tubes. Source: Courtesy of Valeo.

Plates are smoothed, corrugated, or embossed (i.e. with protrusions). The corrugations on the successive plates ensure the support of the plate stack through a large series of contact points (Shah and Sekulić, 2003). Also, corrugations create highly interrupted and tortuous channels, which yield high level of turbulence (Shah and Sekulić, 2003). This high turbulence reduces fouling of the heat exchanger. Also, high turbulence combined with low hydraulic diameters yields high heat transfer coefficients. An example of plate heat exchanger used as an oil cooler (the role of which is explained in Chapter 2) is given in Figure 1.26.

Plates are sealed around their edges by gaskets, welding, or brazing. Gasketed plate heat exchangers are not adapted to corrosive fluids. Moreover, the use of gaskets limits the operating pressures

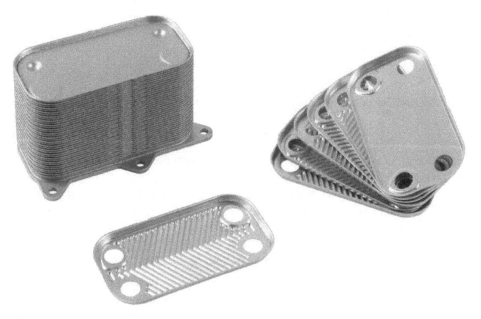

Figure 1.26 Plates from a plate heat exchanger used as coolant-to-oil cooler. Source: Courtesy of Valeo.

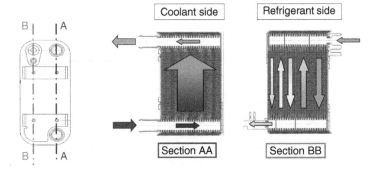

Figure 1.27 Multi-pass configuration of a BPHEX. Source: Adapted from Valeo.

and temperatures. Such limitations are partly overcome with welded or brazed plate heat exchangers. Such heat exchangers are perfectly tight but cannot be opened for servicing.

Usually, in plate heat exchangers, each fluid shows one single pass, and the flow configuration is purely counterflow. Reduced fouling combined with pure counterflow configuration and large heat transfer coefficients allow for a reduction of the heat transfer area with respect to shell-and-tube heat exchangers. Also plate heat exchangers are lighter and more compact.

It should be mentioned that some plate heat exchangers are characterized by hybrid flow configurations. This is illustrated in Figure 1.27, which represents a water-cooled condenser with one single pass on the coolant side and five passes on the refrigerant side. Regarding the latter passes, the first, third, and fifth ones are in the counter-flow configuration. The second and fourth refrigerant passes are in the parallel-flow configuration. In general, increasing the number of passes increases the heat exchanger compactness.

1.8.1.5.3 Plate-Fin Heat Exchangers
Plate-fin heat exchangers are made of a stack of plates with corrugated fins sandwiched in between. There are different types of fins: they could have triangular or rectangular cross-sections, and they could be plain or multilouvered or wavy (Figure 1.28).

Among the plate-fin heat exchangers, one can mention the bar and plate heat exchangers. Such heat exchangers are composed of parting sheets (plates), corrugated fins, and side bars. The fins act as spacers between the plates. Side bars are used to form a flow channel between two plates. Such heat exchangers are mechanically robust. An example of bar and plate heat exchanger used as charge air cooler (CAC) (the role of which is explained in Chapter 2) is given in Figure 1.29.

On one fluid side, the combinations of two plates and two side bars can be replaced by flat extruded tubes. In such configuration of plate-fin heat exchangers, fins are sandwiched between the flat tubes and attached to them by brazing. In automotive applications, liquid or two-phase fluid flows inside the tubes, while air flows outside the tubes on the finned side. The presence of fins compensates for the low convection heat transfer coefficient on the air side by increasing the heat transfer area.

1.8.1.5.4 Tube-Fin Heat Exchangers
In this configuration, one of the fluids flows through the tubes, while the other one flows at the outside of the tubes and is in contact with fins. The latter fluid is usually a gas such as air.

Figure 1.28 Components of a plate-fin heat exchanger used as an evaporator. Source: Courtesy of Valeo.

Different shapes of tubes are used, such as round, oval, or peanut. Either the tubes have individual fins or a matrix of tubes share continuous fins that can be plain, wavy, multilouvered, etc. This will be illustrated in Chapter 2 for radiators. Also the inner surface of the tubes can have fins or corrugations to decrease the thermal resistance of the fluid flowing inside the tubes. This is illustrated in Figure 1.30.

The different technologies of heat exchangers show different compactness. The compactness is defined in terms of heat transfer area per unit envelope volume of the heat exchanger. Compact heat exchangers show a compactness factor larger than 700 m^2 m^{-3} (Incropera and De Witt, 2002).

1.8.2 Energy Balance Across a Heat Exchanger

Before introducing the indicators of the performance of a heat exchanger, it is relevant to establish the energy balance on a heat exchanger. For all types of heat exchangers, the energy balances can be derived from the First Law of Thermodynamics applied to the control volume (Eq. (1.96)). Several assumptions are introduced:

- The heat exchanger is in the steady-flow regime.
- There is no heat transfer between the heat exchanger and the ambient.
- The variations of the kinetic and potential energies of both the cold and hot fluids between the supply and the exhaust of the heat exchanger are negligible.
- Only two fluids travel through the heat exchanger, even though there exist heat exchangers with more than two fluids.
- No work is produced or consumed by the heat exchanger. Note that thermoelectric generators can be seen as heat exchangers producing electricity.

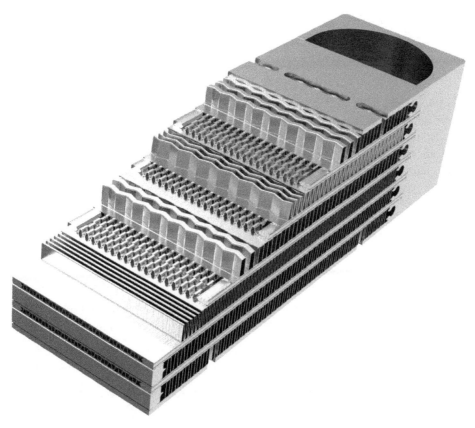

Figure 1.29 Cutaway photograph of a bar and plate heat exchanger used as a charge air cooler. Source: Courtesy of AKG.

Figure 1.30 Cutaway photograph of a (flat) tube-fin heat exchanger used as charge air cooler. Source: Courtesy of Valeo.

Under all these assumptions, Eq. (1.96) applied to the control volume comprising both fluids reduces to

$$\dot{m}_h.(h_{h,su} - h_{h,ex}) = \dot{m}_c.(h_{c,ex} - h_{c,su}) \tag{1.135}$$

where

$\dot{m}_h$ is the hot fluid mass flow rate, [kg s^{-1}]
$\dot{m}_c$ is the cold fluid mass flow rate, [kg s^{-1}]
$h_{h,su}$ is the specific enthalpy of the hot fluid at the heat exchanger supply, [J kg^{-1}]
$h_{h,ex}$ is the specific enthalpy of the hot fluid at the heat exchanger exhaust, [J kg^{-1}]
$h_{c,su}$ is the specific enthalpy of the cold fluid at the heat exchanger supply, [J kg^{-1}]
$h_{c,ex}$ is the specific enthalpy of the cold fluid at the heat exchanger exhaust, [J kg^{-1}].

To express heat transfer rate $\dot{Q}$ [W] between the hot and the cold fluids, Eq. (1.96) must be applied to control volumes comprising one single fluid. This gives

$$\dot{Q} = \dot{m}_h.(h_{h,su} - h_{h,ex}) \tag{1.136}$$

$$\dot{Q} = \dot{m}_c.(h_{c,ex} - h_{c,su}) \tag{1.137}$$

In the absence of phase change and assuming that the specific heats on both fluid sides are constant (and equal to the average value between the supply and exhaust temperatures) during heat transfer, one gets:

$$\dot{Q} = \dot{m}_h \cdot c_{p,h}.(T_{h,su} - T_{h,ex}) = \dot{m}_c \cdot c_{p,c} \cdot (T_{c,ex} - T_{c,su}) \tag{1.138}$$

This equation can also be written in the following way by introducing the capacity flow rate $\dot{C}$, which is the product of the mass flow rate times the specific heat.

$$\dot{Q} = \dot{C}_h \cdot (T_{h,su} - T_{h,ex}) = \dot{C}_c \cdot (T_{c,ex} - T_{c,su}) \tag{1.139}$$

1.8.3 Performance

Performance of heat exchangers can be quantified in terms of energy-related indicators (thermal performance and hydraulic performance), economic indicators (e.g. the cost), compactness indicators (compactness factor, weight, etc.), or reliability. This section only focuses on the energy-related indicators owing to their impact on the energy performance of the thermal management systems.

1.8.3.1 Thermal Performance

The energy balance across the heat exchanger (Eq. (1.135)–(1.137)) does not bring information about its thermal performance. More specifically, the previous energy balance equations cannot predict the heat transfer rate $\dot{Q}$ knowing the capacity flow rates and the supply temperatures of both fluids. Another equation involving an intrinsic characteristic of the heat exchanger is necessary. This characteristic is overall heat transfer conductance AU [W K^{-1}], which is the product of heat transfer area A [m^2] and of the overall heat transfer coefficient U [W m^{-2}K^{-1}]. The conductance is the reciprocal of the overall heat transfer resistance R [K W^{-1}].

The overall heat transfer conductance AU [W K^{-1}] of the heat exchanger can be defined independent of the side of the heat exchanger, which is not the case of the overall heat transfer coefficient U [W m^{-2}K^{-1}]. Considering Figure 1.31, one can write

$$\frac{1}{AU} = \frac{1}{A_h U_h} = \frac{1}{A_c U_c} = R \tag{1.140}$$

Figure 1.31 Desegregation of the heat exchanger conductance into thermal resistances.

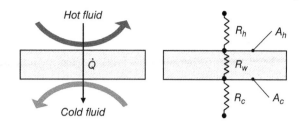

where

A_h is the area of the hot surface of the heat exchanger, [m^2]
A_c is the area of the cold surface of the heat exchanger, [m^2]
U_h is the overall heat transfer coefficient defined on the hot fluid side, [W m^{-2}K^{-1}]
U_c is the overall heat transfer coefficient defined on the cold fluid side, [W m^{-2}K^{-1}].

The overall thermal resistance between the hot and cold fluids can be seen as the association of several resistances in series:

– the convection resistance R_h between the hot fluid and the heat exchanger hot surface,
– the conduction resistance R_w between the hot and cold surfaces through the heat exchanger wall,
– the convection resistance R_c between the cold fluid and the heat exchanger cold surface.

Associating these resistances yields

$$\frac{1}{AU} = R_h + R_w + R_c \tag{1.141}$$

Assuming that both the hot and the cold surfaces of the heat exchanger are finned, Eq. (1.141) can be developed as (Incropera and DeWitt, 2002)

$$\frac{1}{AU} = \frac{1}{(\eta_0 \, A \, h)_h} + R_w + \frac{1}{(\eta_0 \, A \, h)_c} \tag{1.142}$$

where

η_0 is the overall surface efficiency, [−]
h is the convective heat transfer coefficient, [Wm^{-2}K^{-1}].

The overall surface efficiency η_0 can be related to the efficiency of a single fin η_f by

$$\eta_0 = 1 - \frac{A_f}{A}(1 - \eta_f) \tag{1.143}$$

where

A_f is the fin surface area, [m^2]
A is the total surface area (exposed base and fin), [m^2].

Hence, enlarging the heat transfer area (and thus the size of the heat exchanger), increasing the convective heat transfer coefficients on both fluid sides, and limiting the conduction resistance of the wall separating the two fluids can increase the heat exchanger conductance.

Two methods are commonly used for computing the heat transfer rate $\dot{Q}$ as a function of the heat exchanger conductance AU: the log-mean temperature difference method and the epsilon-NTU method. It can be shown that both methods are algebraically equivalent.

1.8.3.1.1 Log-Mean Temperature Difference (LMTD) Method In this method, the heat transfer rate $\dot{Q}$ [W] is expressed by the product of the conductance AU [W] and the log-mean temperature difference ΔT_{lm} [K]. The log-mean temperature difference is the average temperature difference between the hot and cold fluids across the heat exchanger. Therefore,

$$\dot{Q} = A \cdot U \cdot \Delta T_{lm} \tag{1.144}$$

For a counter-flow heat exchanger, the log-mean temperature difference is

$$\Delta T_{lm} = \frac{(T_{h,su} - T_{c,ex}) - (T_{h,ex} - T_{c,su})}{\ln\left(\frac{(T_{h,su} - T_{c,ex})}{(T_{h,ex} - T_{c,su})}\right)} \tag{1.145}$$

For a parallel-flow heat exchanger, the log-mean temperature difference becomes

$$\Delta T_{lm} = \frac{(T_{h,su} - T_{c,su}) - (T_{h,ex} - T_{c,ex})}{\ln\left(\frac{(T_{h,su} - T_{c,su})}{(T_{h,ex} - T_{c,ex})}\right)} \tag{1.146}$$

For a cross-flow heat exchanger or a shell-and-tube heat exchanger with multiple passes, the log-mean temperature difference is computed by applying a correction factor to the log-mean temperature difference obtained as if the heat exchanger was counter-flow (the reader is invited to refer to Incropera and DeWitt (2002) for more information).

1.8.3.1.2 Epsilon-NTU (ε − NTU) Method The *LMTD* method is well adapted to problems where the 4 temperatures are known, for instance, to identify the necessary conductance for a given heat transfer rate. This is the case during the design phase of a heat exchanger. If only two temperatures are known (at least one of each fluid; the supply temperatures for instance), an iterative procedure must be implemented. In that situation, the ε − NTU method is more adapted. Here, the heat transfer rate is computed as a fraction ε of the maximal heat transfer rate. The latter is equal to the minimum capacity flow rate multiplied by the maximal temperature difference.

$$\dot{Q} = \varepsilon \cdot \dot{Q}_{max} = \varepsilon \cdot \dot{C}_{min} \cdot (T_{h,su} - T_{c,su}) \tag{1.147}$$

$$\dot{C}_{min} = \min(\dot{C}_h, \dot{C}_c) \tag{1.148}$$

The effectiveness ε of the heat exchanger can be expressed as a function of the NTU, which is the ratio of the overall thermal conductance and the minimal capacity flow rate. It gives

$$NTU = \frac{AU}{\dot{C}_{min}} \tag{1.149}$$

The effectiveness ε can be expressed as a function of the number of transfer units NTU [−] and the capacity ratio C_r[−] for different flow configurations. Expressions are given in Table 1.5.

Table 1.5 Relation between epsilon, NTU, and Cr for different flow configurations.

Counter-flow	$C_r < 1$	$\varepsilon = \dfrac{1 - \exp(-NTU(1 - C_r))}{1 - C_r \cdot \exp(-NTU(1 - C_r))}$
	$C_r = 1$	$\varepsilon = \dfrac{NTU}{1 + NTU}$
Parallel flow		$\varepsilon = \dfrac{1 - \exp(-NTU(1 + C_r))}{1 + C_r}$
Cross-flow	Both fluids mixed	$\varepsilon = \dfrac{1}{\dfrac{1}{1-\exp(-NTU)} + \dfrac{C_r}{1-\exp(-C_r \cdot NTU)} - \dfrac{1}{NTU}}$
	Both fluids unmixed	$\varepsilon = 1 - \exp\left[\left(\dfrac{1}{C_r}\right) NTU^{0.22} \cdot (\exp(-C_r \cdot NTU^{0.78}) - 1)\right]$
	$\dot{C}_{max}$ mixed, $\dot{C}_{min}$ unmixed	$\varepsilon = \left(\dfrac{1}{C_r}\right) \cdot [1 - \exp(-C_r(1 - \exp(-NTU)))]$
	$\dot{C}_{min}$ mixed, $\dot{C}_{max}$ unmixed	$\varepsilon = 1 - \exp\left(-\dfrac{1}{C_r}(1 - \exp(-C_r \cdot NTU))\right)$

The relation $\varepsilon = f(NTU, C_r, \text{flow configuration})$ is used to compute the heat transfer rate knowing the capacity flow rates and supply temperatures ($T_{h,su}$ and $T_{c,su}$) as well as the conductance AU. Similar relations but under the form of $NTU = g(\varepsilon, C_r, \text{flow configuration})$ can be used in the frame of heat exchanger design problems.

With the heat capacity ratio C_r defined as:

$$C_r = \frac{\dot{C}_{min}}{\dot{C}_{max}} \tag{1.150}$$

Independent of the construction type of heat exchanger, if $C_r = 0$, then the effectiveness reduces to

$$\varepsilon = 1 - \exp(-NTU) \tag{1.151}$$

The previous relation applies to semi-isothermal heat exchangers such as an air-conditioning evaporator if the vapor superheat zone is neglected (this will be described in Chapter 3). In the case of $C_r = 0$, it can be observed that ε tends to 100% when NTU tends to infinity. For other values of C_r, the limiting value of ε depends on flow configuration and C_r. However, independent of flow configuration and C_r, the pinch point temperature (which is the minimal temperature difference between the two fluids at the same location in the heat exchanger) tends to 0 K when NTU tends to infinity.

It must also be stressed that both the *LMTD* and *epsilon-NTU* methods rely on some assumptions, the major ones of which are: the heat exchanger is insulated, the specific heats of the two fluids are constant, and the overall heat transfer coefficient U is constant along the heat exchanger.

1.8.3.2 Hydraulic Performance

Besides the energy performance, the hydraulic performance of the heat exchanger is another performance indicator to consider. The hydraulic performance can be expressed in terms of pressure drops encountered by the different fluids flowing through the heat exchanger. Limiting these pressure drops allows for limiting the energy consumptions of fans and pumps displacing the fluids.

References

Bell, I.H., Wronski, J., Quoilin, S., and Lemort, V. (2014). Pure and pseudo-pure fluid thermophysical property evaluation and the open-source thermophysical property library coolprop. *Industrial and Engineering Chemistry Research* 53 (6): 2498–2508. https://doi.org/10.1021/ie4033999.

Braun, J.E. and Mitchell, J.W. (2012). *Principles of Heating, Ventilation, and Air Conditioning in Buildings*. Wiley.

Çengel, Y.A. and Boles, M.A. (2006). *Thermodynamics, An Engineering Approach*, 5e. McGraw-Hill Higher Education.

Incropera, F.P. and DeWitt, D.P. (2002). *Fundamentals of Heat and Mass Transfer*, 5e. Wiley.

Klein, S. and Nellis, G. (2016). *Thermodynamics*. Cambridge University Press.

Shah, R.K. and Sekulić, D.P. (2003). *Fundamentals of Heat Exchanger Design*. Wiley.

2

Internal Combustion Engine Thermal Management

2.1 Introduction

Internal combustion engine (ICE) thermal management is the whole set of mechanisms to maintain the temperatures of the ICE structure and working fluids (coolant, oil, gases, etc.) within optimized ranges.

ICE thermal management is important for:

- ensuring the reliability of the engine and auxiliary systems
- improving the engine efficiency and thus reducing the fuel consumption
- limiting the formation of pollutant emissions and correctly post-treating them

Advanced thermal management systems must also

- adapt the temperature and flow rate set-points to the operating regime of the engine
- reduce the fan and pump consumptions
- limit the penalty, on the performance and emissions, associated with short-duration journeys

In addition, ICE thermal management should also consider as a constraint the thermal comfort and visibility needs of the cabin occupants. Also, it will be shown in that the introduction of hybrid (and fully electric) powertrain architectures (Chapter 4) and waste heat recovery solutions (end of this chapter) will modify the conventional ICE thermal management systems. Rather than considering the thermal management of the engine alone, a global vehicle thermal management must be considered. This allows maximizing the vehicle energy performance, while meeting the numerous constraints imposed by the vehicle sub-systems.

This chapter is organized as follows. The second section recalls the fundamentals of ICEs, highlighting the key points related to energy efficiency and thermal management analysis. The third section deals with engine cooling and heating. The fourth section, fifth section, and sixth section cover oil cooling, charge air cooling, and exhaust gas recirculation cooling, respectively. The seventh section treats the integration of the different heat exchangers described in this chapter inside the front-end module. The last section provides a short overview of waste heat recovery technologies.

Thermal Energy Management in Vehicles, First Edition. Vincent Lemort, Gérard Olivier, and Georges de Pelsemaeker.
© 2023 John Wiley & Sons Ltd. Published 2023 by John Wiley & Sons Ltd.
Companion website: www.wiley.com/go/lemort/thermal

2.2   Fundamentals of Internal Combustion Engines

2.2.1   Characteristics of the Internal Combustion Engines

In an ICE, the combustion of the fuel occurs inside the engine. The resulting exhaust gases produce work during an expansion process before being expelled outside the engine. Examples of ICEs are the reciprocating engine (described in detail in this chapter) and the gas turbine equipped with an internal combustion chamber.

In an *external combustion engine*, the combustion of fuel is achieved outside the engine. The heat produced by the combustion is transferred to the engine working fluid, different from the combustion gases, to produce work. This typically occurs in a heat exchanger fed on one side by combustion gases and on the other side by the engine working fluid (e.g. air or vapor). Examples of external combustion engines are the Rankine cycle power system (presented at the end of this chapter), the externally fired gas turbine, and the Stirling engine.

Besides internal and external combustion engines, there is a third category of energy converters that can be used for vehicle powertrain, which are the fuel cells. The latter ones represent the category of electrochemical energy converters. They will be described in Chapter 4.

An ICE is called reciprocating or piston engine if the motion is generated by the linear displacement of pistons inside cylinders. Such reciprocating engines typically use a crankshaft — connecting rod mechanism to transform the linear motion of the piston into a rotational motion of the crankshaft. The latter one can, for instance, drive the wheels of a vehicle.

Figure 2.1 features a typical representation of a crankshaft-connecting rod mechanism. Conversely to reciprocating engines, in rotary engines, moving elements show a rotary motion rather than a reciprocating motion. This is in the case of Wankel engines.

Today, the main piston engines used in cars, trucks, buses, and ships are:

o The gasoline engine or spark-ignition engine (SI engine),
o The diesel engine or compression-ignition engine (CI engine).

This book focuses only on the description of these types of reciprocating ICE engines.

In these ICEs, the fuel drawn into the cylinder provides the chemical energy from which only a part is converted, after combustion, into useful mechanical energy available at the crankshaft. The remaining energy is converted and dissipated as thermal energy in the exhaust gases line and in the losses through the wall of the combustion chamber and in mechanical losses (see section 2.2.5).

The main components and geometrical characteristics of a reciprocating engine are shown in Figure 2.1. A reciprocating engine is made up of one or several cylinders inside which pistons are moving back and forth in a reciprocating fashion. Pistons are connected to the crankshaft by means of connecting rods. The crankshaft is located inside the crankcase. At the bottom of the crankcase is located the oil pan.

The volume delimited by the cylinder wall, piston head, and engine head is the cylinder volume or combustion chamber volume.

The engine's moving parts create a rotating motion from the pressure of the combustion gases. These moving parts constitute the assembly piston (with its wrist pin and rings) – connecting rod – crankshaft. The piston is the most solicited component. It transmits the force of the combustion gases and ensures the tightness between the combustion chamber and the crankcase. It must be designed to minimize the friction between the piston and cylinder. It ensures its own cooling by transferring the heat toward the cylinder and must be as light as possible to limit efforts associated with inertia.

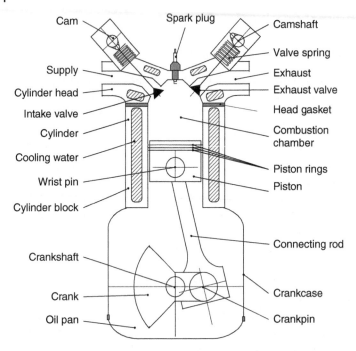

Figure 2.1 Main components of an internal combustion reciprocating engine.

The connecting rod converts the reciprocating motion of the piston into a rotating motion of the crankshaft.

The crankshaft is the final link in the chain of movement conversion before the actual use of the power.

The cylinder block ensures the good operation of the pistons and piston rings, supports the crankshaft and the cylinder head, and ensures the circulation of oil and glycol water. It must have a high thermal conductivity, must be as light as possible, and must be tight.

The head gasket is interposed between the cylinder block and the cylinder head. It is subjected to mechanical constraints (clamping pressure, pressure of the gases, pressure of oil and glycol water, and vibrations), thermal constraints (differential expansion and combustion temperature), and chemical constraints (it must maintain its physical integrity in contact with the fuel, coolant, and oil).

The engine head integrates the combustion chamber in conjunction with the piston upper surface, the supply piping, and the systems to prepare the fuel–air mixture. It ensures air supply to the cylinders and exhaust of the burned gases, integrates the valves control mechanism, and allows for the circulation of the cooling and lubricating fluids.

The valvetrain is the set of engine components, the displacement of which is synchronized with that of the piston, that ensures, at specific times of the cycle, the introduction of an air (and possible air–fuel mixture) mass necessary for the combustion and exhaust of burned gases. It is made up of valves, valve springs, valve spring retainers, rocker arms, pushrods, cam rollers and cams. The rotation of the eccentric camshaft pushes up the pushrods that open the valves through the tilting movement of the rocker arm.

The valves can seal the combustion chamber, but they also allow for the intake and exhaust of the admission fluids and burned gases, respectively.

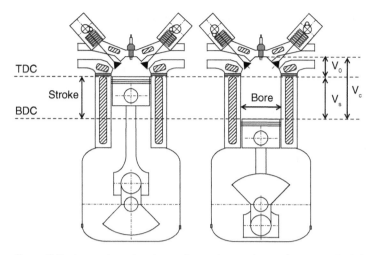

Figure 2.2 Internal combustion reciprocating engine main geometrical characteristics. (a) Piston at the TDC. (b) Piston at the BDC.

During its reciprocating motion inside the cylinder, the piston shows two particular positions. The *Top Dead Center* (TDC) is the top position of the piston (depicted in Figure 2.2 left). The *Bottom Dead Center* (BDC) is the bottom position of the piston (depicted in Figure 2.2 right).

At the TDC, the volume of the cylinder is minimum and equal to the clearance volume V_0 [m³]. At the BDC, the volume is maximum and equal to the maximum cylinder (also called total) volume V_c [m³]. The difference between the cylinder volume and the clearance volume is the volume V_s [m³] swept by the piston. It is also called the displacement volume.

The stroke S [m] is the distance traveled by the piston between the TDC and the BDC. The bore D [m] is the internal diameter of the cylinder. The volume V_s [m³] swept by the piston during one stroke is related to the stroke and bore by

$$V_s = \frac{\pi \cdot D^2}{4} \cdot S \tag{2.1}$$

The mean speed C_p [ms⁻¹] of the piston is given by

$$C_p = 2\,S\,N \tag{2.2}$$

where

S is the piston stroke, [m]
N is the rotational speed of the crankshaft, [Hz]

The total displacement volume V_d [m³] of an engine ("engine displacement") is given by the relation

$$V_d = n_c \cdot V_s \tag{2.3}$$

where

n_c is the number of cylinders of the engine [−]

In practice, the engine displacement V_d is often expressed in cubic centimeters or in liters.

Another important geometrical feature of internal combustion reciprocating engines is the compression ratio r, defined as the ratio of the volume inside the cylinder when the piston is at BDC to the volume inside the cylinder when the piston is at TDC. It is actually a volumetric compression ratio. Hence,

$$r = \frac{V_{BDC}}{V_{TDC}} = \frac{V_0 + V_s}{V_0} = \frac{V_c}{V_c - V_s}$$

(2.4)

2.2.2 Four-Stroke Engine Cycle

Cycles generally described by piston engines are four-stroke cycles or two-stroke cycles. In the automotive industry, most of the engines are four-stroke engines and follow a thermodynamic cycle with four basic steps requiring two revolutions of the crankshaft, i.e. two mechanical cycles, to be completed. Hence, the cycle factor i, defined as the number of thermodynamic cycles described per revolution of the crankshaft, is equal to 0.5.

The general principle of the four strokes for a gasoline engine (i.e. a spark-ignition engine) is represented in Figure 2.3 and described below.

- The gas intake stroke: the piston moves from TDC to the BDC position. The gas intake valve is open and the fuel–air mixture is drawn into the combustion chamber. The intake valve closes at the end of this stroke.
- The gas compression stroke: the intake and exhaust valves are closed and the piston moves up toward the TDC position. The fuel–air mixture is compressed and the cylinder pressure increases.
- The power stroke: when the piston reaches a point just before the TDC position, the spark plug ignites the fuel mixture. Due to combustion, the pressure and temperature increase, and the expansion of gases moves the piston down toward the BDC position transmitting the power to the crankshaft. More work is done on the piston during the expansion stroke than by the piston during the compression stroke.
- The gas exhaust stroke: when the piston reaches the BDC position, the exhaust valve opens allowing the gases to escape out of the cylinder. The piston moves up again from BDC to TDC, and the cylinder pressure decreases. At the end of this stroke, the exhaust valve closes, the intake valve opens, and the cycle is repeated again.

In a four-stroke engine, there is only one power stroke over the four strokes. For single-cylinder engines, this leads to problems of engine vibration. Using multiple cylinders is a way to cope with this issue.

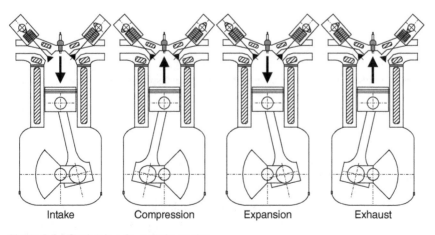

| Intake | Compression | Expansion | Exhaust |

Figure 2.3 Strokes in a four-stroke engine.

In two-stroke engines, the thermodynamic cycle is described only with two strokes — a compression and a power stroke — which makes the engine more compact since the cycle factor is equal to 1 and not 0.5 as for four-stroke engines. Also, valves are replaced by an inlet and outlet opening on the cylinder. These openings are uncovered by the position of the piston. Incomplete expulsion of exhaust gases and partial expulsion of unburned fuel can occur, thereby decreasing the efficiency of the engine. A regain of interest for such compact engines can be explained by the technical improvements implemented to improve their performance.

2.2.3 Combustion Process in the Engines

2.2.3.1 Combustion

Combustion is a chemical reaction between air and fuel (that is oxidized) that delivers the chemical energy stored in the fuel and converts it into thermal energy in exhaust gases. Air and fuels are the reactants and exhaust gases are the products.

Dry air is composed in volume of approximately 20.9% oxygen, approximately 78.1% nitrogen, and rare gases (among others, 0.9% argon). Also, as seen before, atmospheric air contains water vapor. As a simplification, it can be considered that dry air contains 21% in volume of oxygen and 79% of nitrogen. Therefore, each mole of oxygen brings 3.76 (=79/21) of nitrogen.

The fuel is generally composed of mainly carbon, hydrogen, oxygen and sulfur. It can be expressed by the general formula $C_xH_yO_zS$ with x, y, and z representing the number of atoms of carbon, hydrogen, and oxygen, respectively.

The combustion of fuel is stoichiometric when it consumes exactly the amount of air needed to achieve a complete combustion. In the absence of sulfur, the chemical reaction of a stoichiometric combustion can be written as

$$C_xH_yO_z + \left(x + \frac{y}{4} - \frac{z}{2}\right)(O_2 + 3.76\,N_2) \rightarrow x\,CO_2 + \frac{y}{2}H_2O + 3.76\left(x + \frac{y}{4} - \frac{z}{2}\right)N_2 \qquad (2.5)$$

where

O_2 stands for dioxygen
N_2 stands for dinitrogen
CO_2 stands for carbon dioxide
H_2O stands for water

A variable of interest when describing a combustion process is the fuel–air ratio. The latter is defined as the ratio of the fuel mass flow rate $\dot{m}_f$ [kg s^{-1}] to the air mass flow rate $\dot{m}_a$ [kg s^{-1}].

$$f = \frac{\dot{m}_f}{\dot{m}_a} \qquad (2.6)$$

Considering a molar ratio of oxygen to air of 21%, the fuel–air ratio of a stoichiometric combustion is given by

$$\frac{1}{f_{st}} = \frac{\left(\left(x + \frac{y}{4} - \frac{z}{2}\right)/0.21\right) \cdot MM_{air}}{x \cdot MM_C + y \cdot MM_H + z \cdot MM_O} \qquad (2.7)$$

where

MM_{air} is the molar mass of air = 28.97 [kg kmol^{-1}]
MM_C is the molar mass of atomic carbon = 12.01 [kg kmol^{-1}]
MM_H is the molar mass of atomic hydrogen = 1.008 [kg kmol^{-1}]
MM_O is the molar mass of atomic oxygen = 16 [kg kmol^{-1}]

In the actual combustion, more or less air than the stoichiometric amount of air can be provided to the combustion chamber. Fuel–air mixtures with less air and more air than the stoichiometric amount of air are called *rich* and *lean* mixtures, respectively.

Combustion with *less* air than the stoichiometric amount of air will be incomplete. In such combustion, the products contain carbon monoxide CO and, for large deficits of air, unburned fuel or graphite particles C (soot).

Combustion with *more* air than the stoichiometric amount of air yields uncombined oxygen in the products.

The amount of excess air is expressed as a percentage e of the stoichiometric amount of air. That is, the actual fuel–air ratio can be related to the stoichiometric fuel–air ratio by

$$\frac{1}{f} = \frac{1}{f_{st}}(1+e) = \frac{1}{f_{st}} \cdot \frac{1}{\phi} = \frac{\lambda}{f_{st}} \tag{2.8}$$

In the previous equation, ϕ represents the fuel–air equivalence ratio. $\phi < 1$ indicates lean combustion, while $\phi > 1$ indicates rich combustion. The inverse of ϕ is called the relative air–fuel ratio and is denoted by λ (Heywood 1988). The equation of combustion with excess air becomes

$$C_xH_yO_z + (1+e)\left(x + \frac{y}{4} - \frac{z}{2}\right)(O_2 + 3.76\,N_2) \rightarrow x\,CO_2 + \frac{y}{2}H_2O$$
$$+ e\left(x + \frac{y}{4} - \frac{z}{2}\right)O_2 + (1+e)\,3.76\left(x + \frac{y}{4} - \frac{z}{2}\right)N_2 \tag{2.9}$$

For the development of the equation of combustion in rich conditions, the reader should refer to Ferguson and Kirkpatrick (2015).

The engine can work with direct or indirect injection of fuel in the combustion chamber. The air can be supplied at the atmospheric pressure or can be compressed by means of a compressor. The latter mechanism is called forced induction and will be described in section 2.5.1.

The delivered thermal energy depends on the *heating value* (HV) of the fuel, which is defined as the heat quantity released per kilogram of fuel at reference temperature T_{ref} (generally assumed to be 25 °C) completely burned with the air and releasing combustion gases at the reference temperature. Such a chemical reaction occurs in an ideal combustion chamber as depicted in Figure 2.4.

The distinction can be made between the *low heating value (LHV)* and the *high heating value (HHV)*. When defining the LHV, it is assumed that the water in the combustion gases remains in vapor state. Conversely, the quantification of the HHV assumes that the water is condensed to liquid state, yielding a larger thermal energy, since the latent heat of condensation of water is recovered. In engineering practice, since condensation of water is usually prevented in such a way as not to produce acids that could damage pipes and heat exchangers, the *LHV* is generally considered. The distinction between *LHV* and *HHV* is illustrated in Figure 2.4.

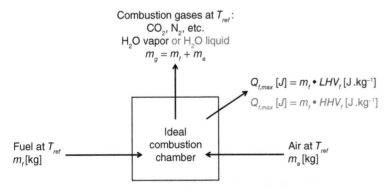

Figure 2.4 Definition of the heating value of the fuel. Source: Adapted from Çengel and Boles (2006).

Table 2.1 Properties of Gasoline and Light Diesel.

	Formula	LHV [MJ kg^{-1}][a)	HHV [MJ kg^{-1}][a)	Density [kg m^{-3}][**	f_{st} [−]
Gasoline (liquid)	$C_nH_{1.87n}$	44.0	47.2	721–785	0.0685
Light Diesel (liquid)	$C_nH_{1.8n}$	43.3	46.1	785–833	0.0690

a) at $T_{ref} = 25\,°C$; **at 20 °C and 1 atm.
Source: Adapted from Çengels and Boles (2006).

The LHV of gasoline and diesel as well as other properties are given in Table 2.1.

The flow process of Figure 2.4 can also be used to write the energy balance on the ideal combustion chamber. That is,

$$\dot{m}_f\, LHV_f(T_{ref}) = \dot{m}_a h_a(T_{ref}) + \dot{m}_f h_f(T_{ref}) - \dot{m}_g h_g(T_{ref}) \tag{2.10}$$

where

T_{ref} is the reference temperature at which the LHV is defined, [°C]
h_a is the specific enthalpy of air, [J kg^{-1}]
h_f is the specific enthalpy of fuel, [J kg^{-1}]
h_g is the specific enthalpy of combustion gases, [J kg^{-1}]

In the case of an incomplete combustion, this energy balance can be modified as

$$\dot{m}_f\, LHV_f(T_{ref}) - \dot{Q}_{incom} = \dot{m}_a h_a(T_{ref}) + \dot{m}_f h_f(T_{ref}) - \dot{m}_g h_g(T_{ref}) \tag{2.11}$$

where

$\dot{Q}_{incomp}$ are the thermal losses associated with incomplete combustion, [W]

The combustion efficiency of an actual combustion can be defined as

$$\eta_c = \frac{\dot{Q}_f}{\dot{Q}_{f,max}} = \frac{\dot{m}_f\, LHV_f - \dot{Q}_{incomp}}{\dot{m}_f\, LHV_f} \tag{2.12}$$

where

$\dot{Q}_f$ is the heat production rate generated by combustion, [W]
$\dot{Q}_{f,max}$ is the maximum heat production rate generated by combustion, [W]

The combustion efficiency depends on the fuel–air ratio: a lack of oxygen, associated with mixture richer than stoichiometric, yields an incomplete combustion and a reduced combustion efficiency (Heywood 1988). Spark-ignition engines show combustion efficiencies ranging from 95% to 98%, and compression ignition engines show combustion efficiencies larger than 98% (Heywood 1988).

Inside the combustion chamber, fuel combustion leads to the heating of the combustion products, reaching temperatures around 800 °C and then to the increase of the pressure inside the cylinder. High temperature and high pressure gases are then expanded delivering mechanical energy that can be recovered at the crankshaft.

Gasoline and diesel engines operate differently. Consequently, the characteristics of engines operating with diesel or with gasoline are different. They are described hereunder.

2.2.3.2 Spark-Ignition Engine (SI Engine)

Gasoline engines, often called SI engines, are used worldwide for passenger cars.

In these engines, fuel and air are usually pre-mixed in the intake system before entering the cylinder to obtain a mixture as homogeneous as possible before initiating the combustion. Mixing can be achieved by means of a carburetor or by injecting fuel in the intake manifold (indirect injection). Alternatively, in direct-injected engines, fuel is injected directly inside the combustion chamber.

The combustion is initiated in one or several areas of the fuel–air mixture by means of an electric spark plug.

2.2.3.3 Compression-Ignition Engine (CI Engine)

Diesel engines, commonly called CI engines, are well known in heavy duty applications and also in passenger cars.

In diesel engines, the process of combustion differs from the gasoline one with carburetor. As shown in Figure 2.5 (right), the air enters through the intake system inside the combustion chamber often at a boosted pressure created by the forced-induction system (see Section 2.5.1). During the compression phase of the cycle, when the piston moves up toward the TDC position, the air pressure increases leading to an increase of its temperature. The fuel is then injected under the form of micro-droplets directly inside the combustion chamber containing high pressure and high temperature air. Pressure and temperature of air are above the ignition point of fuel (Heywood 1988). Fuel is vaporized and mixed with the air to initiate the combustion.

The mixture is generally heterogeneous. Conversely to SI engines, the combustion is not initiated by an external device (spark plug), but by auto-ignition of the fuel made possible by the conditions of pressure and temperature inside the cylinder.

In SI engines, the compression ratio typically ranges from 7 to 12. The upper limit is imposed to prevent any autoignition of the fuel–air mixture, which can be a source of noise and a cause of damage of the engine. Since CI engines compress only air, they are not subject to autoignition and can be operated under larger compression ratios than those of SI engines, in the range of 12 to 24 (Heywood 1988). Also, they can burn less refined and expensive fuels.

2.2.4 Pollutant Emissions

2.2.4.1 Driving Cycles and Pollutant Emissions

Pollutant emissions are limited by regulation in Europe since 1970 with a major step in 1993 imposing catalytic converters on spark-ignition engines (EUR1 emission standard).

Regulations concerning cars pollutant emissions and fuel consumption aim at limiting the latter ones. Regulations describe the procedure to measure pollutant emissions and fuel consumption during a defined chassis dynamometer test realized under specific conditions.

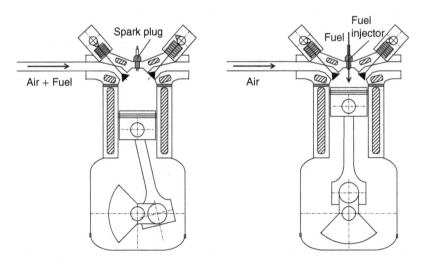

Figure 2.5 Difference between a SI engine (a) and a CI engine (b).

The driving cycle, supposed to represent the typical driving conditions of a car, is defined as the time evolution of the car vehicle speed that needs to be followed during the test. In practice, discrepancies appear between the measured values and actual values. These differences are related to the driving style, vehicle maintenance, driving speed, weather conditions including outdoor temperature, road traffic and conditions, and used accessories (air-conditioning, lighting, etc.).

One of the driving cycles is the *New European Driving Cycle*, NEDC (introduced in 1970), which consists of four repeated Urban Driving Cycles (UDC) of 200 seconds each (with reduced driving speeds and loads) and one Extra-Urban Driving Cycle (EUDC) of 380 seconds (with larger speeds reaching 120 km h^{-1}). Such a driving cycle (represented in Figure 2.28 discussed later in this chapter) is reproduced on a chassis dynamometer. Meanwhile, pollutants are measured.

Since 2017, new driving cycles, called *worldwide harmonized light vehicles test cycles* (WLTC), which are part of the *worldwide harmonized light vehicles test procedures* (WLTP), have been replacing progressively the NEDC cycle. This more representative driving cycles procedure (more representative of real-time driving), represented in Figure 2.6, defines a globally harmonized standard for determining the level of pollutants and CO_2 emissions and fuel consumption from vehicles including more severe tests with the objective to be closer and more consistent with the customer use. This cycle also accounts for auxiliaries such as air-conditioning (which, at 30 °C outdoor temperature, increases the fuel consumption by 1 l 100 km^{-1}) yielding an additional pollutant emission of thermal engine and a decrease of driving range of electric vehicles. There exist different WLTC cycles according to the vehicle category defined as a function of the vehicle power to mass ratio. The average speed on the cycle is 47 km h^{-1}, while it was 20 km h^{-1} for a NEDC. Also, the cycle is longer.

Regulations also include the implementation of the protocol *real drive emissions* (RDE), which is a measurement conducted on real road test.

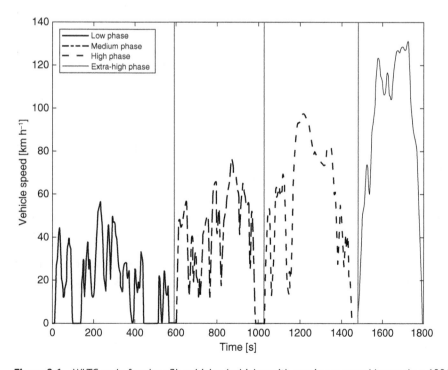

Figure 2.6 WLTC cycle for class 3b vehicles (vehicles with maximum speed larger than 120 km h^{-1}).

Regulations concerning pollutant emissions are more and more stringent and require technological solutions to be developed to respect them.

Ecological incentives also encourage to research new levers of CO_2 emission and customer consumption reduction.

2.2.4.2 Pollutants

The fuel molecule can be defined with the more general formula $C_x H_y O_z S$ composed of a high content of carbon and hydrogen and with oxygen and sulfur. Air is mainly composed of oxygen and nitrogen.

As explained previously, the chemical reaction between the fuel and air provides combustion with the production of different exhaust gases.

If the combustion is complete, the exhaust gases are composed of carbon dioxide CO_2, water H_2O, dinitrogen N_2, and sulfur dioxide SO_2 (mainly produced from the oxidation of sulfur contents if the fuel contains sulfur).

Usually, the combustion is not complete because fuel and air are not mixed in the stoichiometric proportions. It should be mentioned that incomplete combustion can also occur when fuel is burned with more air than the stoichiometric amount of air. Incomplete combustion could actually result from: a lack of mixing between fuel and air, too short time to allow complete combustion, dissociation occurring at high temperatures (Çengel and Boles, 2006), or temperatures not high enough for the reaction to take place.

In the case of an incomplete combustion, the other main products are:

- CO: carbon monoxide, which is formed mainly in rich mixtures
- NO_x: nitrogen oxides, due to the reaction of N_2 with O_2
- HC: unburned and partially burned hydrocarbons
- Particulate matter emissions (soot), containing black carbon (which is pure carbon) among others

These products are pollutants because of their adverse effect on the environment or on people's health.

In gasoline engines, the main pollutant emissions are HC, CO, and NO_x.

In diesel engines, the main pollutant emissions are NO_x (produced in a lean mixture at very high temperature) and soot (produced by a rich mixture at high temperature).

Carbon monoxide (CO) is produced when there is not enough air, and hence oxygen, in the mixture (rich mixture) to form CO_2 or when the temperature is not high enough.

Unburned hydrocarbon (HC) emissions are the fuel quantity that is not burned and goes directly to the exhaust line or that has not burned completely. HC emissions, as CO emissions, are produced when there is not enough oxygen (rich mixture).

NO_x emissions are the result of the oxidation of nitrogen in air at high temperature and produced if there is an excess of air (lean mixture) during combustion.

Particulate emissions, or soot emissions, are produced at high temperature conditions when there is a lack of air in the mixture (rich mixture). In diesel engines, the mixture is heterogeneous, and there are rich zones producing a high quantity of soot and lean zones.

In summary, if the mixture is rich (excess of fuel), the combustion is incomplete and produces HC, CO, and soot. If the mixture is lean (excess of air), the combustion produces NO_x due to an excess of O_2.

2.2.4.3 Trade-off and Technological Levers

It seems to be difficult to reduce in the same way the CO_2 emissions and the other pollutant emissions, particularly NO_x. A trade-off is therefore necessary between both objectives.

For example, the engine efficiency is optimal when the heat source temperature is maximal. But the production of nitrogen oxide in a diesel engine increases with the heat source temperature.

The two main ways to control and reduce the pollutant emissions are:

– the optimization of the combustion by managing the reactant mixture composition and temperature during the combustion to reduce the emissions at the source.
– the treatment of the pollutants by adding after-treatment devices to trap the pollutants or convert them through a chemical reaction.

Regarding reduction of emissions *at the source*, among different technological solutions, the exhaust gas recirculation (EGR) system is a technology mainly used to reduce NO_x at the source. It will be described in Section 2.6. After the combustion, the exhaust gases go through the exhaust line. Part of these gases containing CO_2 are mixed with fresh air and reintroduced into the cylinder. The amount of CO_2 introduced in the mixture inside the engine increases, thereby reducing the relative amount of O_2, increasing the fraction of unreactive gases in the intake mixture, and reducing the combustion temperature. As the temperature decreases, NO_x production decreases but soot increases. EGR is commonly used in diesel engines to reduce NO_x emissions and is increasingly being used in modern gasoline engines with direct injection.

The exhaust gases can also pass through *after-treatment* devices with the objective to convert HC and CO into H_2O and CO_2 by oxidation process, to convert NO_x into H_2O and N_2 by a reduction process, and to filter and burn the particles to give CO_2 and H_2O.

Solid particles present in combustion gases are retained by a porous material filter. The efficiency depends on filter porosity and filtration surface. The accumulation of particles inside the filter will increase the engine back-pressure, negatively impacting the performance of the latter one. It is therefore necessary to regenerate the filter by combustion of the trapped particles. Particles mainly consist of carbon. Their combustion temperature ranges from roughly 500 to 600 °C, temperatures that are seldomly reached in normal operating conditions of diesel engines. In diesel engines, the strategy consists of increasing the temperature of the exhaust gases by acting on engine parameters and through post-injections.

Major technologies of NO_x reduction consist of:

– Catalytic decomposition of NO.
– The storage of NO_x (NO_x-trap catalyst): the catalyst stores and treats NO_x by using reducing gases emitted by the engine during specific operating phases. Nitrogen oxides are therefore indirectly treated by the fuel.
– The selective reduction of NO_x in the presence of hydrocarbons (HC-DeNox)
– The selective reduction of NO_x by urea (SCR: selective catalytic reduction). It reduces NO_x on a catalytic substrate, with ammonia (NH_3) generated from the decomposition of aqueous urea solution (AdBlue) and injected in exhaust gases. This system requires, in addition to the volume of the catalyst in the line, the implementation of a tank (with volume typically ranging from 10 to 15 L) and an AdBlue injection system. The AdBlue consumption is around 1 L/1000 km.

The NO_x-trap is cheaper than the SCR, but its efficiency decreases with the engine load and with the exhaust gas temperature.

NO_x post-treating enlarges the possibilities of engine settings allowing for a better trade-off between CO_2/NO_x (see the section about EGR indicating necessary trade-offs).

With the increase in the efficiency of thermal engines, the available temperatures at the engine exhaust for the after-treatment systems are decreased. This yields a decrease in the performance of the after-treatment systems. To increase this temperature, it may be interesting to insulate and reduce the thermal inertia of the exhaust-line components.

2.2.5 Energy Analysis

Different losses appear along the process of chemical energy conversion into useful mechanical energy inside an ICE. An energy analysis helps to understand the main energy conversion and transfer processes and the associated losses inside an ICE and ultimately their impact on the engine thermal management system.

In a real engine cycle, there are different sources of energy conversion losses that reduce the efficiency from the ideal cycle. Around 55–60% of the energy of the burned fuel is lost. In the combustion chamber, losses associated with exchange between gases and wall constitute a significant part of the chemical energy introduced by the fuel in the engine.

In addition, the conversion of the produced power into useful mechanical power makes some mechanical losses appear that can dissipate around 5–10% of the energy. These mechanical losses are due to the friction between all moving mechanical parts and to the mechanical energy needed to drive the different accessories of the engine. These losses evolve in function of the operating point and the thermal conditions particularly during the transient phase.

It is therefore 60–70% of the energy introduced in the engine that is lost yielding an engine efficiency of around 30–40%.

2.2.5.1 Energy Conversion Processes in Engines

The energy flow chart represented in Figure 2.7 can be used to express the overall energy balance across the engine. Such a flow chart relies on some assumptions and simplifications of the actual processes involved in an ICE. Among others, it is assumed that all processes are in the steady flow regime.

Fuel enters the engine at a mass flow rate $\dot{m}_f$ [kg s^{-1}] and at a specific enthalpy $h_{f,su,eng}$ [J kg^{-1}].

Simultaneously, air enters the engine at a mass flow rate $\dot{m}_a$ [kg s^{-1}] and at a specific enthalpy $h_{a,su,eng}$ [J kg^{-1}]. If a turbocharger is used, air is compressed in a compressor and cooled down in a charge air cooler (CAC). If the latter is air-cooled, a power $\dot{Q}_{CAC}$ is rejected to the ambient.

2.2.5.1.1 Combustion of Fuel

Fuel and air enter the combustion chamber at specific enthalpies of $h_{f,su,cc}$ and $h_{a,su,cc}$, respectively. The engine combustion chamber, delimited by the cylinders' walls and the pistons, can be fictitiously represented by the association of an adiabatic combustion chamber and a heat engine. The adiabatic combustion chamber describes the combustion of fuel, while the heat engine describes the conversion of heat into work. The adiabatic combustion chamber comprises four fictitious heat exchangers. Heat exchangers 1 and 2 cool or heat the reactants to reach the reference temperature T_{ref} at which the LHV of the fuel is defined (see Section 2.2.3.1). That is,

$$\dot{Q}_1 = \dot{m}_f(h_{f,su,cc} - h_{f,ref})$$

$$\dot{Q}_2 = \dot{m}_a(h_{a,su,cc} - h_{a,ref}) \tag{2.13}$$

Heat exchanger 3 represents the heat production rate released by the combustion of fuel at the reference temperature T_{ref}. Combustion allows for the release of the chemical energy contained in the fuel.

$$\dot{Q}_3 = \dot{Q}_f = \eta_C \, \dot{m}_f \, LHV_f \tag{2.14}$$

As shown previously, generally, the combustion is not complete and involves some unburned fuel, the energy of which is lost. The combustion efficiency η_c is typically around 98–99%.

Heat exchanger 4 heats the combustion gases from the reference temperature to the adiabatic combustion temperature.

$$\dot{Q}_4 = \dot{m}_g \, (h_{g,adiab} - h_{g,ref}) = (\dot{m}_a + \dot{m}_f)(h_{g,adiab} - h_{g,ref}) \tag{2.15}$$

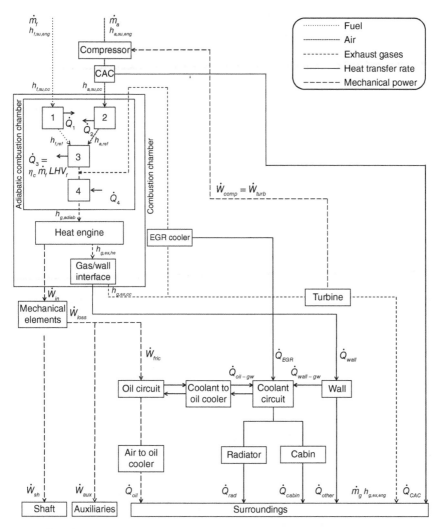

Figure 2.7 Representation of engine energy balance (engine equipped with a turbocharger and a high-pressure EGR).

Since the combustion chamber is described as adiabatic, the net heat production rate provided to the chamber must be null. It yields

$$\dot{Q}_1 + \dot{Q}_2 + \dot{Q}_3 - \dot{Q}_4 = 0 \tag{2.16}$$

The latter equation can be used to compute the enthalpy of the exhaust gases $h_{g,adiab}$ and hence the adiabatic combustion temperature $T_{g,adiab}$.

2.2.5.1.2 Conversion of Heat into Mechanical Work After the fuel burns, the heat produced by combustion ($Q_f = \eta_c\, m_f\, LHV_f$) is partly converted into work through the engine thermodynamic cycle. The work done by the combustion gases on the piston during one cycle is called the indicated work per cylinder and per cycle W_{in}. Such a work can be computed by integrating the area enclosed by the pressure–volume diagram (indicator diagram).

$$W_{in} = \oint p \cdot dV = IMEP \cdot \frac{\pi \cdot D^2}{4} \cdot S = IMEP \cdot V_s \tag{2.17}$$

As developed later, the indicated mean effective pressure (IMEP) is the theoretical constant pressure applied on the piston along the stroke S and giving the indicated work W_{in}.

The rate of work transfer from the combustion gases to the piston is called the indicated power $\dot{W}_{in}$ [W] (Heywood, 1988) and is the complete power developed in the cylinders resulting from the fluid pressure acting on the pistons.

$$\dot{W}_{in} = i\,N\,n_c\,W_{in} = i\,N\,n_c \cdot IMEP \cdot V_s \tag{2.18}$$

where

i is the cycle frequency (the number of thermodynamic cycle per crankshaft revolution), [−]
N is the crankshaft rotational speed, [Hz]
n_c is the number of cylinders, [−]

As mentioned previously, the working cycle frequency (or cycle factor) is equal to 0.5 for a four-stroke engine and to 1 for a two-stroke engine.

Note that the indicated work is defined here as the *net* indicated work and accounts for the pumping work associated with intake and exhaust strokes. The net indicated work is the *gross* indicated work minus the pumping work. An example of measured indicated work is given in Figure 2.8.

The *indicated thermal efficiency* is defined as

$$\eta_{th,in} = \frac{\dot{W}_{in}}{n_c\,\dot{m}_f\,LHV_f} = \frac{W_{in}}{n_c\,m_f LHV_f} \tag{2.19}$$

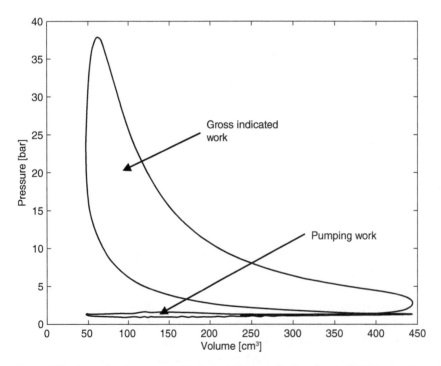

Figure 2.8 Example of a measured indicated diagram. Engine: Toyota SI NA (naturally aspirated) 1.6 4A-GE with T-VIS technology – 120HP at 7200 rpm – 150 N m at 4800 rpm. Testing conditions: throttle of 25%, regime of 4400 rpm. Source: Data provided by Technifutur.

The *indicated fuel conversion efficiency* includes the combustion efficiency and is defined as

$$\eta_{f,in} = \frac{\dot{W}_{in}}{\dot{m}_f \, LHV_f} = \frac{W_{in}}{m_f LHV_f} \tag{2.20}$$

That is, the indicated fuel conversion efficiency, indicated thermal efficiency, and combustion efficiency are related by (Heywood 1988)

$$\eta_{f,in} = \eta_c \, \eta_{th,in} \tag{2.21}$$

The indicated thermal efficiency is inherent to the thermodynamic cycle and quantifies the performance of the system in converting heat into work.

The indicated thermal efficiency depends on the compression ratio. As mentioned earlier, the latter is defined as the ratio of the volume of the combustion chamber at *BDC* to that at *TDC*. To develop simple qualitative relations between the indicated thermal efficiency and the compression ratio, the concept of ideal cycle can be introduced. An ideal cycle is made of internally reversible processes, wherein the combustion is represented by a heat-addition process and the exhaust gas rejection is represented by a heat-release process (Çengel and Boles, 2006). Additionally, the working fluid is assumed to be an ideal gas characterized by constant c_p and c_v values (Heywood, 1988).

The cycle idealizing the spark-ignition engine (Otto cycle) assumes that combustion occurs at a constant volume. For such an ideal cycle, the thermal efficiency is related to the compression ratio r by

$$\eta_{th,otto} = \frac{W_{in}}{Q_{su}} = 1 - \frac{1}{r^{\gamma-1}} \tag{2.22}$$

where

   r is the compression ratio, [−]
   γ is the specific heat ratio of the gas, $(=c_p/c_v)$ [−]

The ideal cycle representing the compression-ignition cycle (Diesel cycle) assumes that the combustion occurs at constant pressure. The ratio of the combustion chamber volume at the end of the combustion to the combustion chamber volume at the beginning of the combustion is the cutoff ratio r_c. The cycle efficiency is related to the compression ratio r and the cutoff ratio r_c by

$$\eta_{th,diesel} = \frac{W_{in}}{Q_{su}} = 1 - \frac{1}{r^{\gamma-1}} \left[\frac{r_c^{\gamma} - 1}{\gamma(r_c - 1)} \right] \tag{2.23}$$

where

   r_c is the cutoff ratio, [−]

The Otto and diesel cycles can be visualized and compared in the P—V diagrams shown in Figure 2.9. Typical values of compression ratios were given previously.

It should be stressed that the thermal efficiency $\eta_{th,in}$ of the actual engine is lower than that of the ideal cycle it represents. However, the performance of ideal cycles behaves qualitatively in the same manner as the actual engines under variation of operating conditions. For instance, increasing the compression ratio yields an increase of the thermal efficiencies of both the actual and ideal cycles.

Despite the fact that the expression in brackets in Eq. (2.23) is larger than 1, the thermal efficiency of diesel engines (CI engines) is higher than that of gasoline engines (SI engines). Actually, as mentioned earlier, diesel engines operate under a larger compression ratio. Also, in the ideal diesel cycle, the specific heat ratio γ is higher because only air is compressed during the compression stroke.

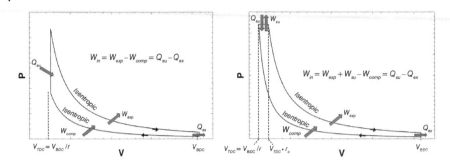

Figure 2.9 Representation of the Otto (a) and Diesel (b) cycles in the P–V diagrams.

In Figure 2.7, the conversion of heat to the indicated work is represented by the heat engine section. Combustion gases coming from the adiabatic chamber enter the heat engine section, where part of the heat is converted into the indicated work. In terms of power, assuming the heat engine to be adiabatic, we can write

$$\dot{W}_{in} = \dot{m}_g \left(h_{g,adiab} - h_{g,ex,he} \right) \tag{2.24}$$

2.2.5.1.3 *Transmission of the Mechanical Work*

Only part of the work done by the gas on the piston will be available in a useful form at the engine shaft. Actually, part of the indicated work is used to overcome friction between the moving elements (for instance, the bearings and the pistons) and to drive engine auxiliaries (pumps, etc.). All these losses are merged into the mechanical losses term W_{loss}. We can write

$$W_{loss} = W_{fric} + W_{aux} \tag{2.25}$$

The *brake power*, also named useful or effective or shaft power, is the energy produced per unit of time and delivered at the outlet of the engine. It is expressed as the torque delivered on the shaft multiplied by the rotational speed of the engine. That is,

$$\dot{W}_{sh} = T_{sh} \cdot \omega = T_{sh} \cdot 2 \cdot \pi \cdot N = T_{sh} \cdot 2 \cdot \pi \cdot \frac{\text{rpm}}{60} \tag{2.26}$$

where

$\dot{W}_{sh}$ is the brake or shaft power, [W]
T_{sh} is the engine torque, [N · m]
ω is the rotational speed, [*rad/s*]
N is the rotational speed, [Hz]

The brake power is determined by measuring the torque T_{sh} and the rotational speed N at the engine shaft. The torque is measured by means of a brake dynamometer.

The brake power is equal to the indicated power diminished by mechanical losses. That is,

$$\dot{W}_{sh} = \dot{W}_{in} - \dot{W}_{loss} \tag{2.27}$$

Mechanical losses range from 5% to 10% of the indicated power $\dot{W}_{in}$ and evolve in function of the operating point and of the thermal conditions especially during the starting phases or at zero load.

These friction losses result from the relative movement between components in different zones, particularly the piston rings–cylinder liner assembly, the crankshaft and output shaft bearings, and the camshaft bearings. The friction losses level mainly depend on the temperature of the oil and the engine metallic masses and on the engine rotational speed. These losses lead to heat dissipation localized mainly on the piston–cylinder assembly and in the oil circuit inducing an overconsumption particularly when the engine is cold. The heat production associated with the friction losses in the engine is not negligible.

During the thermal transient phases, the optimal level of the oil temperature is not reached particularly at the cold start when the oil has a temperature close to the ambient temperature. This is the reason why the friction losses are much higher than during stabilized conditions when the oil temperature is higher. Reducing the warm-up time during the transient phases and maintaining the optimized oil temperature allow for the reduction of the friction losses.

The mechanical efficiency, defined as the ratio of the brake power to the indicated power, can be used to quantify the relative impact of mechanical losses. Hence,

$$\eta_m = \frac{\dot{W}_{sh}}{\dot{W}_{in}} \tag{2.28}$$

The *brake fuel conversion efficiency* is defined as the indicated fuel conversion efficiency multiplied by the mechanical efficiency. That gives

$$\eta_{f,sh} = \eta_{f,in}\,\eta_m \tag{2.29}$$

The brake fuel conversion efficiency is the most global indicator of performance of the engine.

It should be stressed that the aforementioned analysis has assumed the engine to be in the steady-state regime. During the warm-up thermal transient phase, the thermal inertia of the coolant fluid, oil, and metallic masses are important and involve higher thermal losses through the combustion wall.

2.2.5.1.4 Energy Conversion Processes on the Exhaust Gas Line
Gases leaving the combustion chamber flow through the turbine of the turbocharger. The turbine delivers a power $\dot{W}_{turb}$ that is transferred to the compressor.

If no other elements are placed on the exhaust line, the exhaust gases at the outlet of the turbine are rejected to the ambient at an enthalpy $h_{g,exhaust}$.

Part of the energy of the exhaust gases can be valorized through other mechanisms: turbo-compound, thermo-electric generators (TEG), or Rankine cycle (RC) systems. To date, only turbocompound is commercialized, but intense research has been conducted on TEG and RC, especially during the period 2000–2020.

2.2.5.1.5 Transfer and Conversion of the Heat Released at the Combustion Chamber Walls
Exhaust gases from the heat engine section exchange heat with the combustion chamber wall, which allows for the computation of the gas enthalpy at the outlet of the combustion chamber $h_{g,ex,cc}$. Energy balance on the exhaust gases gives

$$\dot{Q}_{wall} = \dot{m}_g\,(h_{g,ex,he} - h_{g,ex,cc}) \tag{2.30}$$

The rate of heat $\dot{Q}_{wall}$ exchanged between the exhaust gases and the walls is injected into the walls. The latter are cooled down by means of the coolant loop $\dot{Q}_{wall-gw}$, and also by natural convection and by radiation from the engine to the ambient ($\dot{Q}_{other}$).

The coolant circuit is mainly cooled by the radiator ($\dot{Q}_{rad}$). Part of the heat of the circuit is also released in the cabin through the heater core ($\dot{Q}_{cabin}$), whose function is to heat the cabin (see Chapter 3). Additionally, the coolant circuit can exchange heat with the oil circuit through the coolant to oil cooler ($\dot{Q}_{oil-gw}$).

The oil circuit can be heated up or cooled down by the coolant through the coolant-to-oil cooler ($\dot{Q}_{coolant,oil}$). The oil circuit is heated by the frictional losses ($\dot{W}_{fric}$). In the configuration shown in

Figure 2.7, it is cooled down by the air through the air-to-oil cooler ($\dot{Q}_{oil}$). As it will be seen in Section 2.4, oil is either cooled down by coolant or by air.

2.2.5.2 Engine Overall Energy Balance

In steady-state regime, the overall energy balance across the engine can be expressed as

$$\dot{m}_f h_{f,su,eng} + \dot{m}_a h_{a,su,eng} = \dot{W}_{sh} + \dot{W}_{aux} + \dot{Q}_{oil} + \dot{Q}_{rad} + \dot{Q}_{cabin} + \dot{Q}_{other} + \dot{Q}_{CAC} + \dot{m}_g h_{g,ex,eng}$$
(2.31)

When writing this energy balance, the thermodynamic system is assumed to comprise the combustion chamber and all the surrounding components as shown in Figure 2.7.

Also, the mass balance across the engine imposes that

$$\dot{m}_g = \dot{m}_a + \dot{m}_f$$
(2.32)

The enthalpy of the fuel entering the engine can be written as

$$h_{f,su,eng} = h_{f,form,ref} + h_{f,su,eng,sens} - h_{f,sens,ref} = h_{f,form,ref} + \Delta h_{f,sens}$$
(2.33)

where

$h_{f,su,eng}$ is the specific enthalpy of fuel entering the engine, [J kg^{-1}]
$h_{f,form,ref}$ is the specific enthalpy of formation of fuel at the standard reference state, [J kg^{-1}]
$h_{f,su,eng,\,sens}$ is the specific sensible enthalpy of fuel entering the engine, [J kg^{-1}]
$h_{f,sens,ref}$ is the specific sensible enthalpy of fuel at the standard reference state, [J kg^{-1}]

The enthalpy of formation at the standard reference state is defined as the energy absorbed or released when a chemical compound is formed from stable chemical elements during a steady-flow process at 25 °C and 1 atm (Çengel and Boles 2006).

In a similar manner, the enthalpy of air entering the engine can be expressed as

$$h_{a,su,eng} = h_{a,form,ref} + h_{a,su,eng,sens} - h_{a,sens,ref} = h_{a,form,ref} + \Delta h_{a,sens}$$
(2.34)

The enthalpy of the exhaust gases can also be expressed as a function of the enthalpy of formation.

$$h_{g,ex,eng} = h_{g,form,ref} + h_{g,ex,eng,sens} - h_{g,sens,ref} = h_{g,form,ref} + \Delta h_{g,sens}$$
(2.35)

Combining Eq. (2.31), Eq. (2.33), Eq. (2.34), and Eq. (2.35), we obtain

$$\dot{m}_f \Delta h_{f,sens} + \dot{m}_a \Delta h_{a,sens} + (\dot{m}_f h_{f,form,ref} + \dot{m}_a h_{a,form,ref} - \dot{m}_g h_{g,form,ref}) = \dot{W}_{sh} + \dot{W}_{aux} + \dot{Q}_{oil}$$
$$+ \dot{Q}_{rad} + \dot{Q}_{cabin} + \dot{Q}_{other} + \dot{Q}_{CAC} + \dot{m}_g \Delta h_{g,sens}$$
(2.36)

The latter equation can be written as

$$\dot{Q}_f + \dot{Q}_a + \eta_C \dot{m}_f LHV_f = \dot{W}_{sh} + \dot{W}_{aux} + \dot{Q}_{oil} + \dot{Q}_{rad} + \dot{Q}_{cabin} + \dot{Q}_{other} + \dot{Q}_{CAC} + \dot{Q}_g$$
(2.37)

$\dot{Q}_f$ and $\dot{Q}_a$ represent the rates of heat transfer provided or recovered to bring the fuel and air from the engine supply state to the reference state.

It can be even more manipulated to make appear the thermal losses associated with incomplete combustion. That is,

$$\dot{Q}_f + \dot{Q}_a + \dot{m}_f LHV_f = \dot{W}_{sh} + \dot{W}_{aux} + \dot{Q}_{oil} + \dot{Q}_{rad} + \dot{Q}_{cabin} + \dot{Q}_{other} + \dot{Q}_{CAC} + \dot{Q}_g + \dot{Q}_{incomp}$$
(2.38)

Assuming that the fuel and air enter the engine at the reference state, Eq. (2.38) reduces to

$$\dot{m}_f LHV_f = \dot{W}_{sh} + \dot{W}_{aux} + \dot{Q}_{oil} + \dot{Q}_{rad} + \dot{Q}_{cabin} + \dot{Q}_{other} + \dot{Q}_{CAC} + \dot{Q}_g + \dot{Q}_{incomp}$$
(2.39)

This equation expresses that the heat generation rate associated with the combustion of fuel $\dot{m}_f LHV_f$ is split into the shaft power $\dot{W}_{sh}$, the power consumed by the auxiliaries $\dot{W}_{aux}$, the heat transfer rate rejected by the oil cooler $\dot{Q}_{oil}$, the heat transfer rate rejected by the radiator $\dot{Q}_{rad}$, the heat transfer rate injected into the cabin $\dot{Q}_{cabin}$, the heat transfer rate dissipated from the engine block to the ambient $\dot{Q}_{other}$, the heat transfer rate rejected through the cooling of the charge air $\dot{Q}_{CAC}$, the heat transfer rate available in the exhaust gases $\dot{Q}_g$, and the heat transfer rate associated with incomplete combustion $\dot{Q}_{incomp}$.

Neglecting the auxiliary losses, cabin heating, other losses, charge air cooling, and incomplete combustion losses, we can write

$$\dot{m}_f LHV_f = \dot{W}_{sh} + \dot{Q}_{oil} + \dot{Q}_{rad} + \dot{Q}_g \tag{2.40}$$

To give orders of magnitude of losses, we can consider that, with a brake fuel conversion efficiency of the engine of 35%, the different losses are shared with the approximate respective values of 5% in the oil circuit ($\dot{Q}_{oil}$), 20% in the coolant circuit ($\dot{Q}_{rad}$), and 40% in the exhaust gases line ($\dot{Q}_g$). The share of the losses depends on the engine operating conditions.

Note that the energy balance can also be expressed upward the coolant circuit. It yields

$$\dot{m}_f LHV_f = \dot{W}_{sh} + \dot{W}_{loss} + \dot{Q}_{EGR} + \dot{Q}_{wall} + \dot{Q}_{CAC} + \dot{Q}_g + \dot{Q}_{incomp} \tag{2.41}$$

During the transient thermal phase, the thermal conditions, particularly the coolant and oil temperatures, are key characteristics in the energy exchanges.

At the beginning of the engine warm-up phase, the coolant temperature is still low. The large temperature gradient between the combustion chamber wall and the coolant increases the energy losses through the wall. Similarly, the low oil temperature increases the friction losses.

Moreover, the thermal inertia of the volume of coolant, oil and metallic masses involve other energy losses while these parts are heated. In steady-state conditions, these thermal inertia effects disappear.

The heat associated with the different losses could be partly recovered. The end of this chapter presents some solutions or technologies to recover heat for heating purpose or to convert heat into mechanical or electrical energy.

2.2.5.3 Engine Overall Energy Performance Indicators

2.2.5.3.1 Overall Engine Efficiency The *brake fuel conversion efficiency* has been introduced earlier. It quantifies the performance of the engine to convert the available chemical energy introduced by the fuel into a useful mechanical energy on the crankshaft available to power the vehicle. This efficiency accounts for combustion efficiency, thermal conversion efficiency, and mechanical losses.

$$\eta_{f,sh} = \frac{\dot{W}_{sh}}{\dot{m}_f LHV_f} = \eta_c \, \eta_{th,in} \, \eta_m \tag{2.42}$$

In technical literature, this efficiency is also called effective efficiency or *overall engine efficiency*.

2.2.5.3.2 Mean Effective Pressures As mentioned earlier, the mean effective pressure (MEP) [Pa] is the theoretical constant pressure applied on the piston along the stroke S [m] and giving the work W [J].

$$W = MEP \cdot \frac{\pi . D^2}{4} \cdot S = MEP \cdot V_s = \oint p \cdot dV \tag{2.43}$$

The power $\dot{W}$ [W] can be expressed as a function of the MEP by

$$\dot{W} = i \cdot N \cdot n_C \cdot MEP \cdot V_s \tag{2.44}$$

The distinction can be made between:

- the brake mean effective pressure (BMEP), computed from the engine shaft power $\dot{W}_{sh}$
- the indicated mean effective pressure (IMEP), computed from the indicated power $\dot{W}_{in}$

Also, the friction mean effective pressure (FMEP) can be defined as

$$FMEP = IMEP - BMEP \tag{2.45}$$

From Eq. (2.44), it can be observed that the BMEP is a relative assessment of the engine performance. This is an indicator of the efficiency of the engine to use the available displacement (Heywood 1988). Hence, we have for instance

$$BMEP_{turbocharged\ engines} > BMEP_{naturally\ aspirated\ engines} \tag{2.46}$$

The BMEP can be used to compare different engines. Alternatively, for a given shaft power, the engine displacement can be evaluated based on typical values of BMEP.

$$V_s = \frac{\dot{W}_{sh}}{i \cdot N \cdot n_C \cdot BMEP} \tag{2.47}$$

Typical values of BMEP can be found in the literature.

2.2.6 Quantification of the Major Heat Transfers in ICEs

2.2.6.1 Heat Transfer Between Gases and Engine Walls

The energy losses associated with the thermal exchanges between the gases and the combustion chamber and cylinder head wall, combined with the thermal energy in the exhaust gas line, represent the main energy losses of the thermodynamic cycle. The other main energy losses are the mechanical losses

The thermal exchanges between the gases and the wall are made up of *convection* by the aerodynamic effects of the gas movement inside the combustion chamber and by *radiation* mainly by the particles in the case of diesel engines.

The thermal exchanges in the combustion chamber take place between the inlet mixture or combustion gases and the combustion chamber wall (cylinder jacket, piston, and cylinder head). They are largely dependent on the aerodynamic conditions inside the combustion chamber, particularly due to the piston movement in the cylinder, to radiation mainly by the particles in the diesel engine case and on the temperature gradient between the gases and the wall. These thermal exchanges also depend on the engine operating point and are important in fully loaded conditions.

The expression of these thermal exchanges introduces a convective term characterized by a convective exchange coefficient h_c, [W $\cdot$ m^{-2} $\cdot$ K^{-1}], and a radiative term:

$$\dot{Q}_{wall} = A_{wall}\, h_c\, (T_g - T_{wall}) + A_{wall}\, K\, \sigma\, \left(T_g^4 - T_{wall}^4\right) \tag{2.48}$$

where

$\dot{Q}_{wall}$ is the heat transfer rate between the gases and the wall, [W]
A_{wall} is the heat transfer area between the gases and the wall, [m^2]
h_c is the convective heat transfer coefficient between the gases and the wall, [W m^{-2}K^{-1}]
T_g is the gas temperature, [K]
T_{wall} is the wall temperature, [K]
σ Stefan constant parameter, 5.67×10^{-8}, [W m^{-2} K^{-4}]

In the radiative heat transfer, we can distinguish radiation from gases and radiation from soot particles. The heat exchange by radiation can generally be neglected in SI engines (Heywood, 1988).

However, this is not the case in diesel engines, where the radiation from soot particles is approximately five times higher than that from combustion gases (Heywood 1988).

In Eq. (2.48), K is a constant. Annand (1963) (cited in Amara (1994)) suggested to use 0.529 for diesel (CI) engines.

With a global thermal exchange coefficient h_g, [W m^{-2} K^{-1}] defined after gathering the convective and radiative terms, the following correlation from Ferguson and Kirkpatrick (2015), originally proposed by Woschni (1967) (and also recalled in Heywood (1988)), is often used. This correlation computes the instantaneous cylinder average global heat transfer coefficient $h_g(\theta)$ as function of the crank angle θ.

$$h_g[W\ m^{-2}K^{-1}] = 3,26 \cdot D^{-0,2}\ P^{0,8}\ T_g^{-0,55}\ C^{0,8} \tag{2.49}$$

where

> D is the cylinder bore, [m]
> P is the gas pressure in the cylinder, [kPa]
> T_g is the gas temperature in the cylinder, [K]
> C is the characteristic speed, [m s^{-1}]

The characteristic speed is given by the following relation, which comprises a term proportional to the mean piston speed and a pressure rise term associated with combustion.

$$C = \left[K_1\ \overline{C_p} + K_2\ \frac{V_c\ T_1}{P_1\ V_1}\ (P - P_0) \right] \tag{2.50}$$

where

> V_c is the volume of one cylinder, [m^3]
> P_1, T_1, and V_1 are the pressure, [bar], temperature, [K] and volume, [m^3] of the combustion chamber, respectively, at a reference instant (gas intake valve closing or compression stroke beginning)
> $\overline{C_p}$ is the mean speed of the piston, [m/s]
> P_0 is the gas pressure in the cylinder for the motored (versus fired) engine, [kPa]
> K_1 and K_2 are specific coefficients given at each phase of the engine cycle, [–] and [m s^{-1}K^{-1}]

Coefficients K_1 and K_2 can be found in Woschni's original paper (1967). Woschni showed that the heat transfer by radiation from the working gas (water vapor and CO_2) is small and can be neglected. However, he accounted for flame radiation, which consists of the radiation by soot particles. Later, Woschni introduced a swirl effect in the coefficient K_1.

The thermal exchanges inside the cylinder head take place between the inlet gases or exhaust gases and the cylinder head wall.

The convection thermal exchanges between the inlet gases or exhaust gases and the cylinder head wall are performed with gas speed higher than that in the cylinder. The heat transfer under high temperature gradient between the exhaust gases and the cylinder head wall is then important and gives high heat flux levels.

2.2.6.2 Heat Transfer Between Coolant and Engine Walls

As explained later in Section 2.3.2, ICEs are typically cooled down by a liquid coolant, which is an aqueous solution of glycol. The coolant flows through water cores inside the engine block and head, called *cooling water jacket*.

Part of the heat transferred by the exhaust gases to the engine walls flows by conduction through the latter ones and reaches the cooling water jacket.

At the level of the cooling water jacket, the heat transfer between the wall and coolant fluid is generally supposed to be convective with turbulent flow because of the complex form of the coolant cores and the poor surface quality conditions of the wall, obtained from iron or aluminum foundry.

The following expression proposed for turbulent flow by Colburn (and modified by Sieder and Tate (1936) to account for large variations of physical properties) is often used:

$$Nu = 0.023 \cdot Re^{0,8} \cdot Pr^{1/3} \cdot \left(\frac{\mu(T_{gw})}{\mu(T_{wall})} \right)^{0,14} \tag{2.51}$$

where

T_{gw} is the local coolant bulk temperature, [°C]
T_{wall} is the wall surface temperature, [°C]
μ is the dynamic viscosity, [Pa · s]

The Nusselt number Nu is the ratio of the heat transfer by convection to the heat transfer by conduction. It quantifies the importance of the former versus the latter.

$$Nu = \frac{h \; D_h}{k} \tag{2.52}$$

where

h is the convective heat transfer coefficient, [W m^{-2}K^{-1}]
D_h is the characteristic length (here the hydraulic diameter), [m]
k is the thermal conductivity of the fluid, [W m^{-1}K^{-1}]

Therefore, knowing the Nusselt number (evaluated by correlations such as the one given hereunder) allows for the convective heat transfer coefficient to be determined.

The Reynolds number Re is the ratio of the inertial forces to viscous forces within a flowing fluid. At low Re ($Re < 2300$ for fully developed flows in pipes), the flow is laminar and viscous forces are dominant. At high Re ($Re > 2900$ for fully developed flows in pipes), the flow is turbulent and inertial forces are dominant. The Reynolds number is defined as

$$Re = \frac{\rho \, C \, D_h}{\mu} \tag{2.53}$$

where

ρ is the fluid density, [kg m^{-3}]
C is the flow speed, [m s^{-1}]

The Nusselt and Reynolds numbers use the hydraulic diameter as the characteristic length defined by

$$D_h = \frac{4 \cdot A_{pass}}{L_m} \tag{2.54}$$

where

A_{pass} is the area of the flow passage, [m^2]
L_m is the length of the wetted perimeter, [m]

The Prandtl number Pr is the ratio of momentum diffusivity to thermal diffusivity. It compares the effect of speed profile and thermal conductivity on the temperature profile within a fluid.

For $Pr \ll 1$, thermal diffusivity dominates momentum diffusivity. The Prandtl number is defined as

$$Pr = \frac{\mu \, c_p}{k} \tag{2.55}$$

where

c_p is the specific heat of the fluid at constant pressure, [J kg^{-1}K^{-1}]

Knowing the convective heat transfer coefficient, the heat transfer rate is computed by:

$$\dot{Q}_{wall-gw} = A_{wall-gw} \cdot h \cdot (T_{wall} - T_{gw}) \tag{2.56}$$

where

$A_{wall-gw}$ is the heat transfer area between the wall and the coolant fluid, [m^2]

During the engine warm-up phase, the temperature gradient between the wall and the coolant fluid is higher than that during the stabilized regime. The thermal inertia of the engine postpones the achievement of the optimal temperature conditions. The energy losses during this phase are then more important and involve an overconsumption and an increase in pollutant emissions.

2.2.6.3 Overall Heat Transfer Between the Gas and Coolant

The correlation proposed by Taylor (1985) and recalled by Ferguson and Kirkpatrick (2015) allows for the calculation of an "overall average" heat transfer coefficient (spatial and crank angle average) implicitly accounting for conduction and radiation.

$$h_o = 10.4 \cdot Re^{0.75} \frac{k_g}{D} \tag{2.57}$$

With the Reynolds number defined as

$$Re = \frac{(\dot{m}_a + \dot{m}_f) \, D}{\frac{\pi D^2}{4} \mu_g} \tag{2.58}$$

Orders of magnitudes of gas conductivity and viscosity k_g and μ_g are 0.06 W m K and 20×10^{-6} Ns m^{-2}, respectively (Fergusson and Kirkpatrick, 2015).

The crank-angle average heat transfer rate from the gas to the coolant can then be computed knowing the average gas temperature and the combustion chamber heat transfer area:

$$\dot{Q}_{g-gw} = h_o \, A \, (\overline{T}_g - T_{gw}) \tag{2.59}$$

2.2.6.4 Heat Transfer with the Surroundings

The heat transfer to the surroundings is achieved, on the one hand, in the underhood by direct thermal exchange with the engine parts and, on the other hand, in the vehicle front-end zone through the air-to-coolant heat exchanger (radiator of the coolant circuit), the air-to-oil cooler, and the charge air cooler.

The direct heat transfer with the engine parts consists of heat directly exchanged between the external surfaces of the cylinder head, cylinder block, and oil sump of the engine and the ambient air of the confined underhood environment. These losses are convective and radiative thermal heat transfers. They have a limited influence on the engine internal temperatures.

The performance and design of the radiator depend mainly on the air and coolant temperatures, the available front surface, the distribution and speed of the air through the surface and the presence of other heat exchangers as condenser, and the charge air cooler located upstream or downstream the radiator. These considerations lead to some trade-offs.

2.3 Engine Cooling and Heating

2.3.1 Purpose of Engine Cooling and Heating

As explained previously and expressed in Eq. (2.41), in ICE, only part of the thermal energy generated by the combustion of fuel ($\dot{m}_f\, LHV_f - \dot{Q}_{incomp}$) is converted into mechanical work. Part of the remaining energy is present in the hot combustion gases at the exhaust line of the engine ($\dot{Q}_g$). Another part ($\dot{Q}_{wall}$) is converted into thermal energy absorbed by the combustion chamber wall. It contributes to warm the engine metal parts, the coolant fluid, and the engine oil before rejection to the air, thus increasing their temperatures during the warm-up phase until the control of the temperatures during the steady-state phase.

To cool down the engine with glycol water, passageways are designed in the block and cylinder head to let the coolant flow. These passageways form the coolant jacket. Regarding the passage in the engine block, two main design configurations are possible: either the coolant passage is formed by both the engine block and the back side of the cylinder liner («wet liner») or it is fully nestled in the cylinder liner («dry liner»). Figure 2.10 features the different levels of temperature inside an ICE. The display configuration is a dry liner one. The wall temperature limit acceptable for the cylinder head is generally around 250 °C with the current aluminum alloys. This maximal allowable temperature of the metallic masses is largely lower than the operating temperature of the combustion gases in the ICE, which is typically higher than 800 °C. Moreover, the increasing trend of the specific power in engine increases itself the thermal and mechanical loads.

The temperature of the engine components and oil must be maintained in an appropriate range.

Very high temperatures and temperature gradients cause very high thermo-mechanical stress, critical fatigue phenomena, and deformation of the engine parts. Among other things, such deformations yield cylinder cracks and cylinder-head-gasket leakages. Hence, the presence of hot spots has to be prevented to ensure a good thermo-mechanical resistance and to limit the metal expansion.

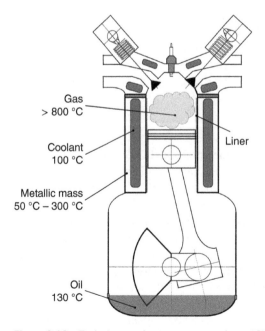

Figure 2.10 Typical operating temperatures in an ICE.

Excessive temperature of oil results in very low viscosity and potentially to its deterioration (aging phenomenon). Excessive operating temperatures can also lead to the boiling of coolant and to abnormal combustion.

On the contrary, very low engine and oil temperatures yield excessive pollutant emissions and frictional losses. This situation typically occurs when the cold engine is started and extends over the engine warm-up phase. One way for reducing the warm-up time involves reducing the thermal inertia of structure and fluids.

While the purpose of a basic engine cooling system is to protect the engine, advanced systems can also play a significant role in reducing fuel consumption and emissions by reducing frictional losses, reducing auxiliary power consumption, and improving combustion (Pang and Brace 2004). Indeed, the ICE thermal management allows improving the engine consumption and emissions by controlling the fluids (coolant, oil) and metallic parts temperature.

Hence, an advanced engine cooling and heating system ensures the optimal thermal level for consumption, emissions, performance, and cabin heating and contributes to optimize the vehicle overall thermal management.

The engine cooling and heating system is more and more complex and must meet constraints due to various geographic applications with different climate specifications, economical and architecture aspects, and ecological standards.

2.3.2 Working Principle of Engine Cooling and Heating Systems

The objectives of the engine cooling and heating system are as follows:

- to evacuate into the external environment (engine surroundings) the heat transferred through the wall by the combustion gases
- to maintain a relatively constant optimal temperature of each engine part, taking into account the thermo-mechanical behavior of the materials
- to accelerate the temperature warm-up during the transient phase

Cooling of the engine can be ensured by air or a liquid coolant. The oil also participates in the cooling. Liquid coolants are generally aqueous solution of ethylene glycol or propylene glycol. Glycol is used to cool down water to negative temperatures without freezing. Glycol water coolants also contain several corrosion inhibitor additives. Other types of coolants can be used. Cooling by a liquid shows advantages over cooling by air, such as better convective coefficient and larger density and specific heat capacity of the liquid coolant. Today, air-cooling systems are mainly used for small-scale engines. Liquid coolant is also an appropriate heat transfer fluid for heating the engine and hence reducing the warm-up time.

As mentioned hereunder, the engine temperature control system also aims at reducing the warm-up time. Actually, increasing the temperatures of the metallic masses, of the coolant, of the oil and of the gases reduces mechanical friction and increases combustion efficiency. As a consequence, the fuel consumption and the pollutant emissions are minimized. Also reducing the warm-up time allows for the optimization of the cabin heating for ensuring the thermal comfort and the visibility. Different mechanisms aiming at decreasing the time duration of the warm-up phase can be considered, including radiator by-pass and exhaust heat recovery systems (EHRS). The latter solution will be described at the end of this chapter.

A conventional architecture of the engine liquid coolant loop is represented in Figure 2.11. Major components of the loop are the thermostat, the radiator with its fan, the pump, an expansion tank (or an overflow tank), and the heater core. All components are connected through a coolant circuit. In normal operation, the coolant flows through the engine where it is heated. Hot coolant

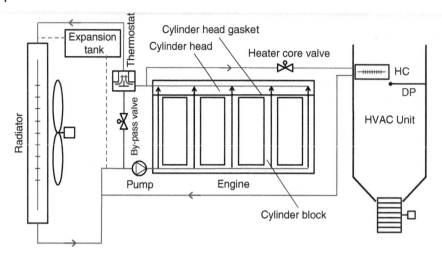

Figure 2.11 Conventional architecture of the engine liquid coolant loop.

is then routed to the radiator located in the front-end module. The radiator – an air-to-coolant heat exchanger – cools down the coolant. The heat transfer rate in the radiator can be increased by means of an axial fan, the rotational speed of which can be controlled. The pump and fan can be driven either mechanically by the engine shaft or independently by means of an electric motor. Often, the pump is mechanically driven, and the fan is electrically driven.

The energy balance on the coolant circuit, viewed as a thermodynamic system, can be expressed by distinguishing thermal energy entering and leaving the circuit. The work of the pump should also be taken into consideration. Thermal energy entering the system consists of:

– the heat transferred by combustion gases to the coolant through the cylinder head and cylinder block
– the frictional losses between the piston rings and the cylinder liner
– the heat exchanged with the engine oil
– the heat provided by an additional heating system, which is sometimes used

The total rate of thermal energy transferred from the ICE to the coolant is noted as $\dot{Q}_{ICE-gw}$.
Thermal energy leaving the system consists of:

– the heat transferred from the coolant to the ventilation air in the heater core (HC in Figure 2.11)
– the heat transferred from the coolant to the cooling air in the radiator
– the heat losses to the ambient in the piping
– the heat transferred to the metallic masses and to the oil during the warm-up phase

In conventional vehicles, the coolant loop also ensures, at least partly, the heating of the air supplied into the cabin. At the outlet of the engine, part of the hot coolant is routed to the heater core that is connected to the heating, ventilation and air-conditioning (HVAC) unit. Hence, the engine cooling system has an impact on the passenger comfort and security. To better control the comfort inside the cabin, a heater core valve (in series with the heater core in the coolant circuit in Figure 2.11) and a small circulating pump can also be used. The management of the coolant loop can contribute to improve the thermal comfort inside the cabin.

Figure 2.12 shows the thermal energy transfer from the engine to the coolant and from the coolant to the air. In the steady-state regime, the rate of heat $\dot{Q}_{ICE-gw}$ [W] transferred from the engine to the coolant is related to the coolant mass flow rate $\dot{m}_{gw,ICE}$ [kg s^{-1}], coolant specific heat c_{gw} [J kg^{-1}K^{-1}]

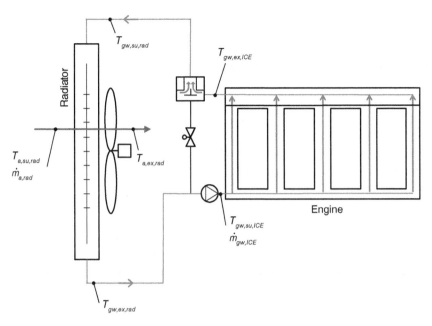

Figure 2.12 Thermal energy transfer between the engine and the coolant and between the coolant and the air.

and supply and exhaust coolant temperatures $T_{gw,su,ICE}$ and $T_{gw,ex,ICE}$ [°C] by

$$\dot{Q}_{ICE-gw} = \dot{m}_{gw,ICE}\, c_{gw}\, (T_{gw,ex,ICE} - T_{gw,su,ICE}) \tag{2.60}$$

If the heating up by the pump and the heat losses from the piping are neglected, if the engine by-pass is not in use (the mass flow rates through the engine and through the radiator are the same), if the heater core is not activated, and if the coolant loop is in the steady-state regime, we can write

$$\dot{Q}_{ICE-gw} = \dot{Q}_{gw-rad} = \dot{m}_{gw,ICE}\, c_{gw}\, (T_{gw,su,rad} - T_{gw,ex,rad}) \tag{2.61}$$

The heat transfer rate in the radiator is computed by the epsilon-NTU method described in Chapter 1.

$$\dot{Q}_{cool,rad} = \varepsilon_{rad}\, \dot{C}_{min,rad}\, (T_{cool,su,rad} - T_{a,su,rad}) \tag{2.62}$$

2.3.3 Circulation of the Coolant through the Engine

The coolant circulates through a set of passages inside the engine block (surrounding the cylinders) and head, called the water jacket. In a basic configuration, the coolant flows successively through the engine block, head gaskets, and engine head. The cooling water jacket allows for the cooling of the hot spots on the cylinder head and the cylinder block zones. The coolant flow rate is distributed among all cylinders.

In a cross-flow configuration, all the cylinders are cooled by fluid at the same temperature. Such configuration reduces the pressure drop in the engine and hence the pump consumption.

In a split-cooling configuration, the engine block and head are separately supplied by the coolant. This allows the warm-up time to be decreased by supplying only the engine head by the coolant during the start-up phase with zero flow in the engine block. After the warm-up phase, the engine block branch and the head branch can be controlled at different temperatures with a thermostat on each branch. The cooling of the cylinder head is improved by means of a lower temperature of the coolant crossing this part.

For a fixed engine rotational speed, the engine heat rate $\dot{Q}_{ICE-gw}$ rejected in the coolant fluid is increasing with the mechanical power transmitted to the crankshaft.

For a fixed mechanical power transmitted to the crankshaft, the engine heat rate $\dot{Q}_{ICE-gw}$ rejected into the coolant fluid is increasing with the engine rotational speed.

Heat transfer rates are increased with a larger heat transfer surface around the combustion chamber, an appropriate metal choice and a limited wall thickness. When reducing the thickness, care should be taken to prevent any risk of coolant boiling. Boiling can locally appear as soon as the wall temperature exceeds the boiling temperature of the fluid and creates damage due to local cavitation erosion. The pressurizing of the coolant circuit allows reducing these risks.

Note that generally, the operating coolant flowrate is defined to avoid a very high temperature of the cylinder head material (temperature to be kept under 230–250 °C).

The coolant circuit is defined to transport and evacuate a high rate of generated heat. The design of the architecture of the circulation, the coolant flow rate, and the temperature and pressure of the circuit depend generally on the trade-off between the risk of boiling, the heat rate level to evacuate, the internal pressure drop that impact the choice of the coolant pump, and the architecture in the underhood. These elements impacts largely the consumption and pollutant emissions.

The internal coolant flow rate in the engine is defined generally considering the thermo-mechanical behavior of the engine structure on the operating points with the maximum mechanical energy production.

The cooling demand for diesel engines is generally higher than that for gasoline engines because of higher thermo-mechanical stresses and a larger part of the generated heat rejected in the coolant circuit.

2.3.4 Radiator

2.3.4.1 Purpose of the Radiator

As mentioned previously, the role of the radiator is to cool down by air the hot coolant coming from the engine.

Radiators can also be used to reject heat from other components that can be cooled by the coolant, such as the water-cooled charge air cooler (WCAC), the water-cooled condenser, the oil cooler, the EGR cooler, the power electronics, the electric motor, and the battery. When multiple components are cooled down by the coolant with different temperature level specifications, several loops at different temperatures can be implemented. Usually, a low-temperature loop and a high-temperature loop are used.

2.3.4.2 Technologies of Radiators

As any heat exchanger, the radiator must show high thermal and hydraulic performance with compromise on the high air permeability for a sufficient air flow rate. Thermal performance depends on the radiator characteristics, the operating conditions (the air and coolant flow rates and temperatures) and the concentration of glycol in the coolant. The radiator must also be as compact as possible. From a mechanical point of view, the radiator must sustain pressure cycling, vibrations, thermal shocks and external and internal corrosion.

Radiators are usually aluminum tube–fin heat exchangers. A typical arrangement of a radiator is shown in Figure 2.13. The radiator is made up of the radiator core, two water tanks (on both sides of the radiator), the gaskets, headers, and side plates. Fins and tubes, which compose the radiator core, are assembled either by mechanical expansion or by brazing. Tubes can be extruded,

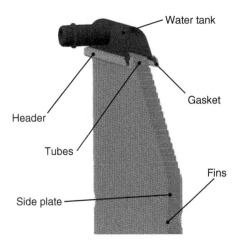

Figure 2.13 Illustration of a tube-and-fin radiator. Source: Courtesy of Valeo.

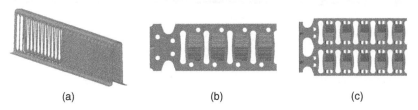

(a)   (b)   (c)

Figure 2.14 Different louvered fins designs for brazed and mechanically assembled radiators. Source: Adapted from Valeo.

folded, or electro-welded. Dimple tubes instead of smooth tubes can possibly be used to increase the convective heat transfer coefficient on the coolant side.

Compared to mechanical assembly, brazed assembly is little more expensive but offers better performance, compactness, and weight.

Figure 2.14 shows different designs of fins for a brazed (a) and for mechanically assembled (b and c) radiators. In (b) and (c) fins, tubes show peanut shape. Fin (b) is associated with a radiator having one single row of tubes, while fin (c) is associated with a radiator having two rows of tubes.

The assembly of the water tank, gasket, header, and tubes is shown in Figure 2.15. The water tank is crimped on the radiator, compressing the gasket.

It will be shown in Section 2.4 that one of the radiator tanks can include another water-cooled heat exchanger, which is the in-tank oil cooler.

2.3.4.3 Flow Configurations in Radiators

Coolant flow configuration in the radiator can be horizontal or vertical (Figure 2.16). The vertical configuration (also named "down flow") is based on a large number of vertical short tubes in parallel. The advantage of such a configuration is the low pressure drop it creates on the coolant side. This configuration is mainly found in coolant loops working at atmospheric pressure.

For horizontal configurations (also named "cross flow"), the distinction can also be done among *I flow* and *U flow* circulations (Figure 2.17). In the former case, the coolant enters on one side and exits on the other side. In the latter case, the coolant enters and exits on the same side and makes two passes through the radiator. A third flow circulation mode is the *Z flow*, where the coolant makes three passes through the heat exchanger. To ensure these flow configurations, partitions are

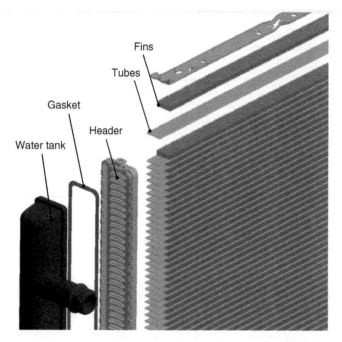

Fins

Tubes

Gasket

Header

Water tank

Figure 2.15 Assembly of water tank, gasket, header, and tubes. Source: Adapted from Valeo.

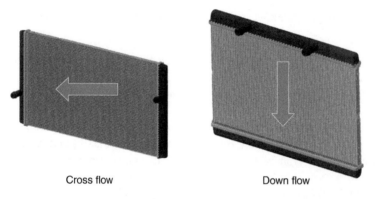

Cross flow

Down flow

Figure 2.16 Horizontal flow ("cross flow") and vertical flow ("down flow") configurations in radiators. Source: Adapted from Valeo.

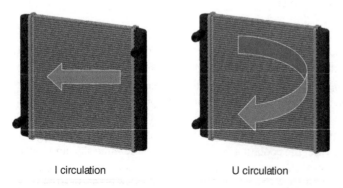

I circulation

U circulation

Figure 2.17 I and U flow configurations in radiators. Source: Adapted from Valeo.

nestled inside the water tanks. I flow circulation offers low coolant pressure drop and low coolant velocity. U flow circulation offers better performance.

Radiators can show a bi- or multi-temperature configuration (Figure 2.18). For instance, after two passes, part of the coolant exits the radiator. The remainder flows through additional passes where it is further cooled down. Colder coolant can be used for instance to cool the charge air cooler. The flow through the additional passes is imposed either by a calibrated supply orifice or by a dedicated small capacity electric pump. In that configuration, the radiator has one inlet and two outlets.

Figure 2.19 gives the performance of a radiator, in terms of rejected heat rate $\dot{Q}_{rad}$, as a function of the coolant volume flow rate $\dot{V}_{gw,rad}$ for different air-side pressure drops. The larger the pressure drop, the larger is the air speed through the radiator, and therefore, the larger is the air flow rate.

Figure 2.18 Illustration of a radiator with a multi-temperature configuration. Source: Adapted from Valeo.

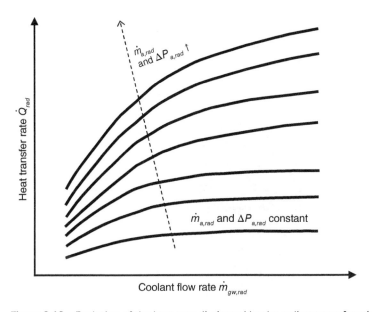

Figure 2.19 Evolution of the heat rate dissipated by the radiator as a function of the coolant mass flow rate for different air-side pressure drops.

This figure indicates that the heat transfer performance on the air-side largely impacts the overall performance. This is due to the low convective heat transfer coefficient and hence large heat transfer resistance on the air side. In case of low air flow rate, the performance "saturates" rapidly with the coolant flow rate.

2.3.5 Expansion Tanks

The temperature of the coolant at the supply of the radiator typically ranges between 100 and 120 °C. When a liquid coolant is heated, it expands. This was explained in Chapter 1 when describing the properties of incompressible fluids. To account for the increase in coolant volume, the radiator is connected to an expansion tank. Moreover, the radiator is equipped with a pressure cap to which is fitted the hose connected to the expansion tank.

The pressure cap is characterized by a rated pressure. When the coolant is heated up, as soon as the pressure inside the radiator reaches the rating pressure, a valve opens letting the coolant flow into the expansion tank. When the coolant inside the radiator is cooled down, a vacuum valve allows coolant to flow back into the radiator.

Besides allowing for the thermal expansion of the coolant due to temperature increase, the expansion tank also serves other functions:

– Filling of the coolant circuit.
– Reserve of liquid volume.
– Degassing the liquid circuit. Gas bubbles formed in the coolant circuit when the vehicle is running are separated and trapped. These gas bubbles come from:
  o air introduced during the filling of the circuit
  o exhaust gases coming from combustion chamber in case of leakage through the head gasket
  o vapor from local boiling in the engine, EGR heat exchanger, turbocharger, etc.
– Absorption of the volume increase due to the formation of vapor in case of local boiling.
– Pressurization of the circuit.
– Pressure limitation: extraction of liquid and vapor in the case of overpressure.

The expansion tank is transparent to read easily the level of liquid inside the tank.

A schematic representation of an expansion tank is given in Figure 2.20. The minimum level of the tank is generally positioned at the highest level of the circuit or above this level.

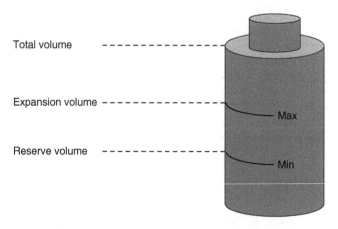

Total volume

Expansion volume

Max

Reserve volume

Min

Figure 2.20 Schematic representation of an expansion tank.

Two main different integrations and usages of the expansion tank in the cooling circuit can be identified:

- The circuit with "cold" tank, where there is no circulation of liquid inside the tank. In that configuration, the tank is connected with the main coolant circuit by one duct.
- The circuit with "hot" tank where the coolant fluid passes through the tank. In such a configuration, the liquid temperature inside the tank is higher, and the pressure in the circuit is higher with respect to cold tank configuration. It is better to limit the cavitation of the coolant pump and local boiling in the engine. This configuration helps to separate and trap the gases from the liquid circuit owing to the circulation flow rate inside the tank.

2.3.6   Thermostat

2.3.6.1   Purpose of the Thermostat
The thermostat is a device that adjusts the coolant flow rates in different branches of the engine coolant loop to control the engine coolant temperature at the engine inlet or outlet. To adjust the cooling rate of the coolant, the thermostat allows more or less coolant to flow through the radiator.

The thermostat is thus a key element of the engine coolant loop, since it controls its operation. As explained hereunder, thermostats can also ensure a quicker warm-up of the engine.

When the coolant temperature is too low, the thermostat prevents any coolant flow to the radiator to limit the engine heat rejected to the air and to accelerate the increase in engine temperature. Therefore, the oil viscosity (and thus the frictional losses) and the combustion chamber temperature decreases and increases quickly, respectively. Consequently, the mechanical and combustion efficiencies also increase quickly.

2.3.6.2   Working Principle of a Thermostat
A thermostat is composed of a cooling temperature sensing element and one or several valves that adjust the flow rates in the different branches of the cooling system including the radiator branch. These valves can be actuated thermally or by an electric engine. Thermally actuated valves rely on the expansion of a material, such as wax, with temperature and on the resulting mechanical force. Two main types of wax thermostats are typically used: conventional and heated. Electrically actuated thermostats represent the third major category of thermostats.

2.3.6.3   Technologies of Thermostats
2.3.6.3.1   Conventional Wax Thermostats   In a basic configuration, the wax-type thermostat is located between the engine and the hose to the radiator. In Figure 2.21, a simple configuration of two-way thermostat is illustrated through a simplified scheme adapted from Hillier and Coombes (2004). The thermostat comprises a cylinder that contains wax. This cylinder slides along a thrust pin. The pin is nestled inside a flexible rubber sleeve. On top of the cylinder is the valve. The pin is attached to the bridge. The thermostat is filled with engine coolant and senses its temperature. When the coolant reaches a temperature close to 80–85 °C, the wax starts to melt. This temperature is named the thermostat rating temperature. Expansion of the wax during the solid–liquid transition forces the cylinder to slide down along the pin. This movement leads to the opening of the thermostat valve, allowing the coolant to flow to the radiator. When the coolant temperature decreases, the wax temperature decreases. Below a given threshold, it starts to solidify. The cylinder is then lifted under the action of the spring.

According to the blend of paraffins used, the phase change occurs in an interval of temperature of roughly 10 K. Once it is completely melted, liquid wax keeps on slightly expanding with

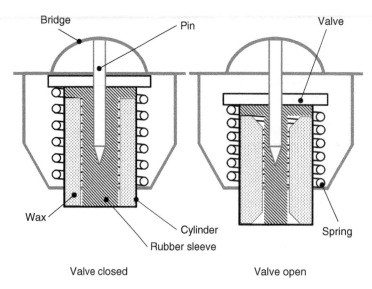

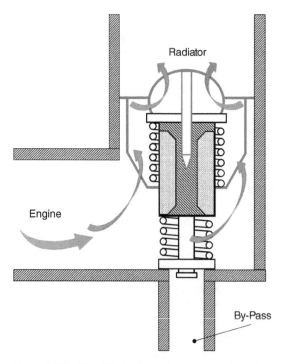

Figure 2.21 Components and working principle of a wax-type thermostat. Source: Adapted from Hillier and Coombes (2004).

temperature. Consequently, the valve opening will be larger for high coolant temperature at the engine outlet, forcing more coolant to flow to the radiator.

To decrease the engine warm-up time, thermostats can be equipped with a bypass valve at the bottom of the thermostat. Decreasing the warm-up period allows for fuel and emission savings. Such a thermostat is illustrated in Figure 2.22 in a position where the by-pass valve is fully closed.

Figure 2.22 Simplified schematic representation of a wax-type thermostat valve with a by-pass valve.

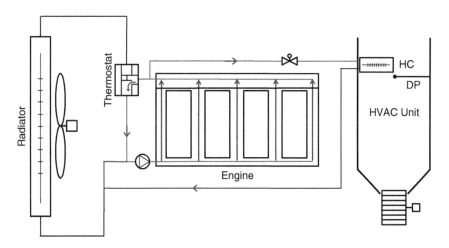

Figure 2.23 Engine cooling circuit equipped with a thermostat with by-pass valve: radiator valve fully closed.

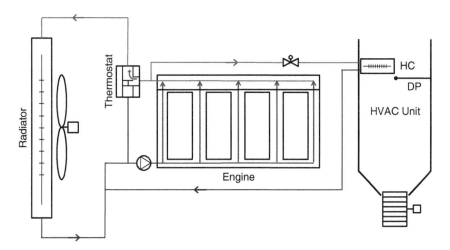

Figure 2.24 Engine cooling circuit equipped with a thermostat with by-pass valve: radiator valve fully open.

When the valve to the radiator is fully closed (Figure 2.23), the by-pass is open allowing for a recirculation of the coolant through the engine. As the temperature of the coolant increases and reaches the optimal engine temperature, the radiator valve starts to open and the by-pass valve closes. Consequently, more coolant flow is routed to the radiator. On highly loaded regimes, when the radiator valve is fully open, the by-pass valve is fully closed (Figure 2.24).

Another configuration of engine cooling circuit is illustrated in Figure 2.25. Here, the thermostat is located at the engine inlet.

Thermostatic valves are characterized by a hysteresis, which is represented in Figure 2.26. For the same opening rate of the valve, the temperature at which the wax solidifies (cooling phase – left curve) is lower than the temperature at which it melts (heating phase – right curve). This hysteresis of a few K is useful for the control of the thermostat.

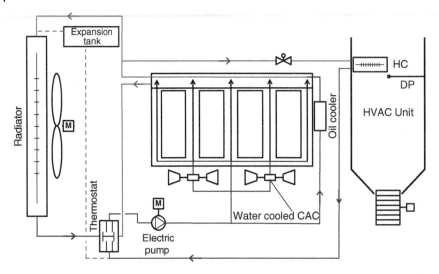

Figure 2.25 Engine cooling circuit equipped with a thermostat located at the engine inlet.

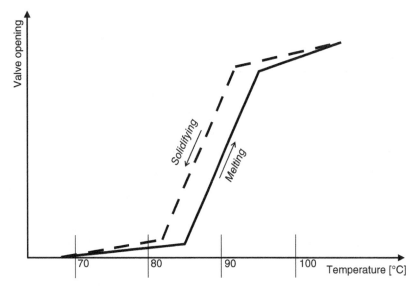

Figure 2.26 Characteristic curves of a wax-type thermostat: visualization of the hysteresis.

2.3.6.3.2 Heated Wax Thermostats The conventional wax-type thermostat presented in the previous section relies on a wax that starts to melt at a predefined engine coolant temperature. In a heated wax thermostat, the wax is heated by both the coolant and by an electrical heat source. Consequently, the opening temperature of the valve can be adjusted externally with a control strategy (engine management system). One of the benefits is the possibility to operate the engine at a higher temperature in part load, thereby reducing the fuel consumption without increasing the risks of durability on the engine. The time response depends on the wax inertia. At high load or during sharp acceleration, it could show a slow response to decrease the coolant temperature to protect the engine (Chanfreau et al., 2003).

2.3.6.3.3 ***Electrically Actuated Thermostats*** As an alternative to wax-type thermostats, electrically actuated thermostats can be used to control the coolant flow rates through the different branches of the coolant loop. Similar to heated-wax thermostats, the engine coolant temperature set point is not permanently set and can be adjusted externally by the engine management system. Such thermostats can offer additional functions to engine thermal control. Electrically actuated thermostats are made up of a valve and an electric motor. Electrically actuated thermostats show much smaller time responses than heated-wax ones (approximately 1 second versus 30–40 seconds). As a drawback, short-time responses can yield thermal shocks combined with fan cycling.

One example of such a valve integrated inside the engine coolant loop is presented in Figure 2.27. This example also shows that the heater core is associated with an auxiliary water pump whose functionality is described hereunder.

The valve comprises four connections: one inlet and three outlets. The valve inlet (*A*) is connected to the engine coolant outlet. The three outlets allow directing the coolant flow to the engine by-pass (*B*), to the radiator (*C*) and to the heater core (*D*).

Different modes of operation can be achieved:

Zero Flow In this mode, there is no coolant circulation inside the engine. When the engine is switched on, the electric valve is closed and prevents any flow through *B*, *C*, and *D* during the first few minutes. As a consequence, the engine warm-up time is reduced. The coolant volume to be heated is reduced, and the temperatures of the coolant and engine metal around the combustion chamber and of the air–fuel mixture are increased faster. This allows the hydrocarbons and CO emissions and the frictional losses to be reduced. However, the oil temperature increase could be slightly longer because of the thermal exchange absence between the coolant and the oil through a coolant-to-oil heat exchanger. Also, the zero-flow phase has to be limited because of risk of boiling in the cylinder head. During zero flow phase, a cabin heating or demisting/defrosting demand cannot use the engine heat. Therefore, such demands could prevent zero flow mode.

Warm-up In summer, this mode only allows the by-pass line to be fed with the coolant (outlet *B* is open). This prevents hot spots in the cylinder head. There is no flow of hot coolant through the heater core, which reduces the coolant quantity to heat. In winter or if there is a need of heat to

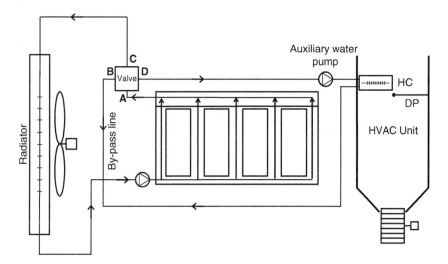

Figure 2.27 Schematic representation of the integration of an electrically actuated thermostat valve inside the engine coolant loop.

prevent any mist formation risk in the cabin, all the coolant is directed to the heater core (only the outlet *D* is open). However, in that case, cabin heating can recover a large part of the engine heat and slow down the engine warm-up.

Temperature Control (After Warm-up Phase) In summer, the valve outlets *B* and *C* are open. There is no flow of hot coolant through the heater core. This allows the blown air temperature not to be increased by a few Kelvins in air-conditioning mode by the hot coolant enclosed in the heater core even though the air does not pass through it (Chanfreau et al., 2003). In winter, all branches are fed by coolant (outlets *B*, *C* and *D* are open).

The engine coolant temperature is imposed to 90 °C at high load and can be increased until 110 °C at part load. Combining the faster warm-up and the higher coolant temperature set point (Figure 2.28) yields a reduction of fuel consumption by roughly 4% on the high-speed portion of the NEDC cycle (Chanfreau et al., 2003).

This electric valve also shows the benefit of decreasing the coolant temperature faster from 110 to 90 °C due to its shorter response time in comparison with a heated wax thermostat (Chanfreau et al., 2003)

Post-cooling When the engine is switched off after high-load application, engine heat soak happens, which increases the coolant temperature, possibly up to boiling. To prevent engine heat soak and coolant boiling, the coolant needs to be cooled down even if the engine is stopped. Natural circulation of coolant to the radiator, due to thermosyphon effect, is limited with wax thermostat and heated wax thermostat, due to the closing of the former and the pressure drop of the latter (Chanfreau et al., 2003). A larger flow through the radiator can be achieved with the (low pressure drop) electric valve yielding a lower fan consumption and actuation time (Chanfreau et al., 2003). Also, the auxiliary water pump can be used to decrease even more the engine cooling time (Chanfreau et al., 2003). In summer, with such a mode, the auxiliary water pump circulates the coolant through the engine, the heater core and the radiator (inlet *A* is open, outlet *D* is open, outlet *C* becomes the valve inlet, outlet *B* is closed; the direction of the flow in the radiator is inversed; the engine pump is deactivated). This yields a smooth engine cooling.

Post-heating In winter, the engine can act as a heat storage and provide heat to the heater core when the engine is off (vehicle stopped or electric mode of a hybrid vehicle). In that case, only the outlet *D* and inlet *A* are open.

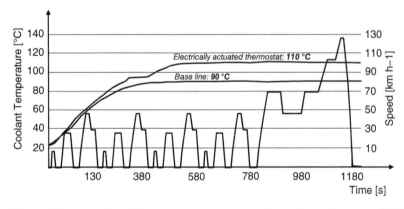

Figure 2.28 Evolution of the coolant temperature at the engine outlet on the NEDC cycle: comparison between a coolant loop equipped with an electrically actuated thermostat and a base line. Source: Adapted from Chanfreau et al., (2003).

2.3.7 Heating Systems

It has been shown previously that the main components that allow for the cooling of the coolant are the engine radiator and to a lesser extent the heater core. To reduce the warm-up time, when the engine is started, it may be useful to heat the coolant for a short period of time. This can be achieved by electric heaters (PTC) or fossil fuel burners or by recovering thermal energy from exhaust gas and injecting it into the coolant. The latter option, named EHRS, will be presented in Section 2.8.1.

2.4 Oil Cooling

2.4.1 Purpose of Oil Cooling and Heating

In vehicles, oil is used both as a lubricant and as a heat transfer fluid. As a lubricant, oil reduces the wear of moving components and friction losses associated with elements in contact. As a heat transfer fluid, oil evacuates the heat. Different oil circuits are used in the vehicle. They are respectively in contact with the engine, the gear box, and the transmission. Engine oil is mainly used to lubricate the moving parts and to reduce frictional losses (crankshaft bearings, piston rings-cylinder liner, distribution system, etc.). Oil also evacuates part of the heat generated by the combustion of fuel and not converted into work and by frictional losses.

Thermal energy entering the engine oil circuit is associated with piston cooling, friction in distribution, friction in crankshaft bearing, piston ring-cylinder liner friction, friction in turboexpander bearings, and heat transfer from the coolant (in a coolant-to-oil heat exchanger during the engine warm-up). Thermal energy leaves the engine oil circuit by heating some of the engine metallic masses (during engine warm-up), by transferring heat to the ambient (in an air-cooled heat exchanger or through natural convection from zones containing oil to the ambient), or by transferring heat to the coolant (in a coolant-to-oil heat exchanger during the engine steady-state regime).

Oil can be in contact with the hottest elements of the engine, since it can be brought to approximately 130 versus 110 °C for the coolant. In the engine oil circuit, oil is filtered through a filter, stored in an oil sump, and pumped with a pump generally mechanically entrained by the engine.

Oil is also employed for cooling and lubricating the gears in the gear box.

Furthermore, oil is used for engine torque transmission and gear shift control in the automatic transmissions.

Temperature of the oil in the different circuits must be controlled: either oil is heated-up or cooled down. The heat transfer rates vary from one circuit to another. Maintaining the appropriate temperature of the oil is important. Too high temperature can yield oil degradation. Also, since the viscosity of oil decreases with the temperature, its lubricating capability is decreased. Too low temperature results in a too high viscosity and thus important friction losses. In that latter perspective, decreasing the time of engine warm-up and oil heating-up is also important.

2.4.2 Working Principle of Oil Cooling and Heating Systems

Oil cooling is partly achieved by convective and radiative heat transfer to the ambient. However, most of the cooling effect is ensured by a dedicated heat exchanger. This heat exchanger can be either cooled down by air (air-to-oil cooler) or by coolant (coolant-to-oil cooler). Oil heating is achieved by preventing any cooling or by a dedicated heat exchanger.

2.4.3 Technologies of Oil Coolers

2.4.3.1 Air-to-Oil Coolers

Air-to-oil coolers are mainly used for cooling the transmission oil. Such coolers are located remotely from the engine inside the front-end module. They are typically plate–fin or tube–fin heat exchangers (Figure 2.29).

2.4.3.2 Coolant-to-Oil Coolers

Regarding the coolant-to-oil coolers, three flow configurations can be considered: either the oil cooler is mounted directly on the engine with no additional piping or it is installed remotely from the engine or it is integrated into one of the radiator tanks.

Figure 2.30 shows the coolant flow circuit associated with the engine-mounted oil cooler configuration.

Engine-mounted and remote coolant-to-oil coolers are typically aluminum plate or plate-and-bar heat exchangers. There are several designs of plate heat exchangers, including donut (Figure 2.31)

Figure 2.29 Air-to-oil cooler, tube–fin technology. Source: Courtesy of Valeo.

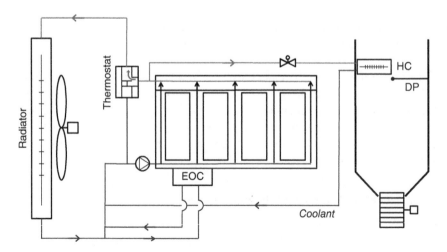

Figure 2.30 Coolant circuit associated with the use of an engine mounted oil cooler (EOC = Engine Oil Cooler).

Figure 2.31 Donut-type coolant-to-oil cooler. Source: Courtesy of Valeo.

Figure 2.32 Stacked plate coolant-to-oil cooler. Source: Courtesy of Valeo.

and stacked plate (Figure 2.32) types. In the construction shown in Figure 2.32 (internal structure shown in Figure 1.26), plates are corrugated; thus, no turbulators are needed. The oil cooler can be equipped with the oil filter within a compact module.

Figure 2.33 shows the coolant and oil flow circuits of a configuration integrating the oil cooler inside the radiator tank. Here the engine oil cooler (EOC) is integrated inside the radiator cold tank to get the largest temperature difference between the engine oil and the coolant.

In-tank heat exchangers are used for both transmission and engine oil cooling. Integrating the oil cooler inside one of the radiator tanks allows for cost reduction.

2.4.4 Oil Temperature Control

Figure 2.34 illustrates an advanced control of the oil temperature. The purpose of the control is to speed-up the warm-up of the oil and to better control its temperature during stabilized operation.

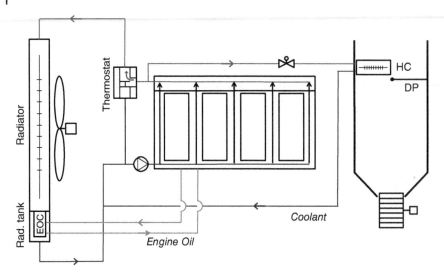

Figure 2.33 Coolant and oil circuits associated with the use of an in-tank oil cooler (EOC = Engine Oil Cooler).

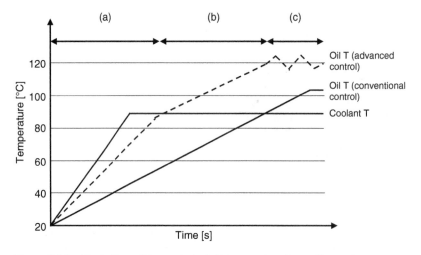

Figure 2.34 Illustration of the control of oil temperature by cooling water.

During phase (*a*), the coolant, whose temperature is higher than that of oil transfers part of its heat to the oil.

During phase (*b*), when the coolant temperature is controlled with the thermostat, the oil is heated by the heat transferred from hot elements it lubricates or cools down. The self-heated oil reaches and exceeds the coolant temperature. There is no active cooling of the oil.

During phase (*c*), if the internal thermal exchanges are not sufficient to control the oil temperature, there is a need for an oil cooler. When oil reaches its temperature set point, the latter is stabilized by means of active cooling by the liquid coolant or by the air. The time evolution of oil temperature achieved by this control strategy is compared to that obtained with a conventional control.

2.5 Charge Air Cooling (CAC)

2.5.1 Purpose of Charge Air Cooling and Forced Induction

Charge air cooling, also named "intercooling," is the process of cooling the compressed air in a forced induction system before being supplied to the cylinder.

Forced induction is the process of increasing the air density by compression and cooling at the engine intake to increase air mass drawn into the system, burn more fuel, and increase the *specific engine power* in [kW per L of displacement]. As a positive result, forced induction engines are thus characterized by a larger specific engine power than naturally aspirated engines. Forced induction is one of the means of downsizing the engines, which consists of reducing the size of the engine for a similar delivered power. Engine downsizing allows for consumption and CO_2 emission reductions. Actually, smaller engines delivering the same power are characterized by a lower weight, lower frictional losses, and shorter warm-up time and operate on a more efficient regime.

2.5.2 Working Principle and Technologies of Forced Induction

Charge air density is first increased by increasing its pressure. This is achieved by means of a single-stage or two-stage compressor. This compressor is used to compress the air. The compressor can be driven either by a turbine (*turbocharger*) or by a drive belt connected to the engine shaft (*supercharger*) or by an electric motor (*electric supercharger*).

2.5.2.1 Turbochargers
The mechanism employing a turbine to drive the compressor is called a turbocharger. The turbine is driven by exhaust gases from the engine (Figure 2.35). Its shaft is connected to the compressor shaft. Part of the exhaust gas can by-pass the turbine through the *waste gate*, which allows the turbine and compressor rotational speed to be controlled. This mechanism is used to control and limit the "boost" pressure at the outlet of the compressor, which protects the engine.

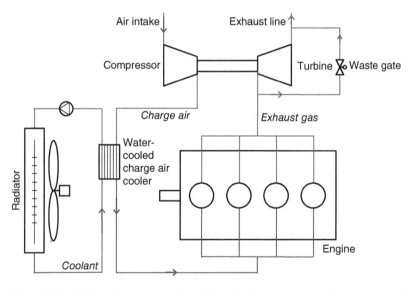

Figure 2.35 Schematic representation of a turbocharger connected to a water-cooled charge air cooler (WCAC).

One of the advantages of the turbocharger is its energy efficiency, since it valorizes the energy of the exhaust gases.

Example 2.1   *Turbocharger*

Performance of a turbocharger has been measured. The turbine is driven by exhaust gas at 590 °C and 1.9 bar absolute. The gas temperature and pressure at the turbine exhaust are 527 °C and 1.136 bar absolute. The charge air enters the compressor at 1 bar 25 °C and leaves it at 99 °C and 1.67 bar. The charge air mass flow rate is 112 kg h^{-1}. The specific heat and isentropic coefficient of exhaust gas are 1212 J (kg K)$^{-1}$ and 1.3, respectively.

Compute: (i) the exhaust gas mass flow rate, (ii) the power consumed by the compressor and the isentropic effectiveness of the turbine and compressor, (iii) the gas flow rate flowing through the waste gate, assuming that fuel is diesel and the equivalence fuel–air ratio is 0.83.

Analysis

In this example, two model outputs are imposed (the compressor and turbine exhaust temperatures) and consequently two parameters are identified (the isentropic effectiveness of the compressor and of the turbine). The power of the turbine is estimated by expressing it as equal to that of the compressor.

Solution

```
"1. INPUTS"
T_g_su_turb=590 [C] "temperature of exhaust gas at the turbine supply"

P_g_su_turb=1,9e5 [Pa] "pressure of exhaust gas at the turbine supply"
P_g_ex_turb=1,136e5 [Pa] "pressure of exhaust gas at the turbine exhaust"
gamma_g=1,3 [-] "specific heat ratio of exhaust gas"
c_p_g=1212 [J/kg-K] "specific heat of exhaust gas"
Phi=0,83 [-] "fuel-air equivalence ratio"
T_a_su_cp=25 [C] "temperature of charge air at the compressor supply"
P_a_su_cp=1e5 [Pa] "pressure of charge air at the compressor supply"
P_a_ex_cp=1,67e5 [Pa] "pressure of charge air at the compressor exhaust"
m_dot_a_cp=112/3600 "[kg/s]" "mass flow rate of charge air"

"2. OUTPUTS"
T_a_ex_cp=99 [C] "temperature of charge air at the compressor exhaust"
T_g_ex_turb=527 [C] "temperature of exhaust gas at the turbine exhaust"
"m_dot_g_turb_kgh=? [kg/h]" "mass flow rate of exhaust gas through the turbine"
"W_dot_cp=?  [W]" "mechanical power of compressor"
"m_dot_g_wg_kgh=? [kg/h]" "mass flow rate of exhaust gas through the waste gate"

"3. PARAMETERS"
"epsilon_s_cp=?  [-]" "isentropic effectiveness of compressor"
"epsilon_s_turb=?  [-]" "isentropic effectiveness of turbine"
"epsilon_s_turb and epsilon_s_cp are 2 model parameters that are identified in
    the present case based on measurement data. They could be imposed to evaluate
    how the turcharger behaves for other operating conditions "

"4. MODEL"
"4.1. Compressor"
h_a_su_cp=enthalpy('air_ha';P=P_a_su_cp;T=T_a_su_cp)
h_a_ex_cp=enthalpy('air_ha';P=P_a_ex_cp;T=T_a_ex_cp)

h_a_ex_s_cp=enthalpy('air_ha';P=P_a_ex_cp;s=s_a_su_cp)
```

```
s_a_su_cp=entropy('air_ha';P=P_a_su_cp;T=T_a_su_cp)
epsilon_s_cp=(h_a_ex_s_cp-h_a_su_cp)/(h_a_ex_cp-h_a_su_cp)  "assuming the
    compressor to be adiabatic"
W_dot_cp=m_dot_a_cp*(h_a_ex_cp-h_a_su_cp)
W_dot_cp_bis=m_dot_a_cp*c_p_a*(T_a_ex_cp-T_a_su_cp) "assuming air to be a perfect
    gas"
c_p_a=CP(air;T=(T_a_su_cp+T_a_ex_cp)/2)

"4.2. Energy balance across the whole turbocharger"
W_dot_turb=W_dot_cp

"4.3. Turbine"
W_dot_turb=m_dot_g_turb*c_p_g*(T_g_su_turb-T_g_ex_turb) "to compute the gas flow
rate through the turbine, assuming it is adiabatic and ideal gas behavior"
m_dot_g_turb_kgh=m_dot_g_turb*3600
epsilon_s_turb=(T_g_su_turb-T_g_ex_turb)/(T_g_su_turb-T_g_ex_s_turb) "assuming
    the turbine to be adiabatic"
(T_g_ex_s_turb+273,15)/(T_g_su_turb+273,15)=(P_g_ex_turb/P_g_su_turb)^((gamma_
    g-1)/gamma_g) "isentropic evolution of ideal gas with constant specific heats,
    see Chapter 1"

"4.4. Waste gate"
1/f_st =((((x+y/4))/0,21)*MM_air)/(x*MM_C+y*MM_H ) "stoichiometric fuel-air
    ratio"
x=1
y=1,8
MM_air=molarmass(air)
MM_C=12,01
MM_H=1,008
f_st=m_dot_f_st/m_dot_a_cp
m_dot_f=Phi*m_dot_f_st "actual fuel flow rate"
m_dot_g_turb_max=m_dot_a_cp+m_dot_f
m_dot_g_turb_max_kgh=m_dot_g_turb_max*3600
m_dot_g_wg_kgh=m_dot_g_turb_max_kgh-m_dot_g_turb_kgh "mass flow rate through the
    waste gate"
```

Results:

```
epsilon_s_cp=0,6341
epsilon_s_turb=0,6521
m_dot_g_turb_kgh=109,3 [kg/h]
m_dot_g_turb_max_kgh=118,4 [kg/h]
m_dot_g_wg_kgh=9,114 [kg/h]
W_dot_cp=2319 [W]
W_dot_cp_bis=2318 [W]
```

2.5.2.2 Superchargers

The system using a belt drive between the air compressor crankshaft and the engine crankshaft is called a supercharger. It is represented in Figure 2.36. A clutch can be used to control the compressor engagement. The compressor by-pass valve allows for the limitation of the boost pressure.

Different types of compressors can be used: roots blowers, G-Laders, screw compressors, and centrifugal compressors.

Superchargers offer several advantages over turbochargers. First, they do not suffer from turbo lag. Turbo lag is the interval of time between the modification of the throttle plate position during acceleration and the response of the turbocharger. Turbo lag creates engine hesitation during

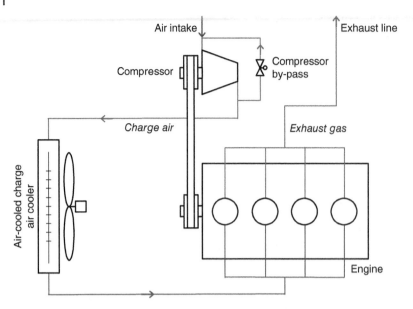

Figure 2.36 Schematic representation of a supercharger connected to an air-cooled charge air cooler (ACAC).

acceleration. Turbo lag is due to the inertia of the turbine. Second, superchargers are able to create more pressure at low engine rotational speed than turbochargers. Third, turbochargers are operating at much higher temperatures than superchargers. However, superchargers are less efficient than turbochargers, since they consume part of the engine shaft energy rather than the energy released in the waste gases.

2.5.2.3 Electric Superchargers

As an alternative to mechanical superchargers described hereunder, the compressor can be driven by an electric motor. The compressor acceleration is thus independent of the engine regime. As another advantage, the compressor works at much lower temperature than the turbocharger and less expensive materials can be used (reference). Different technologies of compressors can be used. An example with a centrifugal compressor is illustrated in Figure 2.37.

Figure 2.37 Electrically driven centrifugal compressor (e-Supercharger). Source: Courtesy of Valeo.

2.5.2.4 Compound Forced Induction

Compound forced induction consists of using different compressor elements, usually two, in the same forced induction system. Different configurations are possible. First, two turbochargers of different sizes can be used, with both the compressors and the turbines associated in series. The smaller turbocharger is used at a low engine rotational speed when the larger turbocharger is not able to reach the boost pressure. Another configuration shown in Figure 2.38 consists of compounding a supercharger and a turbocharger. Compressor elements are usually associated in series. Such a configuration allows taking benefit of the advantages of both technologies of forced induction and compensating for their drawbacks. The supercharger and the turbocharger are used as follows (Volkswagen 2006). The supercharger is engaged when the turbocharger is not able to provide the boost pressure, which happens for low-engine regimes and high-engine loads. At mid-engine regimes, the supercharger can be briefly used to assist the turbocharger in building the required boost pressure during acceleration, for instance. This prevents for any turbo lag. At higher engine regimes, only the turbocharger is used.

2.5.3 Working Principle and Architectures of Charge Air Cooling

Regardless of the technology used to compress the charge air, the latter is cooled down after the compression and before being supplied to the engine intake to ensure a higher density by decreasing the compressed air temperature. Cooling is ensured by the CAC, also called *Intercooler*. In addition to increasing the density of the charge air, charge air cooling (WCAC) is also necessary to prevent the engine from pre-ignition (knock) and allow it to operate under large compression ratios.

Charge air can be cooled down by air (ACAC), water (WCAC), or a refrigerant.

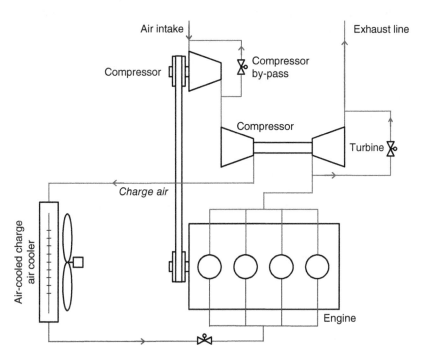

Figure 2.38 Example of compound forced induction with a supercharger and a turbocharger. Source: Adapted from Volkswagen (2006).

2.5.3.1 Charge Air Cooling by Air

Charge air cooling by air is the most classical configuration. As shown in Figure 2.36, at the outlet of the compressor, the charge air flows through the ACAC. The latter is integrated inside the front-end module (see Section 2.7).

2.5.3.2 Charge Air Cooling by Coolant

The water-cooled charge air cooler (WCAC) is cooled down by the coolant of the low temperature loop ($T_{amb} + 15$ K), associated with the low-temperature radiator integrated inside the front-end module. Such a configuration is shown in Figure 2.41.

Water-cooled charge air cooling is becoming more popular. It presents some advantages over charge air cooling by air:

- Water-cooled charge air coolers can be integrated into the engine manifold.
- Shorter air ducts are used, which limits pressure drops.
- The overall forced induction system is more compact.
- Control of air intake temperature is easier, since it is achieved by an electric pump in the coolant circuit and is independent of the outdoor air temperature and vehicle speed. A more stable temperature can be achieved.
- The system shows less turbo-lag.
- This solution needs a low temperature coolant circuit, but is not a constraint (cost constraint among others) if this circuit is shared with other components (e.g. a water cooled condenser).

The water-cooled charge air cooler can be placed either remotely from the engine ("stand-alone configuration") or integrated inside the air intake manifold.

Example 2.2   *Water-cooled Charge Air Cooler*

A turbocharger is equipped with a water-cooled charge air cooler. The coolant is an aqueous solution of glycol water. Specific heat and density of the coolant are 3835 J (kg-K)$^{-1}$ and 1011 kg m^{-3} respectively. The coolant flow rate is 800 L h^{-1}, and the coolant temperature at the CAC supply is 40 °C. The air mass flow rate is 137 g s^{-1}. Air enters the charge air cooler with a temperature of 180 °C and a pressure of 2.8 bar. The pressure at the exhaust of the cooler is 2.75 bar. In the operating conditions considered in this example, the CAC effectiveness is 80%.

Determine: (i) the coolant exhaust temperature, (ii) the charge air exhaust temperature, (iii) the rate of heat exchanged in the CAC, and (iv) the CAC conductance.

Analysis

To solve this heat exchanger problem, the epsilon-NTU method described in Chapter 1 is used. Here, the heat exchanger effectiveness is given. The relation between epsilon and NTU will be used to compute NTU knowing the CAC effectiveness. From NTU, the conductance of the heat exchanger can be determined. It will be assumed that the WCAC is a cross-flow heat exchanger with both fluids unmixed. The library of heat exchangers of EES is used to access the relation between epsilon and NTU for such a heat exchanger.

Solution

```
"1. INPUTS"
T_a_su_CAC=180 [C] "temperature of charge air at CAC supply"
P_a_su_CAC=2,8e5 [Pa] "pressure of charge air at CAC supply"
m_dot_a_CAC=0,137 [kg/s] "mass flow rate of charge air"
P_a_ex_CAC=2,75e5 [Pa] "pressure of charge air at CAC exhaust"
```

```
T_gw_su_CAC=40 [C] "temperature of glycol water at CAC supply"
V_dot_gw_CAC=800e-3/3600 [m^3/s] "volume flow rate of glycol water through the
    CAC"
c_gw=3835 [J/kg-K] "specific heat of glycol water"
rho_gw=1011 [kg/m^3] "density of glycol water"

"2. OUTPUTS"
"T_a_ex_CAC=?? [C]" "temperature of charge air at CAC exhaust"
"T_gw_ex_CAC=?? [C]" "temperature of glycol water at CAC exhaust"
"Q_dot_CAC=?? [W]" "heat transfer rate in the CAC"

"3. PARAMETERS"
epsilon_CAC=0,8

"4. MODEL"
"4.1. CAC heat transfer rate"
v_gw=1/rho_gw
m_dot_gw_CAC=V_dot_gw_CAC/v_gw
C_dot_min_CAC=min(m_dot_a_CAC*c_p_a_CAC;m_dot_gw_CAC*c_gw)
Q_dot_CAC=epsilon_CAC*C_dot_min_CAC*(T_a_su_CAC-T_gw_su_CAC)

"4.2. Energy balance on charge air side"
c_p_a_CAC=CP('air_ha',P=P_a_su_CAC;T=(T_a_su_CAC+T_a_ex_CAC)/2)
Q_dot_CAC=m_dot_a_CAC*c_p_a_CAC*(T_a_su_CAC-T_a_ex_CAC)
"or Q_dot_CAC=m_dot_a_CAC*(h_a_su_CAC-h_a_ex_CAC)"

"4.3. Energy balance on coolant side"
 Q_dot_CAC=m_dot_gw_CAC*c_gw*(T_gw_ex_CAC-T_gw_su_CAC)

"4.4. Conductance of the heat exchanger"
NTU_CAC=HX('crossflow_both_unmixed';epsilon_CAC;m_dot_a_CAC*c_p_a_CAC;m_dot_gw_
    CAC*c_gw;'NTU')
NTU_CAC=AU_CAC/C_dot_min_CAC
```

Results

```
Q_dot_CAC=15578 [W]
T_a_ex_CAC=68 [C]
T_gw_ex_CAC=58,08 [C]
AU_CAC=254 [W/K]
```

2.5.3.3 Charge Air Cooling by Refrigerant

In this less usual configuration, the charge air cooler is cooled down by the refrigerant. The charge air cooler is integrated inside the refrigeration loop of the air-conditioning system. Also, charge air cooling could be achieved by a heat-to-cool refrigeration system, such as a jet-ejector refrigeration system (Galindo et al., 2019).

2.5.4 Technologies of Charge Air Coolers

2.5.4.1 Air-Cooled Charge Air Coolers

Figure 2.39 shows the components constituting a typical air-cooled charge CAC. This heat exchanger shows a tube and fin configuration. Charge air flows through electro-welded or folded

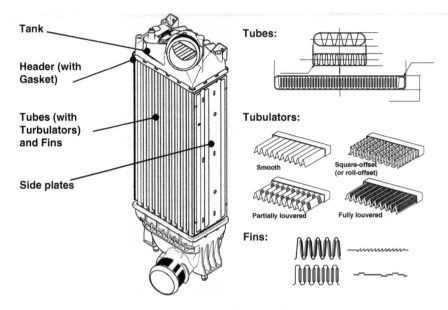

Figure 2.39 Components of an air-cooled CAC. Source: Courtesy of Valeo.

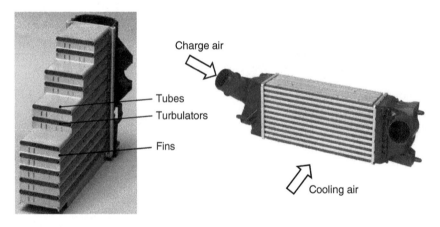

Figure 2.40 View of an air-cooled CAC. Source: Courtesy of Valeo.

tubes equipped with turbulators. On the cooling air side, the heat exchanger comprises fins, which are louvered or roll-offset.

A view of a practical realization of such a heat exchanger for a passenger car application is given in Figure 2.40.

2.5.4.2 Water-Cooled Charge Air Coolers

Figure 2.41 details the components of a water-cooled charge air cooler integrated into the intake air manifold. This heat exchanger has a plate-fin configuration with four passes on the water side. The high compactness of the system is particularly visible on the bottom picture.

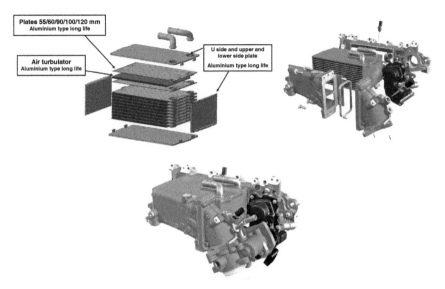

Figure 2.41 Components of a tubular water-cooled CAC. Source: Courtesy of Valeo.

2.6 Exhaust Gas Recirculation (EGR) Cooling

2.6.1 Purpose of EGR and EGR Cooling

The formation of NO and NO_2 (NO_x) is mainly related to the combustion temperature and the oxygen concentration in the intake charge. EGR consists of mixing a fraction of the cylinders exhaust gases with fresh air. The mixture is supplied back to the cylinders. The purpose of the EGR is to reduce the emissions of NO_x by lowering the temperature and the concentration of dioxide O_2 inside the cylinders during combustion. Actually, NO_x are formed by dissociation and recombination of O_2 and N_2 at high temperature.

Exhaust gas recirculation has been initially installed on diesel engines, but it is currently used for both diesel (CI) and spark-ignition (SI) engines. For SI (i.e. gasoline) engines, NO_x are already treated by three-way catalytic converters.

However, for SI engines, cooled EGR recirculation allows for mitigating engine knocking and therefore minimizing the fuel consumption by optimizing the combustion timing (Tornatore et al., 2019). Actually, without EGR, knocking is typically prevented by delaying the ignition, which results in hotter exhaust gases. To stay below the turbine inlet temperature limit, the fuel–air mixture is enriched (fuel acts as a thermal capacity) yielding a fuel consumption increase. Cooled EGR allows to stay closer to stoichiometry ($\lambda = 1$) and to better meet emissions regulations over a larger part of the engine map.

Exhaust Gas Recirculation Cooling is the purpose of cooling the recirculated fraction of the exhaust gases before being re-injected into the cylinders to improve the NO_x emissions reduction by reducing even more the combustion temperature.

Replacing part of fresh air by recirculated gases decreases the oxygen concentration and makes the mixture richer. The major drawback of the use of EGR is the increase in other pollutants (HC, CO, particles, and CO_2) following the richer mixture. Therefore, the EGR rate must be controlled to find the balance between NO_x reduction and other pollutants increase.

2.6.2 EGR Working Principle

The EGR rate, X_{EGR}, is defined as the ratio of the recirculated exhaust gas mass flow rate to the mixture mass flow rate:

$$X_{EGR} = \frac{\dot{m}_{EGR}}{\dot{m}_{EGR} + \dot{m}_a + \dot{m}_f} \qquad (2.63)$$

where

$\dot{m}_{EGR}$ is the mass flow rate of recirculated gas, [kg s^{-1}]
$\dot{m}_a$ is the mass flow rate of intake air, [kg s^{-1}]
$\dot{m}_f$ is the mass flow rate of fuel, [kg s^{-1}]

The fraction of recirculated gas varies from 0% to 55% depending on the engine load and speed. Since the exhaust gas contains less O_2, recirculation decreases the mass concentration of O_2 in the cylinder. Also, exhaust gas contains H_2O and CO_2, which are triatomic molecules; this increases the specific heat of the cylinder contents and therefore decreases the adiabatic combustion temperature.

Before being mixed with fresh charge, the recirculated fraction of the exhaust gases can be cooled down in the *Exhaust Gas Recirculation Cooler*. Cooling down the gases increases the engine volumetric efficiency and further decreases the combustion temperature and NO_x production (Wei et al., 2012). By increasing the density of recirculated exhaust gases, it also prevents to decrease too much the air mass flow rate entering the cylinders and to increase the emissions of particulates for a given EGR rate.

2.6.3 Exhaust Gas Recirculation Architectures

There exist two main architectures of the exhaust gas recirculation: the high-pressure architecture and the low-pressure architecture. These architectures can be combined.

2.6.3.1 High-Pressure EGR

The high-pressure EGR architecture is illustrated in Figure 2.42 in the case of a diesel engine equipped with a turbocharger. Part of the exhaust gas is routed to the engine intake, while the remainder flows to the turbine wheel. Recirculated exhaust gas is cooled down in the EGR cooler before being mixed with fresh air coming from the charge air cooler. The EGR rate is controlled by means of the high pressure EGR valve.

During engine cold start, the by-pass valve is open to increase the engine temperature quickly as well as the temperature of the diesel oxidation catalyst (DOC) in the engine exhaust line (in such a way not to produce very large emissions of HC and CO).

2.6.3.2 Low-Pressure EGR

In the low-pressure EGR configuration, the EGR cooler is located downstream from the turbine. As shown in Figure 2.43, after the DOC and after the diesel particulate filter (DPF), a fraction of the exhaust gas is routed through the EGR cooler. Cooled recirculated gas is mixed with fresh intake air before entering into the compressor. The engine represented in Figure 2.43 is also equipped with a high pressure EGR system that does not include a EGR cooler.

The major interest for such a configuration is to recirculate "clean" gases downstream the DOC and DPF and containing few or no HC. Also, the EGR gases can also be cooled down by the CAC.

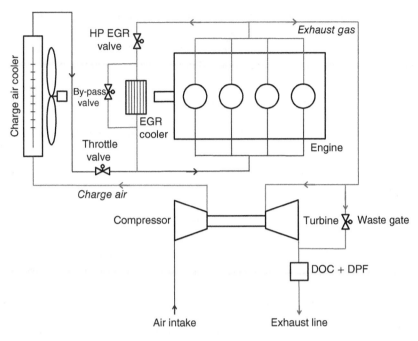

Figure 2.42 Schematic representation of the air/gas circuit of a diesel engine equipped with a high-pressure EGR.

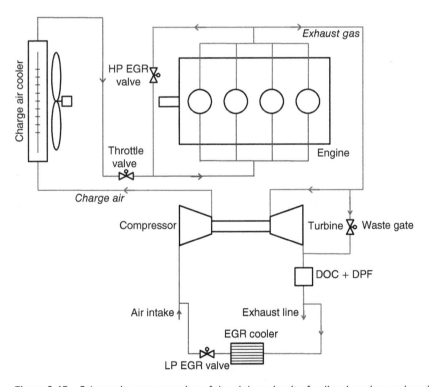

Figure 2.43 Schematic representation of the air/gas circuit of a diesel engine equipped with both a low-pressure and a high-pressure EGR.

The latter allows for reaching lower temperatures at the engine inlet in comparison with a HP EGR configuration. However, the EGR gases contain water vapor, leading to a risk of (acid) condensation in the CAC. The mixture is also more homogeneous. The capacity to decrease the NO_x emission is improved, and the trade-offs with other pollutants emission (PM, CO_2) are better. Finally, in low-pressure EGR systems, the recirculation of exhaust gases does not impact the turbocharger. It should also be mentioned that the packaging of LP EGR is more difficult due to the length of the piping between the DPF and the air intake.

Combining HP EGR with LP EGR allows increasing the EGR rate and using the by-pass of the HP EGR to decrease the engine warm-up time.

2.6.4 Technologies of Exhaust Gas Recirculation Coolers (EGRC)

Exhaust gas recirculation coolers (EGRC) must be able to work at high temperature and to sustain thermal shocks. They should limit the fouling by soot or limit the impact of fouling. They also have to resist to corrosion due to acid condensation. Their thermal effectiveness must be high and the pressure drop encountered by the exhaust gases must be limited. Finally, EGRC must be compact.

EGRC are cooled down by glycol water. Different technologies of heat exchangers can be used:

- Shell and tubes. Tubes can be round (Figure 2.44) or flat oval (Figure 2.45). They can be corrugated to improve the heat exchanger performance and increase the compactness (Figure 2.44).

The flat oval configuration yields better performance, because of lower pressure drop associated with a larger cross-sectional area for the exhaust gas.

- Plate-fin Such heat exchangers, featured in Figures 2.46 and 2.47, are made of a stack of plates and fins. Their large heat transfer area as well as their large cross-sectional area allow to reach a good trade-off between thermal effectiveness and pressure drop. Different types of fins can be used on the gas side. On the coolant side, plates are corrugated. The U-flow configuration allows for an easier packaging and better efficiency of the gas by-pass than the I-flow configuration.
- Tube–fin Performance and compactness are larger than those achieved with the previous technologies (Figure 2.48).

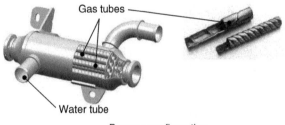

By-pass configuration

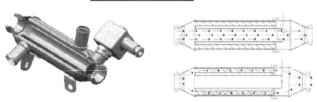

Figure 2.44 Illustration of a shell and round tubes EGRC (length × diameter: 116 × 45 mm; nominal power: 1.21 kW; EGR mass flow rate: 10 g s^{-1}; pressure drop on gas side: 3.4 mbar). Source: Courtesy of Valeo.

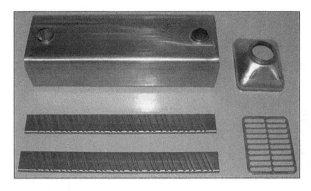

Figure 2.45 Illustration of a shell and flat oval tubes EGRC (width × height × length: 51 × 64 × 200 mm). Source: Courtesy of Valeo.

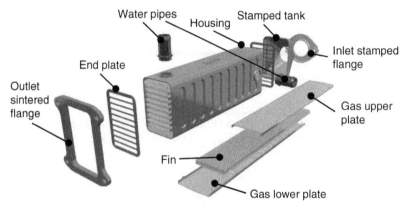

Figure 2.46 Schematic of a plate and fin EGRC – I-flow. Source: Courtesy of Valeo.

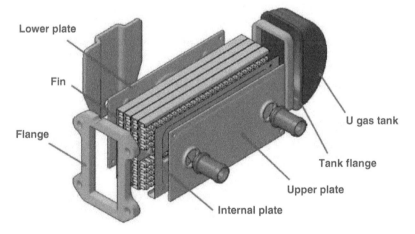

Figure 2.47 Illustration of a plate and fin EGRC – U-flow. Source: Courtesy of Valeo.

The main characteristics of a EGR cooler are as follows:

– Its "permeability", which determines, for a pressure difference between inlet/outlet, the mass flow rate that flows through the heat exchanger.
– Its thermal effectiveness, which defines the capacity of the cooler to cool down the recirculated gases.

Fouling affects the performance of the EGRC. It depends on the strategy considered for EGR settings. Fouling is due to clogging resulting from the interaction of soot with water condensates.

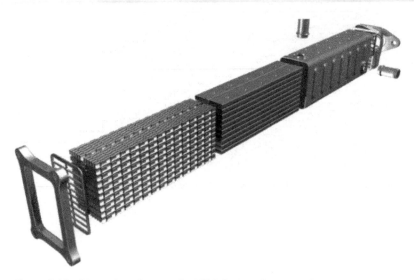

Figure 2.48 Illustration of a tube-fin EGRC. Source: Courtesy of Valeo.

Thermal effectiveness of EGRC typically ranges from 60% to more than 90% according to the technologies of heat exchanger used and the operating conditions. Used materials are usually stainless steel or aluminum.

Example 2.3 *Exhaust Gas Recirculation Cooler*

A plate–fin EGRC is characterized by a conductance that is assumed constant and equal to 49 $W K^{-1}$. Recirculated exhaust gas enters the EGRC at a temperature of 800 °C and an absolute pressure of 3 bar. The coolant is an aqueous solution of ethylene glycol (35% in mass of glycol). The coolant flow rate is 800 $L h^{-1}$ and the coolant temperature at the EGRC supply is 90 °C (Figure 2.49).

Draw the evolution of the thermal effectiveness and heat transfer rate of the EGRC with the EGR mass flow rate ranging from 5 to 30 $g s^{-1}$.

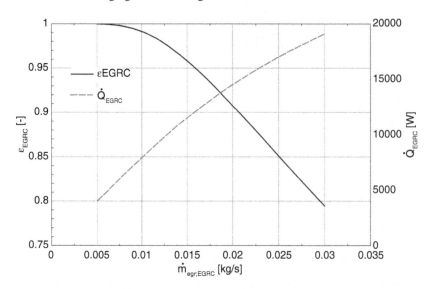

Figure 2.49 Thermal effectiveness and heat transfer rate of the EGRC as a function of EGR mass flow rate.

Additional Data

Specific heat of glycol water (35/65%): 3833 $[\text{J kg}^{-1}\text{K}^{-1}]$
Density of glycol water (35/65%): 1002 $[\text{kg m}^{-3}]$

Solution

```
"1. INPUTS"
T_egr_su_EGRC=880 [C] "temperature of exhaust gas at EGRC supply"
T_gw_su_EGRC=90 [C] "temperature of glycol water at EGRC supply"
P_egr_su_EGRC=3e5 [Pa] "pressure of exhaust gas at EGRC supply"
V_dot_gw_EGRC=800e-3/3600 [m^3/s] "volume flow rate of glycol water through
    the EGRC"
{m_dot_egr_EGRC=0,020 [kg/s]} "exhaust gas mass flow rate through the EGRC"
    "in parametric table"


"2. OUTPUTS"
"T_egr_ex_EGRC=?? [C]" "temperature of recirculated gas at EGRC exhaust"
"T_gw_ex_EGRC=?? [C]" "temperature of glycol water at EGRC exhaust"

"3. PARAMETERS"
AU_EGRC=49 [W/K] "conductance of EGRC"


"4. MODEL"
"4.1. EGRC heat transfer rate"
"c_gw=3833 [J/kg-K]
rho_gw=1002 [m^3/kg]"
"or using EES library"
c_gw=cp(EG;T=T_gw_su_EGRC;C=35)
rho_gw=density(EG;T=T_gw_su_EGRC;C=35)

v_gw=1/rho_gw
m_dot_gw_EGRC=V_dot_gw_EGRC/v_gw
c_p_egr_EGRC=1014
C_dot_egr_EGRC=m_dot_egr_EGRC*c_p_egr_EGRC
C_dot_gw_EGRC=m_dot_gw_EGRC*c_gw
C_dot_min_EGRC=min(C_dot_egr_EGRC;C_dot_gw_EGRC)
C_dot_max_EGRC=max(C_dot_egr_EGRC;C_dot_gw_EGRC)
C_r_EGRC=C_dot_min_EGRC/C_dot_max_EGRC

epsilon_EGRC=(1-exp(-NTU_EGRC*(1-C_r_EGRC)))/(1-C_r_EGRC*exp(-NTU_EGRC*
    (1-C_r_EGRC)))
NTU_EGRC=AU_EGRC/C_dot_min_EGRC
Q_dot_EGRC=epsilon_EGRC*C_dot_min_EGRC*(T_egr_su_EGRC-T_gw_su_EGRC)

"4.2. Energy balance on charge air side"
Q_dot_EGRC=C_dot_egr_EGRC*(T_egr_su_EGRC-T_egr_ex_EGRC)

"4.3. Energy balance on coolant side"
Q_dot_EGRC=C_dot_gw_EGRC*(T_gw_ex_EGRC-T_gw_su_EGRC)
```

2.7   Front-End Module

2.7.1   Purpose of the Front-End Module

The front-end module is located in the front of the vehicle. Its first function is to contain all the heat exchangers supplied with outdoor air. Some of these heat exchangers have been described in this Chapter: radiator, air-cooled charge air cooler and oil cooler. Other heat exchangers hosted by the front-end module will be described later; the air-cooled condenser of the air-conditioning system (Chapter 3), heat exchangers associated with the electrical powertrain (Chapter 4) and heat exchangers associated with waste heat recovery systems (at the end of this chapter).

The design of the front-end module is optimized to decrease the vehicle drag and to ensure the design cooling capacity. The design should also increase its compactness and decrease its mass.

Finally, the front-end module serves other functions. The bumper system ensures the pedestrian protection and the crash management. The front-end module also comprises part of the vehicle lighting system.

2.7.2 Working Principle of the Front-End Module

Air enters into the vehicle underhood through the grilles. It then flows through the different heat exchangers in the front-end module. At low vehicle speeds, airflow through the front-end module results from the action of one or several fans. At higher vehicle speed, airflow is ensured by the ram pressure. This will be explained in detail in this section.

2.7.2.1 Heat Exchangers Configuration

Heat exchangers operating at the lowest fluid temperature are located first on the cooling air flow path. Such configuration yields the best match between the cooling air temperature profile and the temperatures of the different fluids to be cooled in the heat exchangers. Different configurations are represented in Figure 2.50. In a "surfacic" architecture, air flows successively through the condenser, charge air cooler and radiator (and oil-cooler). In a beam architecture, a charge air cooler of smaller frontal area is located in front of the condenser. In a mosaic architecture, a charge air cooler of a smaller frontal area is located aside the condenser on the air path.

Table 2.2 gives typical orders of magnitude of heat transfer rates in the heat exchangers (for a passenger car) as well as inlet and outlet temperatures of the fluids to be cooled down.

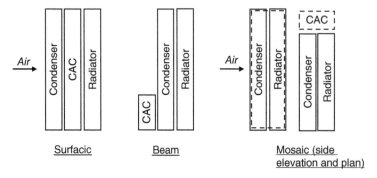

Figure 2.50 Examples of heat exchangers configurations inside the front-end module.

Table 2.2 Typical operating conditions of the different heat exchangers located inside the front-end module (for a passenger car).

	Heat transfer rate [kW]	Fluid inlet temperature [°C]	Fluid outlet temperature [°C]
Radiator	30	90	85
Charge air cooler	8	130	60
Oil cooler	4	130	125
Condenser	8	95	60

2.7.2.2 Aeraulics

Air speed at the inlet of the different heat exchangers is much lower than the vehicle speed. It depends on several vehicle characteristics (such as the size and shape of the grilles, the shape of the vehicle and the geometrical characteristics of the heat exchangers) and vehicle operating conditions (the fan regime and the vehicle speed).

Heat exchangers are cooled down mainly by convection. Cooling airflow passes through the heat exchangers under the action of a pressure gradient. This pressure gradient is created by one or several fans and by the vehicle ram effect. At low vehicle speed, the fan is the major contributor to the airflow through the heat exchangers. The flow path of the air through the front-end module is illustrated in Figure 2.51. Freestream air (∞) enters the front-end module through one or several air inlets or grilles ($0 - 1$). In general, there are upper and lower air inlets. In Figure 2.51, the air inlets are represented by a unique opening ("grille"). Airflow is then routed through the air-conditioning condenser ($1 - 2$) and radiator ($2 - 3$). As mentioned previously, other heat exchangers may be used depending on the types of motorization and thermal management loops implemented onboard. Airflow is then guided through the shroud ($3 - 4$), passes through the fan ($4 - 5$) and the engine bay ($5 - 6$). It finally exits the vehicle through the underhood outlet ($6 - \infty$).

When the airflow reaches the grille, part of its dynamic pressure is converted to static pressure. As described by Ap (1999), the static pressure P_0 at the inlet of the grille can be evaluated by means of

$$P_0 - P_\infty = C_{p,su}\frac{1}{2}\rho C_\infty^2 \tag{2.64}$$

where

P_0 is the static pressure at the inlet of the grille, [Pa]
P_∞ is the freestream static pressure (atmospheric pressure), [Pa]
$C_{p,su}$ is the pressure coefficient at the vehicle inlet, [−]
C_∞ is the vehicle speed, [m s^{-1}]

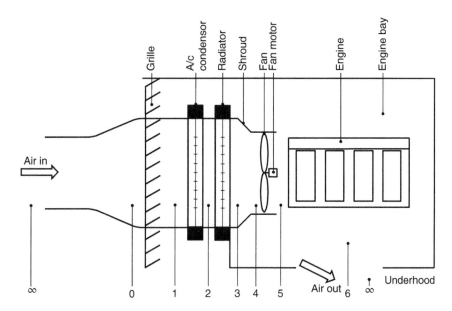

Figure 2.51 Cooling airflow path. Source: Adapted from Schaub and Charles (1980).

The aerodynamic shape of the vehicle determines the value of the pressure coefficient $C_{p,su}$ (Ap 1999).

When airflow leaves the engine bay to the vehicle underhood, it is accelerated, which creates a static pressure decrease. The static pressure P_6 upstream from the underhood outlet can be computed by

$$P_6 - P_\infty = C_{p,ex} \frac{1}{2} \rho C_\infty^2 \qquad (2.65)$$

where

$C_{p,ex}$ is the pressure coefficient at the underhood exhaust, [−]

Ap (1999) proposes to compute the static pressure losses through the different components based on pressure losses coefficients:

$$\Delta P_i = K_i \frac{1}{2} \rho C_i^2 \qquad (2.66)$$

Assuming that the cooling airflow density is constant, the airflow velocity at the component i can be related to the velocity through the radiator by:

$$C_i = \frac{A_r}{A_i} C_r \qquad (2.67)$$

where

C_r is the speed at radiator location, [m s^{-1}]
C_i is the speed at component i location, [m s^{-1}]
A_r is the frontal area of radiator, [m^2]
A_i is the frontal area of component i, [m^2]

When it is activated, the fan introduces a static pressure jump $\Delta P_{fan,s}$ (*fan static pressure*). Since the static pressure upstream and downstream the airflow streamtube are equal and correspond to the ambient pressure, the sum of all the static pressure losses along the streamtube is equal to zero (Verner 2000):

$$P_0 - P_\infty - \sum_i \Delta P_i + \Delta P_{fan,s} - (P_6 - P_\infty) = 0 \iff (P_0 - P_6) + \Delta P_{fan,s} - \sum_i \Delta P_i = 0 \qquad (2.68)$$

Solving the previous equation allows the air velocity through the radiator to be determined (Ap 1999). The ratio between the air velocity through the radiator and the vehicle speed C_r/C_∞, called radiator adaptation coefficient, has been introduced by Ap (1999). Figure 2.52 shows an example of evolution of this coefficient with the vehicle speed when the fan is activated or not.

Eq. (2.68) indicates that the static pressure gradient $P_0 - P_6$ ("ram pressure rise") created by the vehicle motion contributes to compensate the heat exchangers and vehicle pressure drops $\sum_i \Delta P_i$ ("system pressure drop"). For a given air velocity through the radiator, and hence a given cooling airflow volume flow rate, the ram pressure rise can be larger than the system pressure drop. In such a situation, illustrated by point D in Figure 2.53, the fan creates a flow resistance (passive role) and is characterized by a negative pressure jump.

Indeed, in Figure 2.53, the operating point is the intersection of the fan curve (given for a defined rotational speed) and the system curve (sum of the pressure drops of all heat exchangers in series as a function of the air-flow rate). The ram effect shifts down the system curve.

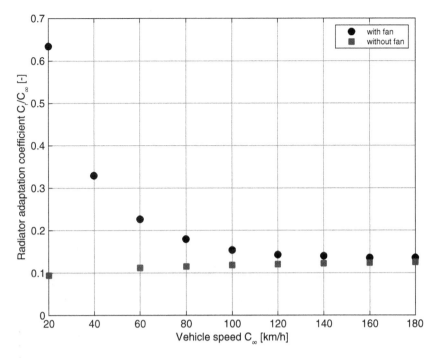

Figure 2.52 Example of air speed as a function of the vehicle speed (with and without fan operation).

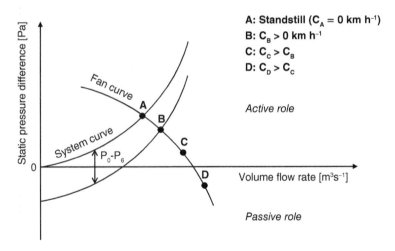

Figure 2.53 Coupling of the fan and system of heat exchangers.

Example 2.4 *Modeling of a Truck Front-end Module*

A truck front-end module is modeled with the elements that can be seen in Figure 2.54. The air from the outside, which is at the atmospheric pressure (P_∞) and at a speed C_∞ (considered as the speed of the truck, i.e. no wind speed considered), is flowing through the front-end module to supply the heat exchangers.

The pressure coefficient $C_{p,su}$ at the vehicle inlet is equal to 0.4. The acceleration of the air stream at the underhood outlet is not described.

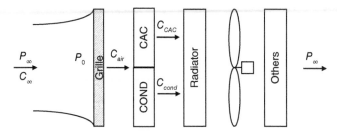

Figure 2.54 Elements of the truck front-end module.

The following simplified fan characteristic curve is proposed:

$$\Delta P_{fan}[Pa] = -113\,\dot{V}_{air} + 850\frac{N_{fan}}{N_{ref}}$$

with $\dot{V}_{air}$ [m³ h⁻¹] the air volume flow rate, N_{ref} = 2100 RPM, N_{fan} the fan speed [RPM].

Let us assume that there is neither leakage nor by-pass of the element found in the front-end module (i.e., the mass flow rate of the components in series must be the same). The air density is given by ρ_{air} = 1.2 kg m⁻³. The pressure losses coefficients of the elements are given by K_{grille} = 1, K_{rad} = 5, K_{CAC} = 2, $K_{AC,cond}$ = 4, K_{other} = 3. The cross-section area of the cooling package is given by A = 0.75 [m²] (as for every component in the front-end module except for the CAC and the A/C condenser that have areas of $A/2$). It is assumed that the pressure at the outlet of the front-end module is the atmospheric pressure (P_∞).

Questions:

1. What is the air mass flow rate crossing the front-end module if the fan is not considered and the vehicle speed is 90 km h⁻¹.
2. What is the air mass flow rate crossing the CAC for these conditions?
3. What is the air mass flow rate when the fan rotates at a speed of 1500 RPM when the vehicle is stopped?

Solution

```
"1. INPUTS"
N_fan = 1500 [RPM] "Fan speed"
C_infinity =" 0"25 [m/s] "Vehicle speed" "to be set to 0 for question 3"
P_infinity=1e5 [Pa] "Atmospheric pressure"
rho_air = 1,2 [kg/m^3] "Air density"

"2. OUTPUTS"
"m_dot_air, m_dot_air_CAC and m_dot_air_AC_cond [kg/s]"

"3. PARAMETERS"
A = 0,75 [m^2] "Cooling package cross section area"
C_p_su = 0,4 [-] "Pressure coefficient at grille inlet"
k_grille = 1 [-]  "Pressure loss coefficient in the grille"
k_rad = 5 [-]  "Pressure loss coefficient in the radiator"
k_CAC = 2 [-] "Pressure loss coefficient in the charge air cooler"
k_AC_cond = 4 [-] "Pressure loss coefficient in the condenser"
k_other = 3 [-] "Pressure loss coefficient in other components"
```

```
"4. MODEL"

"4.1. Conversion of dynamic pressure in static pressure at the grille inlet"
DELTA_P_stream = C_p_su*1/2*rho_air*K "Computation of the pressure losses/gains"
K=if(C_infinity;C_air;-C_air^2;-C_air^2;C_infinity^2) "to simulate the case
    where the vehicle speed is null, then the air is accelerated from 0 to C_air"
DELTA_P_stream =P_0-P_infinity

"4.2. Fan model"
N_ref = 2100 [RPM]
DELTA_P_fan = -113*V_dot_air + 850*N_fan/N_ref
V_dot_air = A*C_air

"4.2. Pressure drop through the different components"
DELTA_P_grille = - k_grille*rho_air*C_air^2
DELTA_P_rad =  -k_rad*rho_air*C_air^2
DELTA_P_CAC =  - k_CAC*rho_air*C_CAC^2
DELTA_P_AC_cond = - k_AC_cond*rho_air*C_AC_cond^2
DELTA_P_other = - k_other*rho_air*C_air^2


"4.3. Description of 2 heat exchangers in parallel"
DELTA_P_CAC = DELTA_P_AC_cond "Since the heat exchangers are in parallel, they
    have the same inlet and outlet pressures"
A*C_air = A_AC_cond*C_AC_cond + A_CAC*C_CAC "mass conservation"
A_AC_cond = A/2 "AC condenser cross section area"
A_CAC = A/2 "Charge air cooler cross section area"

"4.4. Pressure balance over the entire stream tube"
DELTA_P_stream + DELTA_P_grille + DELTA_P_rad + DELTA_P_CAC + DELTA_P_other" +
    DELTA_P_fan" = 0 "the last term of the sum is commented to answer question 1"

"4.5. Mass flow rates"
m_dot_air = A*rho_air*C_air
m_dot_air_CAC = A_CAC * C_CAC*rho_air
m_dot_air_AC_cond = A_AC_cond* C_AC_cond*rho_air
```

Results

Questions 1 and 2	Question 3
C_air=3,262 [m/s]	C_air=4,192 [m/s]
C_CAC=3,822 [m/s]	C_CAC=4,911 [m/s]
C_infinity=25 [m/s]	C_infinity=0 [m/s]
m_dot_air=2,936 [kg/s]	m_dot_air=3,773 [kg/s]
m_dot_air_AC_cond=1,216 [kg/s]	m_dot_air_AC_cond=1,563 [kg/s]
m_dot_air_CAC=1,72 [kg/s]	m_dot_air_CAC=2,21 [kg/s]

2.7.3 Technologies of Components in the Front-End Module

2.7.3.1 Fan System

The fan system typically comprises one or two fans to generate the air flow when that generated by the ram pressure is not large enough. Fan activation can be necessary for high engine loads, high ambient temperature, and low vehicle speed. Practically, the fan activation is controlled by measurements: coolant temperature, condensing pressure in the A/C loop (see Chapter 3), and ambient temperature. The fan system also comprises a shroud to support the fan and to guide the air stream. Note that dual-fan systems are mainly associated with larger engines.

Fans are driven mechanically, electrically, or hydraulically:

- *Mechanically* driven fans are connected to the engine crankshaft by means of a belt and pulley. A fan clutch is typically used to disengage the fan when there is no cooling demand, for instance, during engine warm-up. Viscous fan clutches are commonly used. Such clutches are made up of a drive disk and a driven housing that is connected to the fan. Between the drive disk and the driven housing are a set of grooves that can be filled with a viscous fluid. For high air temperatures at the outlet of the radiator, a bimetal strip controls a valve that allows the fluid to fill the grooves, leading to a strong coupling and a fan speed close to the engine speed (Scott and Xie 2007). For low air speeds, fluid is removed from the grooves, leading to a weaker coupling and a reduced fan speed (Scott and Xie 2007). Instead of using a bimetal strip, the fan engagement can be electronically controlled by using information from different sensors (ambient air temperature, air-conditioning condensing pressure, and oil and coolant temperatures).
- *Electrically* driven fans use a motor, the speed of which is controlled in order to adjust the air flow rate through the heat exchangers. Most of the electrically driven fans are equipped with brushed type DC motors. However, brushless DC motors are more and more used. The latter ones show some advantages over brushed motors; for equivalent powers, they are more compact and lighter, more reliable and durable, and more efficient (80–85% versus 60–75%).

  For passenger cars, the power of the fan electric motor typically ranges from 100 to 600 W. Electrically driven fans can be speed controlled. The fan is actually sized for extreme conditions, but most of the time, a lower air flow is requested. By decreasing the fan speed, the noise and the energy consumption of the fan system can be reduced. There exist different speed control mechanisms including the use of an electrical resistance to decrease the voltage across the motor (bi-speed control) and voltage control by a pulse with modulation (PWM) signal (continuous adaptation of the speed).
- *Hydraulically* driven fans are driven by a hydraulic motor, for instance, a gear motor, supplied with pressurized oil. Hydraulically driven fans allow for speed control by adjusting the oil flow rate through the motor (Jambukar et al., 2014).

Fan system can be located upstream ("pusher configuration") or downstream ("puller configuration") the heat exchangers. Downstream location is better for protection of the fan and homogeneous air distribution, but the fan environment is hotter.

Current developments aim at increasing the fan system efficiency (improvement of both the motor and the fan blades), increasing the compactness, decreasing the weight and reducing the noise.

2.7.3.2 Active Grille Shutters

Grilles can also be equipped with active shutters, the position of which can be controlled as a function of the vehicle cooling needs. Both the upper and lower grilles can be equipped with shutters. Closing the shutters allows the vehicle drag (the Cx coefficient) to be reduced leading to fuel savings. It also allows for faster warm up ("heat encapsulation"), which reduces the pollutant emissions and fuel consumption. Finally, the grilles also contribute to reduce the noise generated by the vehicle (by a few dB).

The air grilles are sized on the worst conditions, corresponding to low vehicle speed, high engine load, and high ambient temperature. However, most of the time, the air flow through the front-end module can be reduced (Figure 2.55).

Figure 2.55 Active Grille Shutters. Source: Courtesy of Valeo.

2.8 Engine Waste Heat Recovery

As it has been explained in Section 2.2.5, only a fraction of the heat released by the combustion of fuel is converted in mechanical energy necessary to drive the vehicle and run auxiliaries (pumps, alternator, and fans). The remaining fraction is dissipated outside the vehicle, mainly through the exhaust gases and through the engine cooling loop radiator. For vehicles equipped with EGR, part of the heat available in the exhaust gas is rejected in the EGR cooling loop. Another heat source available onboard vehicle is the heat rejected in the CAC system.

In general, waste heat is not recovered except for cabin heating. However, valorizing this heat could allow increasing the overall energy performance of the vehicle. In this section, some waste heat recovery and valorization techniques, commercialized or under investigation, are described.

2.8.1 Exhaust Heat Recovery System (EHRS)

The EHRS is a heat exchanger that allows for the heat transfer between the exhaust gas and the engine coolant and/or oil. As represented in Figure 2.56, the engine coolant and the exhaust gas flow in a counterflow configuration through the EHRS. By means of such heat exchanger, higher temperatures are reached at the heater core supply. The EHRS is also equipped with a by-pass valve that can be opened to let the exhaust gas flow outside of the heat exchanger. This bypass valve can be opened for instance as soon as the coolant reaches its temperature set point (Lee et al., 2011). The EHRS is located after the catalytic converter to avoid impact on the emissions (Lee et al., 2011). More advanced configurations allow for the additional heating of lubricating or transmission oil.

The EHRS is used to decrease the warm-up time of the engine coolant and oil (engine oil or lubricating oil). This shorter warm-up yields the following benefits:

- The comfort inside the cabin, which is ensured by the heater core on the coolant circuit, is achieved quickly.
- The engine fuel consumption and emissions are reduced because of the reduced friction.
- For a hybrid vehicle, when the heating system is on, the electric mode is more often used, since the coolant circuit temperature reaches the set point (55–65 °C) required by the heater core more quickly (Barrieu 2011).

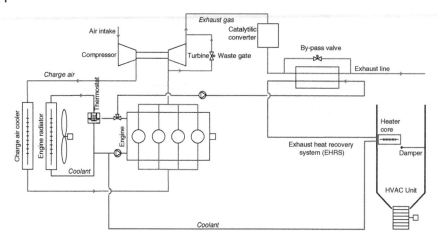

Figure 2.56 Schematic representation of the connection of the EHRS with the engine coolant, heater core, and exhaust gas.

Also, the EHRS can be used to compensate for the deficit of heating capacity of the coolant loop, which is encountered with highly efficient engines (Diehl et al., 2001).

The EHRS is among the simplest technical solutions to recover the thermal energy from the exhaust and is already installed in commercialized vehicles. A few percent of fuel savings can be achieved. The lower the outdoor temperature, the more interesting is this solution (Lee et al., 2001).

2.8.2 (Organic) Rankine Cycle Power Systems

The Rankine Cycle heat engine is another waste heat recovery technique that has been largely investigated during the first two decades of the 2000s, for automotive applications and then for long-haul trucks (Guillaume and Lemort 2019). A Rankine cycle heat engine comprises four major components: an evaporator, an expansion machine, a condenser and a pump. Some Rankine cycle systems comprise a fifth component, which is the recuperator. A working fluid flows successively in all four components in a cyclic way. This working fluid can be water or an organic compound (which contains carbon). At the outlet of the condenser, the working fluid is in liquid state at the low pressure of the system. The pump rises the pressure of the liquid. The high-pressure liquid is then routed to the evaporator where it is heated, vaporized and possibly superheated. The evaporator heat load is provided by engine waste heat. At the outlet of the evaporator, the high-pressure vapor is expanded in an expansion machine, which produces mechanical power. This expansion machine can drive a generator, thereby producing electricity. At the outlet of the expansion machine, the low-pressure vapor enters the condenser. The fluid is cooled down to saturation and condensed, and the resulting liquid is eventually sub-cooled.

When the fluid exits the expansion machine in vapor state, a recuperator could be introduced in the system. It allows the high-pressure liquid to be pre-heated before entering the evaporator (by cooling down the low-pressure vapor from the expansion machine exhaust to the condenser supply), yielding an increase of the thermal efficiency of the system.

In the example of Figure 2.57, the heat source consists of both the EGR cooling and the exhaust gas. Even though the EGR gases are at a higher temperature than the exhaust gas, the cold working fluid at the pump outlet first cools down the recirculated gases. This ensures the recirculated gases to be cooled enough not to impact the depollution system performance. The condenser is water cooled and the glycol–water coolant loop is connected to a dedicated radiator in the front-end module. The expansion machine drives a generator. This is an alternative to a mechanical coupling to the engine shaft. Producing electricity rather than mechanical power offers different advantages: the speed of the expansion machine can be controlled (optimizing the performance of the cycle and

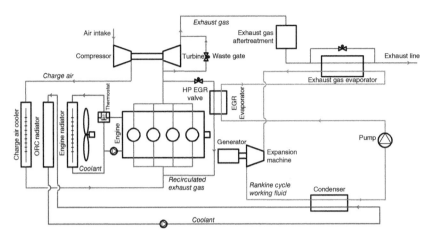

Figure 2.57 Waste heat recovery from both EGR and exhaust gas: Series configuration of the EGR and tailpipe evaporator, dedicated condenser coolant loop, and expander connected to a generator.

of the expansion machine), the Rankine cycle system can keep on working during the motoring phase, the management of the electricity production (in association with batteries) can contribute to the optimization of the performance of a hybrid vehicle. The example of Figure 2.57 presents waste-heat recovery from exhaust gas. Alternatively, or in a combined way, waste heat can be recovered from the engine coolant loop. The latter heat source is at a much lower temperature, but less transient (except during engine warm-up period), which can simplify the design and the control of the RC system. It should also be mentioned that recovering heat from exhaust gases introduces constraints on the maximal back pressure created on the exhaust line by the heat exchanger.

To ensure a large degree of subcooling (for preventing the pump to cavitate), a liquid receiver followed by a subcooler can be placed between the condenser and the pump. In this case, in steady-state regime, the fluid at the outlet of the condenser is in saturated liquid state. The condenser can also be externally pressurized, for instance, by compressed air through a flexible membrane (Galuppo et al., 2021).

Criteria for selection of the working fluid include the thermodynamic performance of the cycle, the cost of the ORC system, pressure levels (pressure not too high in the evaporator and over-atmospheric pressure in the condenser), environmental impact, security (flammability, toxicity), and degradation temperature. The expansion machine can be a displacement expander or a turbomachine. The selection of the working fluid, expansion machine, and cycle architecture must be conducted in parallel through a multi-criteria optimization under constraints. The optimization of the ORC design must account for different engine operating points and aim at minimizing the fuel consumption on a driving cycle. Examples of ORC design optimization for truck applications are given by Bettoja et al. (2016) and Guilllaume and Lemort (2019).

Example 2.5   *Organic Rankine Cycle (ORC)*

An ORC power system has been designed for valorizing the waste heat from the exhaust gases of a truck engine. The ORC condenser is cooled with glycol–water (specific heat of 3700 J $(kg-K)^{-1}$). The expansion machine is a scroll expander of $60\,cm^3$ operating at 2000 rpm. Its isentropic effectiveness is assumed equal to 70% and that of the pump equal to 60%. The working fluid is R1233zd(E). The evaporator and condenser pinch points are equal to 15 and 10 K, respectively. The liquid subcooling at the condenser exhaust and the vapor superheat at the evaporator exhaust are equal to 5 K. The exhaust gas mass flow rate and temperature at the evaporator supply are

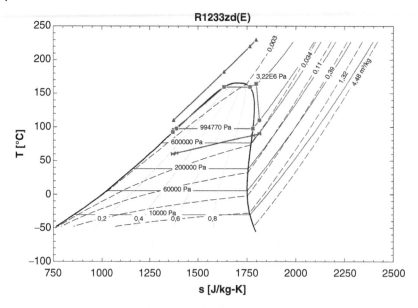

Figure 2.58 Representation of the ORC in the T–s diagram. Green: working fluid. Blue: coolant. Red: exhaust gas. Temperature pinch points are visible.

$0.5\,\text{kg}\,\text{s}^{-1}$ and $230\,°\text{C}$. The coolant mass flow rate and temperature at the condenser supply are $0.65\,\text{kg}\,\text{s}^{-1}$ and $60\,°\text{C}$, respectively.

Compute the efficiency of the cycle, the power delivered by the expander, the power consumed by the pump, the net delivered power and the heat transfer rates in the evaporator and condenser. Represent the cycle in a T–s diagram (Figure 2.58).

Solution

Only the evaporator and expander models are given hereunder. The evaporator model is a three-zone model characterized by a pinch point. When describing the evaporator, guess values on the working fluid temperature at the evaporator supply (pump exhaust) and on the working fluid mass flow rate are necessary. These guesses are removed when describing the expander and pump, respectively.

```
"4. MODEL"
"4.1. Evaporator"
"M_dot_wf_ev=0,35 [C] " "guess value before description of the
   expander"
T_wf_su_ev="70" "guess value"T_wf_ex_pp "the guess value is
   necessary until the pump is described"
"4.1.1. Liquid zone"
"4.1.1.1. Working fluid side"
h_wf_su_ev=enthalpy(fluid$;P=P_ev;T=T_wf_su_ev) "enthalpy of the
   working fluid at the supply of the evaporator"
h_wf_su_ev_tp=enthalpy(fluid$;P=P_ev;x=0) "enthalpy of the working
   fluid at the supply of the two-phase zone of the evaporator"
Q_dot_ev_l=M_dot_wf_ev*(h_wf_su_ev_tp-h_wf_su_ev) "heat transfer
   rate in the liquid zone of the evaporator"
```

```
"4.1.1.2. Secondary fluid side"
Q_dot_ev_l=M_dot_sf_ev*c_p_sf_ev*(T_sf_su_ev_l-T_sf_ex_ev)
"4.1.2. Two-phase zone"
"4.1.2.1. Working fluid side"
h_wf_su_ev_v=enthalpy(fluid$;P=P_ev;x=1) "enthalpy of the work-
ing fluid at the supply of the vapor zone of the evaporator"
Q_dot_ev_tp=M_dot_wf_ev*(h_wf_su_ev_v-h_wf_su_ev_tp) "heat trans-
fer rate in the two-phase zone of the evaporator"
"4.1.2.2. Secondary fluid side"
Q_dot_ev_tp=M_dot_sf_ev*c_p_sf_ev*(T_sf_su_ev_tp-T_sf_su_ev_l)
"4.1.3. Vapor zone"
"4.1.3.1. Working fluid side"
h_wf_ex_ev=enthalpy(fluid$;P=P_ev;T=T_wf_ex_ev) "enthalpy of the
   working fluid at the exhaust of the evaporator"
T_wf_ex_ev=T_ev+DELTAT_wf_ex_ev "temperature of the working fluid
   at the exhaust of the evaporator"
T_ev=temperature(fluid$;P=P_ev;x=1) "evaporating temperature"
Q_dot_ev_v=M_dot_wf_ev*(h_wf_ex_ev-h_wf_su_ev_v) "heat transfer rate
   in the vapor zone of the evaporator"
"4.1.3.2. Secondary fluid side"
Q_dot_ev_v=M_dot_sf_ev*c_p_sf_ev*(T_sf_su_ev-T_sf_su_ev_tp)
"4.1.4. total heat transfer rate"
Q_dot_ev=Q_dot_ev_l+Q_dot_ev_tp+Q_dot_ev_v "heat transfer rate in
   the evaporator"
"4.1.5. Temperature pinch point"
pinch_ev=min(T_sf[1]-T_wf[1];T_sf[2]-T_wf[2];T_sf[3]-T_wf[3];
   T_sf[4]-T_wf[4]) "temperature pinch point in the evaporator"

"4.2. Expander"
M_dot_wf_exp=M_dot_wf_ev
T_wf_su_exp=T_wf_ex_ev
"P_cd=2e5" "guess value"
 "4.2.1. Supply"
h_wf_su_exp=enthalpy(fluid$;P=P_ev;T=T_wf_su_exp) "enthalpy of the
   working fluid at the supply of the expander"
s_wf_su_exp=entropy(fluid$;P=P_ev;T=T_wf_su_exp) "entropy of the
   working fluid at the supply of the expander"
v_wf_su_exp=volume(fluid$;P=P_ev;T=T_wf_su_exp) "specific
   volume of the working fluid at the supply of the expander"
"4.2.2. Displaced mass flow"
M_dot_wf_exp=V_dot_s_exp/v_wf_su_exp "expander displaced mass flow
   rate"
V_dot_s_exp=N_exp/60*V_s_exp "expander displaced volume flow rate"
"4.2.3. Power"
h_wf_ex_exp_s=enthalpy(fluid$;P=P_cd;s=s_wf_su_exp) "enthalpy of
   the working fluid at the exhaust of the expander for an
   isentropic expansion"
w_exp_s=h_wf_su_exp-h_wf_ex_exp_s "isentropic work"
```

```
epsilon_s_exp=w_exp/w_exp_s
w_exp=h_wf_su_exp-h_wf_ex_exp "actual work"
W_dot_exp=M_dot_wf_exp*w_exp "expander power"
```

```
"4.5. Overall performance"
eta=(W_dot_exp-W_dot_pp)/Q_dot_ev "Rankine cycle efficiency"
W_dot_net=W_dot_exp-W_dot_pp "Net power delivered by the Rankine
    cycle power system"
residual=Q_dot_ev+W_dot_pp-Q_dot_cd-W_dot_exp "energy residual
    (must be close to zero)"
```

Results (for question 1 and 2)

```
eta=0,06218 [-] P_cd=994770 [Pa]
P_ev=3,220E+06 [Pa]
Q_dot_cd=72789 [W]
Q_dot_cd_l=3062 [W]
Q_dot_cd_tp=63961 [W]
Q_dot_cd_v=5766 [W]
Q_dot_ev=77615 [W]
Q_dot_ev_l=46441 [W]
Q_dot_ev_tp=24859 [W]
Q_dot_ev_v=6315 [W]
T_cd=97,87 [C]
T_ev=159,5 [C]
W_dot_exp=6348 [W]
W_dot_net=4826 [W]
W_dot_pp=1522 [W]
```

2.8.3   Other Investigated Technologies

Other waste heat valorization technologies include turbocompound (mechanical or electric energy production), thermoelectric generators (electricity production), sorption cycles (cooling production), and jet ejector cycles (cooling production). The cooling effect produced by the latter two technologies can be used for either cabin air-conditioning or engine-forced induction. The interest of the two former technologies versus Rankine cycle power systems has been discussed by Legros et al. (2014).

References

Amara, S. (1994). Elaboration et validation d'un modèle de transferts thermiques instationnaires gaz-paroi dans la chambre de combustion d'un moteur. PhD thesis from Ecole Centrale de Lyon.

Annand, W.J.D. (1963). Heat transfer in the cylinder of a reciprocating internal engine. *Proceedings of the Institution of Mechanical Engineers* 177 (36).

Ap, N.S. (1999). A simple engine cooling system simulation model. *SAE Technical Paper Series*, 1999-01-0237.

Barrieu, E. (2011). EHRS impact on engine warm up and fuel economy. In: Directions in Engine-Efficiency and Emissions Research Conference (Deer 2011), October 3–6, 2011, Detroit, USA.

Bettoja, F., Perosino, A., Lemort, V., Guillaume, L., and Reiche, T. (2016). NoWaste: waste heat re-use for greener truck. In *Proceedings of 6th Transport Research Arena*. doi:https://doi.org/10.1016/j.trpro.2016.05.456

Çengel, Y.A. and Boles, M.A. (2006). *Thermodynamics, an Engineering Approach*, 5e. McGraw-Hill Higher Education.

Chanfreau M., B. Gessier, A. Farkh, and Geels, P. Y. (2003). The need for an electrical water valve in a THErmal Management Intelligent System (THEMIS™). SAE Technical Paper Series 2003-01-0274.

Diehl, P., Haubner, F., Klopstein, S., and Koch, F. (2001). Exhaust heat recovery system for modern cars, SAE Technical Paper 2001-01-1020, https://doi.org/10.4271/2001-01-1020.

Ferguson, C.R. and Kirkpatrick, A.T. (2015). *Internal Combustion Engines: Applied Thermosciences*. Wiley. *ProQuest Ebook Central* https://ebookcentral.proquest.com/lib/unideliege/detail.action?docID=4039918.

Galindo, J., V. Dolz, A. Tiseira, A. Ponce-Mora, Thermodynamic analysis and optimization of a jet ejector refrigeration cycle used to cool down the intake air in an IC engine, *International Journal of Refrigeration*103 (2019) 253–263. doi:https://doi.org/10.1016/j.ijrefrig.2019.04.019.

Galuppo, F., Reiche, T., Lemort, V. et al. (2021). Organic Rankine cycle based waste heat recovery modeling and control of the low pressure side using direct condensation and dedicated fans. *Energy* 216: 119074.

Guillaume, L. and Lemort, V. (2019). Comparison of different ORC typologies for heavy-duty trucks by means of a thermo-economic optimization. *Energy* 182: 706–728. https://doi.org/10.1016/j.energy.2019.05.195.

Heywood, J.B. (1988). *Internal Combustion Engine Fundamentals*, Series in Mechanical Engineering. McGraw-Hill.

Hillier, V.A.W. and Coombes, P. (2004). *Hillier's Fundamentals of Motor Vehicle Technology. Book 1*, 5e. Nelson Thornes Ltd.

Jambukar, V., Aher, V., and Wakchaure, V. (2014). Hydraulic system for fan drive application to cool engine effectively. *International Journal of Engineering and Technical Research (IJETR)* 2 (7): 4–9.

Lee, J., Ohn, H., Choi, J., Kim, S. et al. (2011). Development of effective exhaust gas heat recovery system for a hybrid electric vehicle, SAE Technical Paper 2011-01-1171https://doi.org/10.4271/2011-01-1171.

Legros, A., Guillaume, L., Diny, M. et al. (2014). Comparison and impact of waste heat recovery technologies on passenger car fuel consumption on a normalized driving cycle. *Energies* 7 (8): 5273–5290.

Pang, H.H. and Brace, C.J. (2004). Review of engine cooling technologies for modern engines. *Proceedings of the Institution of Mechanical Engineers Part D Journal of Automobile Engineering* 218: 1209–1215.

Schaub, U.W., and Charles, H.N. (1980). Ram Air Effects on the Air Side Cooling System Performance of a Typical North American Passenger Car. SAE paper n° 800032.

Scott, T. and Xie, Z. (2007). Viscous fan drive model for robust cooling air flow simulation, SAE Technical Paper 2007-01-0595. https://doi.org/10.4271/2007-01-0595.

Sieder, E. N., and Tate, E. C. (1936). Heat transfer and pressure drop of liquids in tubes. Industrial & Engineering Chemistry Research 28: 1429–1436.

Taylor, C.F. (1985). *The Internal Combustion Engine in Theory and Practice*. Cambridge, Massachusetts: MIT Press.

Tornatore, C., Bozza, F., De Bellis, V. et al. (2019). Experimental and numerical study on the influence of cooled EGR on knock tendency, performance and emissions of a downsized spark-ignition engine. *Energy, Volume* 172: 968–976. ISSN 0360-5442.

Verner, D.G. (2000). General ram Correlations for automobiles. Master Thesis, Texas Tech University.

Volkswagen, A.G. (2006). Programme autodidactique 359. Le moteur TSI de 1,4l à double suralimentation. Conception et fonctionnement. 000.2811.73.40 Définition technique 03.2006

Wei, H., Zhu, T., Shu, G. et al. (2012). Gasoline engine exhaust gas recirculation – a review. *Applied Energy* 99 (2012): 534–544.

Woschni, G. (1967) A universally application equation of the instantaneous heat transfer coefficient in the internal combustion engine. *SAE Paper 670931.*

3

Cabin Climate Control

3.1 Introduction

Cabin climate control, also called mobile air conditioning (MAC), is the process of controlling the temperature, humidity, and velocity of the air and indoor air quality (IAQ) inside the cabin. Cabin climate control not only ensures the thermal comfort of occupants but also contributes to their security.

As it will be shown in this chapter, the thermal comfort is function of the temperature, the humidity, the solar radiation through the glazing, the velocity of air inside the cabin, and the temperatures of the surfaces surrounding the occupants and exchanging heat with the latter ones through conduction and radiation. A comfortable cabin climate can be achieved by controlling some of these variables.

The security of occupants is improved by limiting the formation of mist and frost on the glazed surfaces (hence ensuring the visibility of the driver) and maintaining the concentration of CO_2 emitted by occupants at an acceptable level. Filtration of entering outdoor air and cabin air also limits the concentration of pollutants inside the cabin and their adverse effect on the occupants' health.

The cabin climate control increases the fuel consumption of vehicles with internal combustion engines and decreases the driving range of electric vehicles. Hence, it is important to maximize the energy performance of the cabin climate control systems, taking some criteria and constraints into consideration: reliability, cost, weight, and size.

This chapter first addresses thermal comfort. Then, in the third section, the energy balance over a cabin will be introduced to derive cooling and heating loads. Distribution and emission of cooling and heating to the cabin will be described in the fourth section. Cooling and heating production systems will be presented in the fifth and sixth sections, respectively. The last two sections will cover local cooling and heating systems and thermal energy storage.

3.2 Thermal Comfort

When driving or occupying a car, we can expect indoor climatic conditions at least as comfortable as outdoor conditions and even more comfortable. There are multiple aspects that contribute to the comfort of the passengers inside a vehicle cabin.

First of all, there is thermal comfort, which depends, on the one hand, on environmental parameters, which are air and surface temperatures, air humidity, solar and surface radiations and air displacement velocity inside the cabin and, on the other hand, on occupant-related parameters, which are their activity or metabolism and their clothing.

Thermal Energy Management in Vehicles, First Edition. Vincent Lemort, Gérard Olivier, and Georges de Pelsemaeker.
© 2023 John Wiley & Sons Ltd. Published 2023 by John Wiley & Sons Ltd.
Companion website: www.wiley.com/go/lemort/thermal

Second, acceptable IAQ should be achieved, which requires the use of high-efficiency filters often integrated in the heating and air conditioning system to treat the polluted air coming from the outdoor environment or generated inside the cabin. Generally, the gas filter treats the gaseous pollutants such as volatile organic compound (VOC), SO_2, NO_2 and ozone, while the particles filter is able to treat particles as pollen and soot. In addition sometimes, a gas sensor measuring the variation in the outdoor pollution (CO and NO) rates acts to manage the outdoor air entrance flap to prevent polluted outdoor air to enter into the vehicle. The system also ensures a minimum amount of fresh air supplied to the cabin to regenerate the oxygen absorbed by the occupants and to evacuate the humidity and carbon dioxide generated by the occupants. The more polluted the outdoor air is, the more relevant the control of the IAQ is. Actually, the car air treatment will be more and more seen as a protection against the polluted environment. Air filtration also allows for the protection of the evaporator and heater core in the heating, ventilation, and air-conditioning (HVAC) unit (described later in this chapter) and for the removal of odors. It should be mentioned that flagrances can also be used to improve the comfort of occupants.

Third, visibility is an important criterion not only for comfort but also for security: visibility to the outdoor environment should be high enough. For instance, mist on the windshield and the other parts of the glazing will largely alter visibility. The mist can be due to the increased cabin humidity resulting from moisture generated by the occupants and to the low temperature of the components (glazing, etc.). Also, the luminosity inside the cabin should be acceptable.

Fourth, an acceptable acoustic environment should be achieved inside the cabin. Many efforts have been realized to decrease the acoustic level contribution due to the engine compartment or coming from the outdoor environment. Therefore, it is important to limit the contribution of the components involved in the heating and air conditioning system, such as the fan, the compressor, the ducts, etc.

Other aspects involved in overall comfort are the ergonomics of the dashboard and seats (shape, presence of tactile screens, etc.) and driving assistance devices (GPS, etc.).

3.2.1 Definition of Thermal Comfort

Comfort inside the cabin has thus a multisensorial character. However, all the different contributions to comfort can be controlled independently even if there exist interactions between some of the control systems. For example, the air quality control system, which is integrated inside the heating and air conditioning system, generates air pressure drop that tends to decrease the air flow rate when the particle filter is clogged and participates to the flap control strategy.

This chapter will focus only on the thermal comfort and IAQ.

According to ASHRAE (2005a), *"thermal comfort is that condition of mind that expresses satisfaction with the thermal environment."*

Contrary to the situation in a building, the environment inside a vehicle cabin is not uniform, which makes the quantification and the control of thermal comfort complex (Ghosh et al., 2012). Section 3.2.2 will describe the physiological phenomena that contribute to the thermal comfort and how they are related to the cabin environment.

3.2.2 Human Thermo-Physiology

The sensation of thermal comfort or discomfort results from some major physiological variables (core and skin temperatures, metabolism, sweating flow rate, etc.), which are influenced by the climate surrounding the vehicle occupant (temperature, mean radiant temperature, humidity, and air velocity) and by personal parameters (clothing and activity). The thermal sensation is function of these six parameters, but varies between two distinct persons. The relationship between physiological variables and climatic variables can be obtained by expressing heat and mass transfer equations between the body, seen as a thermal system, and its surroundings. These equations are detailed in the following paragraphs.

3.2.2.1 Homeothermy

In opposite to ectothermic animals that gain heat from their environment, humans are an endothermic species who are able to internally produce heat. Moreover, humans are considered a homeothermic species, which means that their internal temperature is relatively constant even despite a great variation of the external environment temperature. The control of the internal body temperature to keep it constant, which defines homeothermy, is necessary to maintain the vital functions. A difference with the usual range of this body temperature often means a diseased state. In steady state, one can distinguish on the one side a thermally stable core composed of organs and of the head and on the other side a skin with variable temperature directly dependent on the external environment. Also, when the overall body surface is considered, it appears that the skin temperature is not homogeneous. The differences between the body segment temperatures are higher in cold conditions than in hot conditions.

Practically, we can distinguish:

- The homeothermic body core, represented by an internal temperature that is maintained close to 37°C and slightly increases with activity.
- The poikilothermic body surface, represented by skin temperature. In comfort conditions, its temperature ranges from 32 to 34°C and decreases with the activity while the sweat flow rate increases.

Homeothermy is maintained by thermoregulation. To maintain constant the internal temperature, the metabolism acts to balance the internal heat production and the energy transferred to the surroundings. The human body holds sensorial sensors on the skin, which send the information of hot or cold states toward the hypothalamus and the central nervous system. The hypothalamus reacts by activating the mechanism of regulation adapted to the situation. The internal production of heat, which depends on the activity or metabolism, has to compensate the energy transfer to the surroundings.

There exist two types of thermoregulation mechanisms:

- Physical thermoregulation:
  - Sweating in warm environment or when the activity is increased, which permits to cool by evaporation the skin surface
  - Variation in the blood flow rate to the body periphery to control the sensible heat dissipated to the surroundings (vasodilatation of skin blood vessels in hot warm environment or when the activity is increased, which increases the blood flow and the thermal exchanges/vasoconstriction of skin blood vessels in cold environment, which blocks the circulation at the extremities, particularly at the fingers, in order to concentrate the blood flow on the vital organs of the core)
- Chemical: variation in the body internal energy production in cold environment by means of shivering generated with an increased muscular tonus.

These mechanisms consist of increasing the metabolism heat production and reducing the heat loss towards the surroundings in cold conditions or on the contrary of increasing the heat loss towards the surroundings in hot conditions.

The cutaneous temperature is considered as a good indicator of thermal comfort. The thermo-physiological system will adapt the exchanges to keep it constant around 32–34°C (as function of the activity).

3.2.2.2 Body Energy Balance

Combustion in body cells produces a metabolic rate $\dot{E}_m$ [W], only a fraction of which is converted into mechanical power $\dot{W}$. A great part of the initial chemical energy is thus converted in thermal energy. The metabolic rate per unit area of skin surface $\dot{E}_m/A_{sk}$ is sometimes expressed in *met*, 1 met being equal to $58.2\,W\,m^{-2}$ of skin surface (it corresponds to the activity level of a seated person at rest such as a passenger in a cabin). Different metabolic rates during vehicle driving are given in Table 3.1 for a driver. For a passenger, the metabolic rate is close to 1 met.

Table 3.1 Metabolic rate during vehicle driving (ISO 14505-2:2007).

Activity level	Metabolic rate $\dot{E}_m/A_{sk}$ [W m^{-2}]
Driving a vehicle on a surfaced road	70
Driving a vehicle on a rough road	80
Driving a vehicle on a 4 × 4 road	90

The area of the skin surface A_{sk} can be evaluated as a function of the body mass m [kg] and height H [m] by Dubois' relationship (Dubois and Dubois, 1916) given in Eq. (3.1).

$$A_{sk} = 0.203\, m^{0.425} H^{0.725} \tag{3.1}$$

The difference $\dot{Q}_m = \dot{E}_m - \dot{W}$ [W] is the heat rate generated by the body and has to be dissipated to the external environment. The mechanisms of energy transfer between the body and its surroundings are as follows:

- sensible heat transfer between the skin and the environment, $\dot{Q}_{sk}$
- sensible heat transfer associated with respiration, $\dot{H}_{r,sens}$
- latent heat transfer associated with moisture evaporation during respiration, $\dot{H}_{r,lat}$
- latent heat transfer associated with sweat evaporation at the skin surface, $\dot{H}_{sweat}$
- latent heat transfer associated with evaporation of water diffused through the skin, $\dot{H}_{diff}$

The heat flow rate that can be dissipated (heat loss) by the human body is the sum of all hereunder-mentioned contributions.

$$\dot{Q}_{loss} = \dot{Q}_{sk} + \dot{H}_{sweat} + \dot{H}_{diff} + \dot{H}_{r,sens} + \dot{H}_{r,lat} = \dot{Q}_{sk} + \dot{H}_{sweat} + \dot{H}_{diff} + \dot{H}_r \tag{3.2}$$

In steady-state regime (no variation in the energy content of the body), the heat flow rate $\dot{Q}_m$ to be dissipated is equal to the heat flow rate $\dot{Q}_{loss}$ that can be dissipated by the human body. This steady-state regime is ensured by the thermoregulatory mechanisms within the physiological limits.

The metabolic heat rate $\dot{E}_m$ can be evaluated by measuring the inhaled flow rate of oxygen and the production of carbon dioxide (ASHRAE, 2005a). The body mechanical efficiency, defined as the ratio of the mechanical power $\dot{W}$ to the metabolic rate $\dot{E}_m$, is most of the time close to 0 and rarely larger than 0.10 (ASHRAE, 2005a). For car occupants, it can be assumed that the mechanical power is negligible. This is also a conservative assumption of the evaluation of the cooling load, since it means that the whole metabolic rate is converted into heat (ASHRAE, 2005a).

The different heat transfer mechanisms between the body and its surroundings are described more in detail in the following Sections 3.2.2.3, 3.2.2.4, and 3.2.2.5.

3.2.2.3 Skin Sensible Losses

The sensible heat loss from the skin $\dot{Q}_{sk}$ [W] is transported through the clothing before being transferred to the environment. The sensible heat transfer rate from the skin through the clothing, per unit area of skin surface, to the environment can be computed by

$$\dot{Q}_{sk} = \dot{Q}_{cl} = \frac{A_{sk}(T_{sk} - T_{cl})}{R_{cl}} \tag{3.3}$$

where

A_{sk} is the skin surface area, $[\text{m}^2]$
T_{sk} is the skin temperature, $[^\circ\text{C}]$
T_{cl} is the clothing outer surface temperature, $[^\circ\text{C}]$
R_{cl} is the clothing thermal resistance, $[\text{m}^2\,\text{K}\,\text{W}^{-1}]$

The thermal resistance of the clothing R_{cl} is defined as the thermal resistance of a uniform insulation layer that would cover the entire body and would result in the same heat transfer as the actual clothing. It is expressed in *Clo*, *1 Clo* being equal to $0.155\ \text{m}^2\,\text{K}\,\text{W}^{-1}$. The standard indoor clothing for wintertime is characterized by a thermal resistance of 1 Clo. The standard indoor clothing for summertime is characterized by a thermal resistance of 0.5 Clo.

Heat is then transferred from the clothing to the surroundings by convection and radiation:

$$\dot{Q}_{cl} = \dot{Q}_c + \dot{Q}_r \tag{3.4}$$

The terms in Eq. (3.4) are developed in Sections 3.2.3.1 and 3.2.3.2. They respectively depend on the dry bulb temperature T and mean radiant temperature T_w. The latter will be defined in Eq. (3.14). It is function of the temperatures of the surfaces surrounding the occupant.

3.2.2.4 Skin Latent Losses

The enthalpy flow rate $\dot{H}_{sweat}$ [W] associated with the vaporization of water produced by sweating can be assessed by the following experimental correlation established in comfortable conditions (ASHRAE, 2005a)

$$\dot{H}_{sweat} = A_{sk} \max\left(0, 0.42\left(\frac{\dot{Q}_m}{A_{sk}} - 58.15\right)\right) \tag{3.5}$$

The enthalpy flow rate associated with the natural diffusion of water through the skin is given by

$$\dot{H}_{diff} = \dot{m}_{diff}\, h_{w,sk,s} \tag{3.6}$$

where

$\dot{m}_{diff}$ is the water flow rate diffused through the skin, $[\text{kg s}^{-1}]$
$h_{w,sk,s}$ is the specific enthalpy of water at saturation and skin temperature, $[\text{J kg}^{-1}]$

The diffused water flow rate can be computed based on the difference of water partial pressures underneath the skin and inside the cabin. The meaning of the water partial pressure (vapor pressure) has been introduced in Chapter 1. Therefore,

$$\dot{m}_{diff} = A_{sk}\, K_{diff}\, (P_{w,sk,s} - P_{w,a}) \tag{3.7}$$

where

K_{diff} is the skin diffusion coefficient, equal to 1.3×10^{-9} [kg s^{-1}m^{-2} Pa^{-1}]
$P_{w,sk,s}$ is the saturation pressure of water at skin temperature, [Pa]
$P_{w,a}$ is the partial pressure of water in the cabin, [Pa]

The previous equations involve skin temperature. In comfortable conditions, the latter is given by (ASHRAE, 2005a)

$$T_{sk}[^\circ C] = 35.7 - 0.0275 \frac{\dot{Q}_m}{A_{sk}} \tag{3.8}$$

3.2.2.5 Respiratory Losses

Respiratory losses can be split into sensible and latent contributions. Actually, air is inhaled under cabin conditions, and exhaled air is slightly oversaturated: its temperature is 34°C, while the saturation temperature is 35.8°C. As a consequence, sensible heat and moisture are transferred to the inspired air before being exhaled. The respiration mass flow rate $\dot{m}_r$ [kg s^{-1}] can be correlated to the exhaled CO_2 mass flow rate considering that the molar fraction of CO_2 in the expired air is 5%:

$$\dot{m}_r = \frac{\dot{m}_{CO_2}}{0.05} \frac{MM_{air}}{MM_{CO_2}} \tag{3.9}$$

where

MM_{air} is the molar mass of air [kg kmol^{-1}]
MM_{CO2} is the molar mass of CO_2 [kg kmol^{-1}]

The expired CO_2 mass flow rate can be itself related to the metabolic power by

$$\dot{m}_{CO_2} = 1.05 \times 10^{-7} \dot{E}_m \tag{3.10}$$

The net humid air enthalpy flow rate transferred by the body to the environment, which is the respiratory losses $\dot{H}_r$ [W], is given by

$$\dot{H}_r = \dot{m}_r (h_{r,ex} - h_{r,su}) \tag{3.11}$$

The sensible contribution can be computed as

$$\dot{H}_{r,sens} = \dot{m}_r c_{p,r} (T_{r,ex} - T_{r,su}) \tag{3.12}$$

The latent contribution corresponds to

$$\dot{H}_{r,lat} = \dot{H}_r - \dot{H}_{r,sens} \tag{3.13}$$

where

$c_{p,r}$ is the specific heat at constant pressure of humid air, [J kg^{-1}K^{-1}]
$T_{r,su}$ is the temperature of the inhaled air (cabin condition), [°C]
$T_{r,ex}$ is the temperature of the exhaled air, [°C]

3.2.2.6 Criteria to Meet to Achieve Thermal Comfort

The body thermal equilibrium (locally and globally) is the first condition to be fulfilled for ensuring thermal comfort. However, body thermal equilibrium is not sufficient to provide alone the sensation of thermal comfort. In steady-state regime, Fanger (1982) proposed to relate the comfort to two physiological variables: the mean skin temperature and the heat transfer rate associated to sweat evaporation. Fanger showed that for a given metabolic power, there exists a combination of values of these two physiological variables that provides thermal comfort. The mean skin temperature decreases with the activity level (Eq. (3.8)), while the sweat flow rate increases with the activity (Eq. (3.5)). Both evolutions tend to increase the heat flow rate from the body core to the environment (ASHRAE, 2005a) and consequently to cool down the body.

As a consequence, the thermal comfort results from the combination of the metabolic rate, clothing insulation, dry temperature, mean radiant temperature, humidity content, and air velocity.

3.2.3 Description of Vehicle Indoor Climate

As presented in Section 3.2.2.6, the thermal comfort is a function of the cabin indoor climate. Actually, the latter influences the rates of the different energy transfer modes between the occupant and his surrounding. The indoor climate can be described by the following climatic variables (the energy transfer modes in which they are involved are mentioned in brackets):

- The cabin dry-bulb temperature $T_{a,cab}$ [°C] (convection)
- The pulsed air dry-bulb temperature $T_{a,su,cab}$ [°C] (convection)
- The temperatures of the surrounding surfaces T_j [°C] (radiation): dashboard, glazing, roof, etc.
- The temperatures of the surfaces in contact with the occupants T_k [°C] (conduction): seat at the level of body base and back, steering wheel, floor, etc.
- The cabin specific humidity ω [kg kg^{-1}] (water mass transfer)
- The ambient pressure P_a [Pa] (water mass transfer)
- The transmitted solar radiation τI_s [W m^{-2}] (radiation)
- The air velocity C_a, [m s^{-1}] (convection)

The cabin environment is highly nonuniform. Because of the direction of solar radiation, of the heating and air conditioning system location generally inside the dashboard in the front zone of the cabin, and of natural convection (which tends to create vertical temperature gradients), temperature gradients are important. Also, air velocities and radiant environment vary from one location to another in the cabin. Hence, it should be mentioned that multizonal models are necessary to accurately describe the indoor climate.

Some of the climatic variables in the cabin listed above can be combined as environmental parameters to define "equivalent" temperatures that are convenient for the thermal comfort assessment.

3.2.3.1 Mean Radiant Temperature

One of these "equivalent" temperatures is the mean radiant temperature T_w (or MRT), which can be computed knowing the surface temperatures. T_w is defined as "*the temperature of an imaginary isothermal black enclosure in which an occupant would exchange the same amount of heat by radiation as in the actual nonuniform environment*" (ASHRAE, 2003b). It can be computed with

$$(T_w + 273.15)^4 = \sum_{j=1}^{N} F_{cl,j} (T_j + 273.15)^4 \tag{3.14}$$

where

T_w is the mean radiant temperature, [°C]
$F_{cl,j}$ is the angle factor between the human subject and the environment, [−]
T_j is the temperature of surface j among the N surfaces, [°C]

Hence, the heat transfer rate by radiation between the occupant and his environment, involved in Eq. (3.4), can be computed by

$$\dot{Q}_r = A_{cl} F h_r (T_{cl} - T_w) \tag{3.15}$$

where

A_{cl} is the surface area of the clothed body, [m^2]
F is the correction factor to compute the effective radiation area of the body, [−]. Its value is around 0.7 for a seated person (ASHRAE, 2005a)
h_r is linearized radiative heat transfer coefficient, [W m^{-2} K^{-1}]

The surface area of the clothed body can be related to the skin surface area and to the clothing thermal resistance R_{cl}. For $R_{cl} > 0.5$ Clo, the following relationship can be used (ASHRAE, 2005):

$$A_{cl} = \left(1.05 + 0.1 \frac{R_{cl}}{0.155}\right) A_{sk} \qquad (3.16)$$

The linearized radiative heat transfer coefficient h_r is given by

$$h_r = \varepsilon_{cl}\, \sigma\, 4 \left(\frac{T_{cl} + 273.15 + T_w + 273.15}{2}\right)^3 \qquad (3.17)$$

where

ε_{cl} is the average emissivity of clothing or body surface, [−]
σ is the Stefan-Boltzmann constant, 5.67×10^{-8} [W m^{-2} K^{-4}]

The definition of the mean radiant temperature given above can be corrected to take into account the effect of a concentrated radiant heat source of rate $\dot{Q}_{rad}$, such as the solar radiation. The corrected mean radiant temperature $T_{w,corr}$ is given by

$$\left(T_{w,corr} + 273.15\right)^4 = (T_w + 273.15)^4 + \frac{\dot{Q}_{rad}}{F\, A_{cl}\, \varepsilon_{cl}\, \sigma} \qquad (3.18)$$

3.2.3.2 Operative Temperature

The operative temperature (or resultant temperature) T_o is "*the temperature of a uniform isothermal black enclosure in which the occupant would exchange the same amount of heat by radiation and convection as in the actual nonuniform environment*" (ASHRAE, 2003). In this imaginary enclosure, the air temperature is equal to the mean radiant temperature, and the air velocity is equal to that in the actual environment. By definition, the heat transfer rate between the clothed human subject and his environment can be expressed as

$$\dot{Q}_{cl} = \dot{Q}_r + \dot{Q}_c = A_{cl} (F \cdot h_r + h_c) \cdot (T_{cl} - T_o) \qquad (3.19)$$

The heat transfer rate by radiation has been developed in Eq. (3.15). The heat transfer rate by convection is given by:

$$\dot{Q}_c = A_{cl}\, h_c\, (T_{cl} - T) \qquad (3.20)$$

In Eq. (3.20), T is the temperature of air surrounding the occupants.

As a first approximation, the convective heat transfer coefficient is supposed to be the largest coefficient between the coefficient $h_{c,n}$ in natural convection and the coefficient $h_{c,f}$ in forced convection. These coefficients can be approximated by (ASHRAE, 2005a)

$$h_{c,n} = 2.38 \left(|T_{cl} - T|\right)^{0.25} \qquad (3.21)$$

$$h_{c,f} = 12.1\, C_a^{0.5} \qquad (3.22)$$

where

C_a is the air velocity, [m s^{-1}]

Figure 3.1 Definition of the operative temperature.

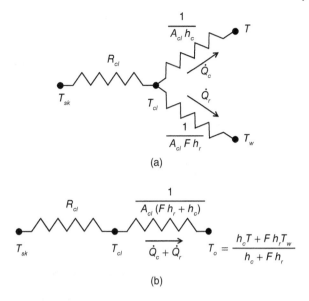

Combining Eqs. (3.15), (3.19), and (3.20), the operative temperature T_o can be expressed as function of the air temperature and mean radiant temperature by

$$T_o = \frac{h_c T + F h_r T_w}{h_c + F h_r} \tag{3.23}$$

The operative temperature takes into account the effects of the air temperature and velocity (convection) and the surrounding surfaces temperatures (radiation).

Schematically, the operative temperature can be represented by the equivalence between the thermal networks (a) and (b) in Figure 3.1.

A simplified expression of the operative temperature states that it is a weighted average of the air temperature and mean radiant temperature, which gives

$$T_o = a\, T + (1 - a)\, T_w \tag{3.24}$$

The weighting factors are function of the air velocity C_a [m s^{-1}]:

$$a = 0.5 + 0.25\, C_a \tag{3.25}$$

For air speed lower than 4 m s^{-1}, a tends to 0.5, and the operative temperature can be expressed as the average of the air temperature and mean radiant temperature:

$$T_o = \frac{T + T_w}{2} \tag{3.26}$$

The operative temperature is useful for the evaluation of the "dry" heat losses (or "non-evaporative" heat losses) of the human subject.

3.2.3.3 Equivalent Temperature

The operative temperature is often considered in buildings for the evaluation of thermal comfort. It is not suitable for vehicles, because the air velocity C_a is much higher in vehicle cabins than in buildings. The equivalent temperature is preferred, since it accounts for the fictitious decrease in the air temperature due to air movement and the additional cooling effect it produces on the skin.

The equivalent temperature T_{eq} [°C] is defined as the temperature of a uniform isothermal black enclosure "*with mean radiant temperature equal to air temperature and zero air velocity, in*

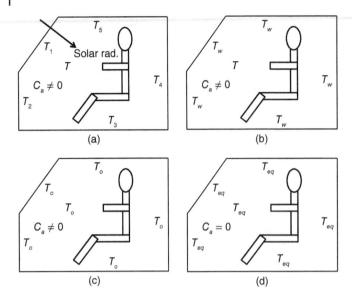

Figure 3.2 (a): Actual environment; (b): Definition of the mean radiant temperature; (c): Definition of the operative temperature; (d): Definition of the equivalent temperature.

which a person exchanges the same heat loss by convection and radiation as in the actual conditions" (ISO14505-3, 2006) (Figure 3.2).

It can be related to the heat transfer from the clothing by

$$\dot{Q}_{cl} = A_{cl} \cdot (h_{c,0} + F \cdot h_r)(T_{cl} - T_{eq}) \tag{3.27}$$

It must be noted that the previous formula can be easily modified to account for conductive heat transfer between the human subject and surrounding surfaces (for instance, the seats) (Holmér, 2005).

The equivalent temperature can be expressed as a function of the operative temperature by means of

$$T_{eq} = T_0 - \frac{h_c - h_{c,0}}{h_{c,0} + F h_r}(T_{cl} - T_0) \tag{3.28}$$

For low air velocities ($C_a < 0.1$ m s^{-1}), the equivalent temperature is equal to the operative temperature (SAE J2234). For larger air velocities ($C_a > 0.1$ m s^{-1}), the equivalent temperature can be computed by Eq. (3.29) developed by Madsen et al. (1984). This equation does not involve the clothing temperature, but the clothing resistance.

$$T_{eq} = 0.55\,T + 0.45\,T_w + \frac{0.24 - 0.75\,\sqrt{C_a}}{1 + R_{cl}}\,(36.5 - T) \tag{3.29}$$

where

T is the air temperature, [°C]
T_w is the mean radiant temperature, [°C]
C_a is the air velocity, [m s^{-1}]
R_{cl} is the clothing thermal resistance, [Clo]

The latter equation indicates that the equivalent temperature is a function of the clothing resistance: for a given air velocity C_a, the cooling effect is larger on a nude skin (SAE J2234, 1993).

3.2.3.4 Local Equivalent Temperature

One major difference with building indoor climate is the large nonuniformity of the environment surrounding the human subject. Experienced temperature, air velocity, and mean radiant temperature vary from one part of the body to another. This thermal asymmetry imposes to represent locally the climatic environment.

The measurement of temperature and air velocity in several points of the cabin allows to evaluate the performance of the heating and air conditioning system and of air distribution but not the human perception of the thermal environment. This perception can be assessed by the use of local and whole body equivalent temperatures.

As suggested by ISO 14505-2 (2006), a local equivalent temperature can be defined as the temperature of a uniform isothermal black enclosure, "*with mean radiant temperature equal to air temperature and zero air velocity, in which a defined zone of the human body surface exchanges the same heat loss by convection and radiation as in the actual conditions.*" Local equivalent temperature can be measured by means of a flat surface heated sensor or an ellipsoid shaped sensor or by means of a thermal manikin.

The *flat surface sensor* consists of a metal surface electrically heated at a constant heat rate. Using the measurement of the surface temperature and heating power, the sensor provides a directional measurement of the local equivalent temperature.

$$T_{eq,local} = T_{cl,local} - \frac{\dot{Q}_{cl,local}}{A_{sensor} \cdot (h_{c,0} + h_r)} \tag{3.30}$$

The *ellipsoidal sensor* measures the omnidirectional local equivalent temperature. Dimensions of the ellipsoid are selected to reproduce the same ratio between convection and radiation as that for the actual human body. The temperature of the heated sensor must represent the mean clothing surface temperature of the portion of the human being that the sensor is simulating (SAE J2234).

The use of a *thermal manikin* is described hereunder. Measurements of surface temperature and heat transfer rate at the zone level can be used to evaluate the local equivalent temperature.

3.2.3.5 Whole Body Equivalent Temperature

An equivalent temperature representative of the whole body can be defined by dividing the body surface into K zones individually heated. These zones are characterized by different surface temperatures. They experience different local climates and consequently exchange different heat rates with their surroundings. In this heterogeneous environment, this local measurement is important.

$$T_{eq,whole} = T_{cl,whole} - \frac{\dot{Q}_{cl,whole}}{A_{cl} \cdot (h_{c,0} + h_r)} \tag{3.31}$$

with

$$T_{cl,whole} = \frac{1}{A_{cl}} \sum_{i=1}^{K} T_{cl,i} A_{cl,i} \tag{3.32}$$

$$\dot{Q}_{cl,whole} = \frac{1}{A_{cl}} \sum_{i=1}^{K} \dot{Q}_{cl,i} A_{cl,i} \tag{3.33}$$

where

$T_{cl,i}$ is the surface temperature of the ith zone, [K]

$A_{cl,i}$ is the surface area of ith zone, [K]

$\dot{Q}_{cl,i}$ is the heat transfer rate by convection and radiation between the ith zone and the environment, [W]

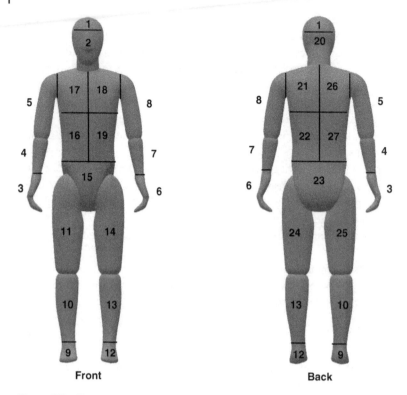

Front   Back

Figure 3.3 Example of a human manikin divided into 27 heated zones. Source: Courtesy of P.T. Teknik.

As shown in Figure 3.3, the whole-body equivalent temperature can be measured by means of a human-shaped man-sized manikin whose surface is divided into K zones heated independently with electric wire to simulate the thermal behavior of the person. The manikin surface is made of heat conductive material, to ensure uniform heat loss, with paint simulating the radiative properties of a human being. The manikin can be equipped with clothes representative of the season (Figure 3.4). Surface temperatures close to the actual surface temperatures of the human subject

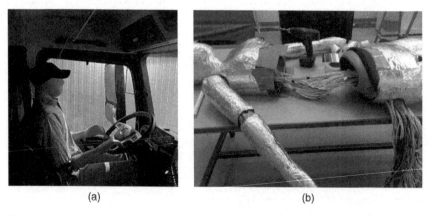

(a)   (b)

Figure 3.4 (a) Practical realization of a thermal manikin. Source: Courtesy of P.T. Teknik. (b) Assembly process of a thermal manikin. Source: Courtesy of University of Rennes, Laboratoire Génie Civil Génie Mécanique.

are reproduced. The thermal manikin is considered as a heat flowmeter. It simulates the sensible heat exchanges between the human body and the environment to which it is exposed. The electrical power necessary to maintain the controlled skin temperature at the different locations is measured and converted in local equivalent temperatures taking into account different environment parameters and the clothing. The humidity in the air and the activity of the manikin are not included in the measurement. The calculated whole-body equivalent temperatures provided by the manikin is the unique value representing the overall parameters and the thermal sensation of the human being inside an environment. The manikin is used to deliver a piece of information about the comfort, but this information has to be interpreted.

Unfortunately, the manikin is often not able to sweat when the conditions are hot, for example, when it is reached by direct solar radiation and/or during the cool-down phase associated with high cabin temperature. In such conditions, the skin temperature cannot be controlled. A solution would consist of supplying the manikin with water and generating vaporization if the skin temperature tends to increase in order to control it. However, in those hot conditions, the thermal manikin can be used to compare different situations.

The manikin has a number of advantages:

- It has the morphology of a person, allowing to evaluate the comfort of each part of the body.
- It detects the asymmetry due to the air flow or due to the radiation (solar radiation, cold surfaces, etc.).
- It uses a unique value, the equivalent whole-body temperature (even if local values are of interest), to evaluate the level of comfort.
- It is more sensitive than a human tester to the variations in the environment.
- It permits repeatable tests and is not influenced by the physiologic and psychologic circumstances of the human tester.

But it has some limitations:

- It does not sweat: it cannot reproduce exactly the mechanisms of thermal regulation of humans and avoid the accumulation of heat. For example, when the temperature of the environment is higher than that of the body or under the effect of solar radiation or for the parts of the body in contact with the seat, the accumulated heat is not rejected by means of the sweating effect. However, in this case, the thermal manikin gives only a piece of information about overheating in some parts of the body. It should be mentioned that some manikins are equipped with sweating functionality. For those manikins, artificial sweating glands are spread over the body to ensure a good coverage of the latter by sweat.
- Because of its thermal inertia, it is necessary to heat it before the beginning of the test in order to obtain the representative temperatures.

The thermal manikin is used to satisfy several objectives, among which:

- Tune the air distribution in the cabin
- Tune the air conditioning regulation
- Compare different technologies of glazing (tinted, over tinted, reflective)
- Compare different situations
- Understand physical phenomena

A large number of flat heated surfaces covering a nonheated manikin can also be used (ISO14505-2, 2006). The more the number of zones, the more realistic is the measured whole body equivalent temperature. More information about the measurement of this temperature is given in standard ISO 14505-2 (2006).

Figure 3.5 Example of use of omnidirectional sensors. Source: Courtesy of P.T. Teknik.

Omnidirectional sensors (ellipsoid-shaped sensors) can also be employed. The size and shape of those sensors are selected to reproduce the ratio between convective and radiative coefficients. Also, the paint color is selected to reproduce radiative properties of the car occupant. As shown in Figure 3.5, at least 6 sensors positioned in relevant locations must be installed on a structure simulating a seated human subject (ISO14505-2, 2006). The measured whole-body equivalent temperature is the average of the temperature measurements, weighted by the sensors surface areas.

If only three ellipsoid-shaped sensors are used (head, abdomen, and feet), the whole-body equivalent temperature can be approximated by (SAE J2234, 1993)

$$T_{eq,whole} = 0.1\, T_{eq}(head) + 0.7\, T_{eq}(abdomen) + 0.2\, T_{eq}(feet) \tag{3.34}$$

Finally, a numerical thermal manikin is a simulation tool that proposes a more detailed approach of the risk of discomfort based on the local skin temperature differences compared with an ideal map. It is adapted to calculate the heat flow exchanged inside the human body and with the environment and takes into account the mechanisms of human thermoregulation. It is a complementary approach, which permits to predict the thermal comfort at the beginning of the project phase before the realization of a physical prototype of the vehicle, including the heating, ventilation and air-conditioning system.

3.2.3.6 Control of Vehicle Indoor Climate

As it has just be explained, thermal comfort can be obtained by controlling the climate inside the cabin. In practice, approaching thermal comfort means reaching the whole-body equivalent temperature set point. To reach this set point, heating or cooling (including latent cooling, corresponding to dehumidification) powers must be provided to the cabin. These powers correspond to the heating and cooling loads, the computation of which is described in Section 3.3. Heating and cooling loads are covered by thermal energy emitters.

The most common emission system consists of pulsing the ventilation air at the right temperature, humidity and mass flow rate to reach the cabin climate set point. This is realized by means of the HVAC unit, the operating principle of which is detailed in Section 3.4.1.

Local comfort allows for reaching the right climate around the occupants without heating or cooling the whole cabin. This allows for energy savings. Local comfort can be achieved by heated seats and steering wheels, radiant panels and jet heating and cooling. This will be discussed in Section 3.7.

3.2.3.7 Transient Evolution of the Indoor Climate

Most of the vehicles driving cycles are of short duration. In such situations, there is no sufficient time for the cabin to enter into the steady-state regime. Expectations of the cabin passengers in terms of thermal comfort vary according to the expected duration of the journey. As a consequence, when describing the indoor climate, evaluating thermal comfort and the A/C system performance, the distinction must be done between transient period ("convergence" period) and steady-state regime. However, after reaching the steady-state phase, perturbations can yield dynamic evolutions: driving under a passing cloud or through a tunnel, variable vehicle speed, etc.

In transient conditions, the distinction is made among the following different phases:

- *Soaking* is a long period of time during which the vehicle is stopped in parking until reaching cabin steady-state temperature. Since the A/C system and heating system are switched off and the vehicle speed is null, the steady-state cabin temperature results from the outdoor temperature and solar radiation. In summer, with windows in closed position, cabin air temperatures higher than 60°C could be easily reached due to greenhouse effect, and some material could reach temperatures higher than 80°C, for example, on dashboard surface. Temperature in the soaked vehicle could be effectively reduced by using strategically located air inlets and outlets for natural-convection-induced flow ("parked car ventilation") (Rugh and Farrington, 2008). Introducing such inlets and outlets is quite complex. Forced ventilation can also be considered using, for instance, the HVAC unit blower. However, the cooling capacity associated with ventilation is limited in comparison with solar gains. In practice, the cabin air temperature and the material temperature (for instance, the dashboard) can be decreased by approximately 15 K and a few K, respectively. In winter, the cabin temperature can be as low as the outdoor temperature.
- *Pull-down* (or cool-down) is the period of time during which the temperature of a soaked cabin is decreased down to the set point temperature by means of the A/C system turned on.
- *Warm-up* is the period of time during which the temperature of a soaked cabin is increased up to the set point temperature by means of the heating system turned on.

Depending on the external conditions and the vehicle driving, the periods of pull down and warm-up can easily reach 20 minutes before reaching the set point temperature.

3.2.3.8 Air Stratification

As mentioned before, the thermal conditions in a cabin are heterogeneous and are source of local discomfort even if the body global energy balance target is reached.

The local discomfort can be the result of:

- Vertical gradient of air temperature (air temperature stratification), which involves low downstairs temperatures.
- Proximity to the cold or hot blown air outlets and sensation of air movement
- Proximity to the cold or hot surfaces (for instance, the glazing), which involves asymmetric radiative heat transfer.
- Contact with cold or hot surfaces (for instance, the seat is equivalent to a clothing and can procure discomfort, the steering wheel could be very hot or cold depending on the environmental conditions).

In comfortable conditions, the temperature at the feet level should be slightly higher than that at the upper body position.

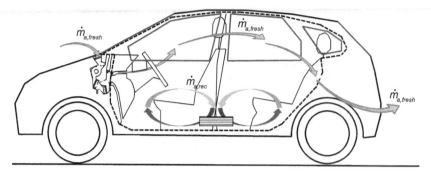

Figure 3.6 Split A/C system.

Also, a split A/C system contributes to a heterogeneous environment. Such a system creates a higher recirculation rate at the feet level, as illustrated in Figure 3.6. Dry fresh air is supplied to the upper zone, while air is recirculated and heated in the lower zone. Thus, the relative humidity becomes higher in the lower zone than in the upper zone. Such a system allows for reduction of fresh air by 20% without decreasing comfort and increasing risk of mist formation on the windshield. It yields lower thermal loads, which is an option to increase the autonomy of electric vehicles.

3.2.4 Evaluation of Thermal Comfort

ISO 14505-3 (2006) defines three types of methods for assessing thermal comfort: the objective methods, the behavioral methods and the subjective methods. Objective methods rely on the measurement of objectives indicators, such as the skin temperature or the skin wetness, which can be correlated with the thermal comfort. Behavioral methods are based on the observation of the occupant behavior, such as modifications of postures or movement patterns (ISO 14505-3, 2006). The subjective methods make use of a subjective scale, such as a thermal sensation scale.

The subjective methods are the most often used. Similarly to what is done for assessing the thermal comfort in buildings, the predicted mean vote (PMV) of the cabin occupants can be computed based on measurements or model predictions of the vehicle climate. Alternatively, and to cope with the limitations of the latter approach, human subject trials can be conducted.

3.2.4.1 PMV Approach

To assess the whole-body thermal comfort, a fictitious thermal load L [W] is defined. It corresponds to the difference between the heat generation rate $\dot{Q}_m$ and the heat losses rate $\dot{Q}_{loss,comfort}$ in the fictitious case where the human subject would be in comfortable conditions.

$$L = \dot{Q}_m - \dot{Q}_{loss,comfort} \tag{3.35}$$

In the previous equation, skin sensible losses can be evaluated using the measured whole-body equivalent temperature.

A thermal stress is associated with this fictitious thermal load. A discrete vote V, with values ranging from -3 to $+3$, corresponds to this stress. The ASHRAE thermal sensation scale is given in Table 3.2.

For a large sample of subjects, a predicted mean vote (PMV) can be defined. It is expressed as (Fanger, 1970):

$$PMV = \left(0.303\, e^{-0.037\frac{\dot{Q}_m}{A_{sk}}} + 0.0275 \right) \frac{1}{A_{sk}} L \tag{3.36}$$

Table 3.2 Thermal sensation scale.

+3	Hot
+2	Warm
+1	Slightly warm
0	Neutral
−1	Slightly cool
−2	Cool
−3	Cold

A comfortable cabin environment is associated with a PMV value of zero. In practice, PMV values comprised between −0.5 and 0.5 are acceptable.

For a given PMV, a predicted percent dissatisfied (*PPD*) can be expressed. A dissatisfied person is anyone whose discrete vote is different from −1, 0 or 1 (ASHRAE, 2005a). The evolution of the PPD with the PMV is given in Figure 3.7. This curve indicates that even for a PMV of 0, 5% of the people would be dissatisfied.

For a given metabolic heat generation rate and clothing resistance, a whole-body equivalent temperature can be identified that yields thermal comfort. Typical values are given in Table 3.3.

Actually, comfort must be ensured for the body as a whole. The air temperature and velocity around the body must deliver a comfortable whole body thermal sensation. However, local discomfort must be prevented. Local discomfort results from unacceptable local equivalent temperature due to asymmetric thermal radiation, draft or important temperature stratification or contact with a component.

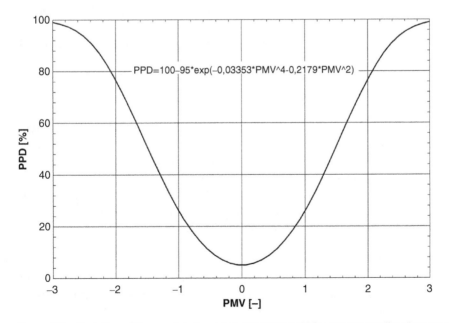

Figure 3.7 Evolution of the predicted percent dissatisfied (PPD) with the predicted mean vote (PMV). Adapted from Fanger 1982.

Table 3.3 Recommended whole body equivalent temperature as a function of metabolic rate and clothing resistance (ISO 14505-2, 2006).

Conditions	Whole body equivalent temperature [°C]
70 W m^{-2} and 0.5 Clo	25.3
90 W m^{-2} and 0.5 Clo	23.8
70 W m^{-2} and 1.0 Clo	22.1
90 W m^{-2} and 1.0 Clo	20.0

3.2.4.2 Human Subject Trials

Human subject trials consist of submitting a questionnaire to a sample of subjects that should be representative of the population of users of the vehicle. Standard ISO 14505-3 (2006) recommends at least eight subjects in the sample. The questionnaire aims at assessing not only the whole-body thermal comfort but also the thermal comfort of some parts of the body. Questionnaires make use of scales such as thermal sensation scale, "uncomfortable" scale, or stickiness scale (ISO 14505-3, 2006).

To perform such tests, the vehicle is put in climatic chambers or some regions of the world with suitable climates.

3.3 Cabin Thermal Loads

The cabin thermal loads are the cooling and heating demands for thermal comfort aspect and visibility aspect (glazing demisting and defrosting) that must be covered by the HVAC system of the vehicle.

As explained previously, there are two different phases. During the transient phase (convergence), which follows the HVAC system activation after the long-term parking of the vehicle, the cooling or heating energy is used to cool or heat the cabin air and components until the thermal comfort is obtained. During the steady-state phase, the thermal comfort is obtained but has to be maintained.

As it will be shown later, the cooling load comprises a sensible contribution and a latent contribution. The former is related to the temperature control inside the cabin, while the latter is related to the control of the moisture content or to the melting of frost from the glazing. As shown previously, temperature, moisture, and frost affect the occupants' thermal comfort and visibility. The cabin moisture content must be controlled to limit mist and frost formation on the interior and exterior faces of the windshield and other windows and ensure the driver's visibility.

Besides temperature and humidity, the CO_2 content inside the cabin must also be controlled by bringing fresh air to the occupants. Large concentrations of CO_2 could actually yield driver drowsiness (Seo and Choi, 2013).

The loads also vary with time. Actually, for passenger cars, most of the time, the driver achieves short trips, and the fraction of time associated with the cabin transient regime is far from being negligible. The desired indoor climate should be reached within a reasonable time to limit the situation where thermal comfort is not achieved before the end of the trip. That is why, a large heating/cooling capacity is needed for situations where the vehicle is parked outside. Also, during the transient regime with extreme environment conditions, a large fraction of the thermal capacity, especially in summer, is necessary for the evacuation of thermal energy stored in the components and in the air during parking. Compared to the design for a room of 40 m^3 in a building with installed heating and cooling capacities of around 2.5–3.0 kW, a conventional car of 3 m^3 needs around 5–7 kW of nominal heating and cooling capacities.

Of course, the loads vary with the environment conditions. In extreme conditions, the load can be very high, and the cooling/heating needs become the first source of energy consumption, which can exceed the power needed to run the vehicle, for example, in a urban cycle. Hence, for a pure electric vehicle or a hybrid vehicle running in the electric mode, the cabin thermal load has to be reduced even in cold conditions to minimize the impact on the driving range.

For security and comfort reasons, it is thus important to satisfy the vehicle thermal load and to design and size the HVAC system accordingly. In this section, we will describe the energy transfer mechanisms involved in the determination of the cabin thermal load and establish mass and energy balances that serve as the basis of cabin lumped-parameter models. Such models are convenient for being coupled to HVAC models in a global vehicle thermal system approach. The global system model could be used to investigate the control of the HVAC components or to evaluate the impact of the indoor climate control on the vehicle performance.

3.3.1 Outdoor Climate

3.3.1.1 Solar Radiation

Solar heat gains depend on the solar irradiance received by the glazed surfaces and on the optical properties of these surfaces. We will first describe the solar spectrum and then detail how the terrestrial solar irradiation can be computed.

3.3.1.1.1 Solar Spectrum Solar energy consists of the electromagnetic radiation emitted by the sun. When this radiation is received by a surface, it can be converted into thermal energy or electricity (with photovoltaic cells). The *solar irradiance* I_s [W m^{-2}] is defined as the power received by a surface per unit surface area. The *spectral solar irradiance* $I_{s,\lambda}$ [W m^{-2} μm^{-1}] is the irradiance per unit of wavelength interval $d\lambda$ about λ and per unit of surface area. On top of the atmosphere, the extraterrestrial solar irradiance on a surface perpendicular to the sun's rays varies with the day of the year (and thus with the distance between the sun and the earth) between 1332 [W m^{-2}] and 1413 [W m^{-2}] (ASHRAE, 2005c).

The spectrum of the extraterrestrial solar irradiance is similar to that of a black body of which temperature is close to 5760 K. Seeing the sun as a black body at a temperature T of 5760 [K], according to Wien's law (λ_m. $T = 2898$ [μm K]) as seen in Chapter 1, the maximum spectral solar irradiance corresponds to a wavelength $\lambda_m = 0.50$ [μm] (ASHRAE, 2005c). The solar spectrum extends from wavelengths of 0.15 to 3.8 [μm]; this interval comprises 99% of the solar irradiance.

The spectrum of solar radiation can be divided in different regions

- ultraviolet region: wavelengths between 0.2 and 0.4 μm
- visible region (region that can be seen by the human eye): wavelengths ranged between 0.4 and 0.7 μm
- near infrared region: wavelengths larger than 0.7 μm

At the sea level, approximately 6% of the radiation energy is comprised in the ultraviolet region, approximately 50% in the visible region, and approximately 40% in the infrared region (Moan, 2001).

When the solar radiation passes through the atmosphere, it undergoes two phenomena: diffusion and absorption. Diffusion results from reflection and scattering of solar radiation by gas molecules (oxygen, nitrogen, ozone, water vapor, etc.), dust and water droplets. Absorption of extraterrestrial radiation by gases in the atmosphere (O_2, H_2O, O_3, CO_2, aerosols, and particles) results into attenuation bands in the spectrum of solar irradiance. As a consequence to these phenomena, the maximum direct solar irradiance on a surface of the earth perpendicular to the sun's rays is around 1000 [$W\,m^{-2}$].

3.3.1.1.2 Calculation of the Terrestrial Solar Irradiance

A portion of the short-wavelength radiation scattered in the atmosphere by N_2, O_2, dust, and other particles actually reaches the earth's surface and is called diffuse radiation (ASHRAE, 2005c).

The total short-wavelength irradiance $I_{s,tot}$ on a surface of the earth is the sum of the direct irradiance I_D, the diffuse sky irradiance I_d and the irradiance reflected by the surrounding surfaces I_r. The latter radiation can be limited to the ground-reflected radiation.

Hence, the total irradiance is given by

$$I_{s,tot} = I_D + I_d + I_r \tag{3.37}$$

where

$I_{s,tot}$ is the total solar irradiance on the surface, [$W\,m^{-2}$]
I_D is the direct solar irradiance on the surface, [$W\,m^{-2}$]
I_d is the diffuse irradiance on the surface, [$W\,m^{-2}$]
I_r is the reflected irradiance on the surface, [$W\,m^{-2}$]

Direct Solar Irradiance The direct solar irradiance I_D is the irradiance of the direct normal radiation I_{DN} corrected by the cosine of the angle θ between the direction of the solar beam radiation and the normal to the surface (named "angle of incidence") of tilt angle Σ. That is,

$$I_D = I_{DN} \cos \theta \tag{3.38}$$

For instance, if the direct normal irradiance is 1000 W m^{-2} and the angle of incidence $\theta = 60°$, then the direct irradiance is 500 W m^{-2} (Figure 3.8).

The direct normal irradiance I_{DN} is a function of the position of the sun in the sky, i.e. of the solar altitude β (defined in Figure 3.9). It is given by

$$I_{DN} = Ae^{-B/\sin \beta} \tag{3.39}$$

Figure 3.8 Definition of the angle of incidence.

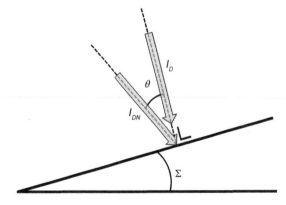

Figure 3.9 Definition of the main angles used in the calculation of the direct solar irradiance. Source: Reproduced and adapted from ASHRAE (2005c).

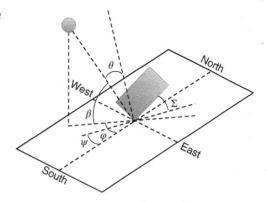

where

A is the apparent extraterrestrial irradiance, [W m^{-2}]

B is the atmospheric extinction coefficient, [−]

Monthly values of constants A and B are given in Chapter 33 of ASHRAE Applications Handbook (ASHRAE, 2003a).

The position of the sun in the sky can be defined by two angles: the solar altitude β and the solar azimuth φ, depicted in Figure 3.9.

The cosine of the incidence angle $\cos\theta$ can be expressed as function of the solar altitude β, the surface tilt angle Σ, the surface azimuth ψ, and the solar azimuth φ by

$$\cos\theta = \cos\beta\cos(\varphi - \psi)\sin\Sigma + \sin\beta\cos\Sigma \tag{3.40}$$

Diffuse Irradiance The diffuse irradiance I_{dh} on a horizontal surface is a fraction of the direct normal irradiance I_{DN}.

$$I_{dh} = C\,I_{DN} \tag{3.41}$$

Monthly values of constant C are given in Chapter 31 of ASHRAE Fundamentals Handbook (ASHRAE, 2005c).

The diffuse irradiance I_d on a tilted surface is given by

$$I_d = I_{dh}\frac{1 + \cos\Sigma}{2} \tag{3.42}$$

where

I_{dh} is the diffuse irradiance on a horizontal surface, [W m^{-2}]

Ground-Reflected Irradiance The ground-reflected short-wave irradiance I_r can be expressed as a function of the total irradiance on a horizontal surface I_{th}:

$$I_r = \rho_g I_{th}\frac{1 - \cos\Sigma}{2} \tag{3.43}$$

where

ρ_g is the albedo (reflectivity or reflectance of the ground), [−]

I_{th} is the total solar irradiance on a horizontal surface, [W m^{-2}]

The total irradiance on a horizontal surface I_{th} is thus

$$I_{th} = I_{DN}\sin\beta + I_{dh} \tag{3.44}$$

3.3.1.2 Atmospheric Radiation

The distinction must be done between the short-wave radiation from the sun and the long-wave radiation from the atmosphere. The latter radiation is emitted by water vapor, carbon dioxide, and ozone, and 90% of the radiation is associated with the 90 first meters over the earth surface (Chapter 33 of the ASHRAE Applications Handbook (ASHRAE, 2003a)).

Rather than considering the emissivity of the atmosphere, it is more convenient to define the atmospheric radiation as the radiation emitted by a black body ($\varepsilon = 1$) at a temperature T_{sky}, named "sky temperature". That is

$$q_{atm} = \varepsilon_{atm}\, \sigma\, T_{atm}^4 = \sigma\, T_{sky}^4 \tag{3.45}$$

Any vehicle outdoor surface (body or glazing) open to the sky also emits long-wave radiation. That is, the net power exchanged between each of the surface and the sky by infrared radiation can be expressed as

$$\dot{Q}_r = A_{surf}\left(\varepsilon_{surf,r}\, \sigma\, T_{surf}^4 - \alpha_{surf,r}\, \sigma T_{sky}^4\right) = A_{surf}\, \varepsilon_{surf,r}\, \sigma\, \left(T_{surf}^4 - T_{sky}^4\right) = A_{surf}\, h_r\, (T_{surf} - T_{sky}) \tag{3.46}$$

where

A_{surf} is the vehicle outdoor surface area, [m^2]
h_r is the linearized radiative heat transfer coefficient, [Wm^{-2}K^{-1}]
$\varepsilon_{surf,r}$ is the vehicle surface longwave emissivity, [−]
T_{surf} is the vehicle outdoor surface temperature, [K]

In Eq. (3.46), the Kirchhoff's relation (Chapter 1) has been used. There exist different correlations to compute the sky temperature. It is proposed to consider the following one (Duffie and Beckman, 1980)

$$T_{sky}[\text{K}] = T[\text{K}]\left(0.8 + \frac{T_{dp}[\text{K}] - 273}{250}\right)^{1/4} \tag{3.47}$$

3.3.2 Energy Transfer Mechanisms Involved in a Vehicle Cabin

The cabin can be defined as the environment occupied by the driver and the passengers. It is represented by the dashed line in Figure 3.10. Seeing the cabin as a thermodynamic system, it is delimited

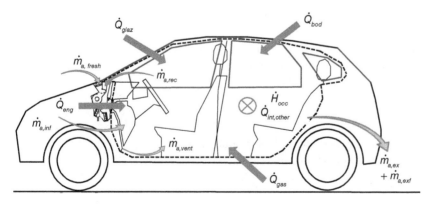

Figure 3.10 Main energy transfer mechanisms between the cabin and its surroundings.

by the windshield, other glazed surfaces, and the car body. The cabin comprises air, as well as internal masses such as the seats and the dashboard.

Glazing highly contributes to the energy exchanges between the cabin indoor and outdoor. Energy enters into the cabin through the glazing by the combined effects of solar heat gain and temperature-driven heat transfer. In total, the rate of heat exchanged through the glazing is $\dot{Q}_{glaz}$, and its expression will be developed in Section 3.3.4.

The cabin body is defined here as the opaque walls delimiting the cabin. The heat transfer rate $\dot{Q}_{bod}$ results from the temperature difference between both sides of the walls. Unlike the glazing, the opaque walls do not transmit solar radiation. However, these walls absorb a fraction of solar radiation, which contributes to heat the cabin. This will be taken into account by artificially correcting the outdoor temperature. On the indoor side of the cabin wall, heat is transferred by convection to the indoor air and by long-wave radiation with the surfaces of the internal masses.

The cabin is an open system, since it exchanges air with the outdoor. Air exchange is achieved intentionally by the ventilation system and unintentionally by infiltration and exfiltration. The fresh air mass flow rate $\dot{m}_{a,fresh}$ entering the cabin through the ventilation system can be mixed with a given amount of recirculated air $\dot{m}_{a,rec}$ from the cabin. The total mass flow rate $\dot{m}_{a,vent}$ is then pulsed into the cabin. The mechanism and the purpose of recirculation will be described in Section 3.4.2. The air that is pulsed into the cabin can be heated up, cooled down, and dehumidified. These operations are achieved by the HVAC unit, which will be described in detail in Section 3.4. Infiltrations and exfiltrations are unintentional airflows through opening submitted to outdoor-indoor pressure gradients that are positive and negative, respectively. In Figure 3.10, the mass flow rates associated with infiltrations and exfiltrations are denoted $\dot{m}_{a,inf}$ and $\dot{m}_{a,exf}$. Infiltrations will be discussed in Section 3.3.6. Air enters the cabin through infiltration and by means of the fan present inside the HVAC unit. Air leaves the cabin through exfiltration and through the air exhausters. The latter ones are usually located at the rear of the vehicle behind the bumper and are made up as a moving rubber membrane. When the cabin indoor pressure is larger than the pressure downstream the membrane (when the blower is on to let fresh air entering the cabin), the membrane uncovers an orifice letting the air leaving the vehicle. When the vehicle is at a standstill (with the blower off) or when the cabin air is totally recirculated, the membrane does not move, preventing any exhaust gas from the engine to enter the vehicle.

The cabin control volume also comprises internal heat sources. Occupants reject heat in both sensible and latent forms. The sensible heat transfer rate $\dot{Q}_{occ,sens}$ contributes to heat the cabin. The latent heat transfer rate is related to a given amount of water vapor $\dot{m}_{w,occ}$ (see Figure 3.15) injected inside the cabin. The cabin may comprise other heat generation sources such as electric heaters. Heat transfer rate associated with these sources is denoted $\dot{Q}_{int,other}$ in Figure 3.10.

Heat could also be transferred into the cabin from high temperature surrounding sources such as the engine or the exhaust gas pipe. They are denoted $\dot{Q}_{eng}$ and $\dot{Q}_{gas}$, respectively.

The cabin comprises internal masses (of temperature $T_{m,cab}$) such as the dashboard and the seats. The thermal inertia associated with these masses highly impacts the evolution of the cabin indoor temperature during transient evolutions such as pull-down and heating-up. These masses exchange heat by convection with the indoor air and by long-wave radiation with the glazing and the cabin body indoor surfaces.

The temperature field inside the cabin (as seen previously, the temperature is not homogeneous) results from these energy transfer mechanisms. The humidity and CO_2 contents of the cabin result from other balances, which will be described later in this chapter.

These different energy transfer mechanisms define the heating and cooling loads, which depend largely on energy transfer through the cabin body and glazing and on the air mass exchange with the outdoor environment. In addition, the cooling capacity also has to take into account the energy associated with the drying of the air (latent energy), which occurs in parallel with the cooling of the air (sensible energy). Depending on the conditions, the latent energy could represent near half of the total cooling energy. These loads define the thermal capacity to install.

3.3.3 Heat Transfer Through the Cabin Body

The cabin body is made up of all opaque walls delimiting the cabin (roof, doors, etc.). These walls are made up of sheets of steel, lining, and plastic. The energy transfer through each wall of the cabin body can be described by the schematics in Figure 3.11. At the outdoor surface of the cabin body, energy is exchanged by convection and radiation. Also, the outdoor surface absorbs a fraction of the total solar radiation. Thermal energy is then transferred by conduction through the n different layers of materials of the cabin body. These layers are characterized by heat capacities, so that heat can be charged or discharged from the materials constituting the body. At the indoor surface, energy is transferred with the cabin indoor environment by convection and radiation. These different energy transfers are described more in detail in Sections 3.3.3.1, 3.3.3.2, and 3.3.3.3.

3.3.3.1 Heat Transfers at the Cabin Body Outdoor Surface

The energy balance on the cabin outdoor surface is given by Eq. (3.48).

$$\frac{\dot{Q}_{bod,out}}{A_{bod}} = h_{out,c}(T_{a,out} - T_{bod,out}) + h_{out,r}(T_{sky} - T_{bod,out}) + \alpha_{bod,s} I_{s,tot} \tag{3.48}$$

where

$\dot{Q}_{bod,out}$ is the rate of heat transfer from the outdoor environment to the cabin outdoor surface, [W]

A_{bod} is the body outdoor surface area, [m^2]

$h_{out,c}$ is the convective heat transfer coefficient on the outdoor surface of the vehicle body, [W m^{-2} K^{-1}]

$T_{a,out}$ is the outdoor temperature, which approximates the operative temperature seen by the surface, [°C]

$T_{bod,out}$ is the temperature of the cabin body outdoor surface, [°C]

$h_{out,r}$ is the linearized radiative heat transfer coefficient on the outdoor surface of the vehicle body, [W m^{-2} K^{-1}]

T_{sky} is the sky temperature, [°C]

$\alpha_{bod,s}$ is the solar absorptivity (or absorptance) of the cabin body, [−]

$I_{s,tot}$ is the total solar irradiance, [W m^{-2}]

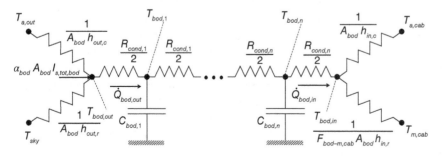

Figure 3.11 Energy transfers through the cabin body.

The convective heat transfer coefficient $h_{out,c}$ at the outdoor surface depends on the vehicle speed.

The radiative heat transfer rate $h_{out,r}(T_{sky} - T_{bod,out})$ given by Eq. (3.48) assumes that the cabin body sees a radiative environment at the sky temperature. This assumption is valid for most of the walls of the cabin body (Breque, 2017).

The definition of the absorptivity will be developed in Section 3.3.4 about the heat transfer through the glazing.

After some algebra, the energy balance given in Eq. (3.48) can be written as

$$\frac{\dot{Q}_{bod,out}}{A_{bod}} = h_{out}(T_{a,out} - T_{bod,out}) + \alpha_{bod,s} I_s - \varepsilon_{bod,r} \Delta I_{IR} \tag{3.49}$$

where $\varepsilon_{bod,r}$ is the vehicle body surface longwave emissivity, [−]

where $h_{out} = h_{out,c} + h_{out,r}$ is the global heat transfer coefficient on the outdoor surface of the cabin body combining convection and longwave radiation.

The expression ΔI_{IR} is a correction to be brought on the calculation of the net long-wave radiation exchanged between the surface and the atmosphere due to the fact the latter is at a temperature T_{sky} and not at $T_{a,out}$.

$$\Delta I_{IR} = \frac{\Delta \dot{Q}_{IR}}{A_{bod}} = \sigma \left(T_{a,out}^4 - T_{sky}^4 \right) = \frac{h_{out,r}}{\varepsilon_{bod,r}}(T_{a,out} - T_{sky}) \tag{3.50}$$

It is common to simplify the expression of the energy balance on an outdoor surface by introducing the sol–air temperature $T_{sa,out}$:

$$T_{sa,out} = T_{a,out} + \frac{1}{h_{out}} \left(\alpha_{bod} I_s - \varepsilon_{bod,r} \Delta I_{IR} \right) \tag{3.51}$$

The sol–air temperature can be seen as a correction on the external temperature in order to take into account:

– the long-wave radiation exchanged between the vehicle body and the atmosphere
– the fraction of the solar radiation the latter absorbs.

That is, the rate of heat transfer at the cabin outdoor surface can be written as

$$\frac{\dot{Q}_{bod,out}}{A_{bod}} = h_{out}(T_{sa,out} - T_{bod,out}) \tag{3.52}$$

3.3.3.2 Heat Transfer and Storage Through the Cabin Body Materials

In Figure 3.11, the cabin body wall is supposed to be made of n layers of different materials. Each layer is characterized by a heat capacity C_j [J K^{-1}]. This capacity is the product of the mass m_j [kg] of the layer by its specific heat c_j [J kg^{-1} K^{-1}]. Assuming that each layer is uniform, the energy balance on each layer can be written as

$$C_j \frac{dT_j}{dt} = \frac{k_j A_{bod}}{\frac{e_j}{2}}(T_{j-1} - T_j) - \frac{k_j A_{bod}}{\frac{e_j}{2}}(T_j - T_{j+1}) \tag{3.53}$$

where

k_j is the thermal conductivity of layer j, [W m^{-1} K^{-1}]
e_j is the thickness of layer j, [m]

3.3.3.3 Heat Transfers at the Cabin Body Indoor Surface

At the indoor surface, the energy balance can be written as

$$\frac{\dot{Q}_{bod,in}}{A_{bod}} = h_{in,c}(T_{bod,in} - T_{a,cab}) + F_{bod-m,cab}\, h_{in,r}(T_{bod,in} - T_{m,cab}) \tag{3.54}$$

where

$\dot{Q}_{bod,in}$ is the rate of heat transfer from the cabin indoor surface to the indoor environment, [W]

$h_{in,c}$ is the convective heat transfer coefficient on the indoor surface of the vehicle body, [W m^{-2} K^{-1}]

$T_{bod,in}$ is the temperature of the cabin body indoor surface, [°C]

$T_{a,cab}$ is the cabin indoor temperature at the vicinity of the surface, [°C]

$F_{bod-m,cab}$ is the view factor between the cabin indoor surface and the internal mass, [−]

$h_{in,r}$ is the linearized radiative heat transfer coefficient on the indoor surface of the vehicle body, [W m^{-2} K^{-1}]

$T_{m,cab}$ is the internal mass temperature, [°C]

The previous equation states that each cabin indoor surface exchanges heat by radiation with the internal masses (seats, steering wheel, etc.). The latter ones are assumed to be lumped into one single node of temperature $T_{m,cab}$.

It should be stressed that in Eq. (3.54), the radiative heat transfer component is expressed with the linearized heat transfer coefficient. Another expression would be

$$\dot{Q}_{bod,in,r} = F_{bod-m,cab}\, A_{bod}\, \varepsilon_{bod,r}\, \sigma\, \left(T_{bod,in}^4 - T_{m,cab}^4\right) \tag{3.55}$$

Equation (3.55) assumes that the internal masses are a black body, and hence, its emissivity is unity. This assumption is justified by the dark colors usually used and leading to large IR emissivities (0.8–0.95) and by the complex shapes of these internal masses that tend to "trap" the radiation (Breque, 2017).

3.3.3.4 Heat Transfer Through the Cabin Body in the Steady-State Regime

In the steady-state regime, there is no heat accumulation or depletion in the different materials constituting the body. The overall energy balance on the body can be expressed as

$$\dot{Q}_{bod,in} = \dot{Q}_{bod,out} = \dot{Q}_{bod} = \left(\sum_j \frac{e_j}{k_j A_{bod}}\right)^{-1} (T_{bod,out} - T_{bod,in}) \tag{3.56}$$

Introducing the sol–air temperature and defining the indoor operative temperature (see Eq. (3.23)), the previous energy balance can be written as

$$\dot{Q}_{bod} = \left(\frac{1}{A_{bod}h_{out}} + \sum_j \frac{e_j}{k_j A_{bod}} + \frac{1}{A_{bod}(h_{in,c} + F_{bod-m,cab}h_{in,r})}\right)^{-1} (T_{sa,out} - T_{o,in})$$

$$= A U_{bod}(T_{sa,out} - T_{o,in}) \tag{3.57}$$

3.3.4 Heat Transfer Through the Glazing

Glazing represents a large fraction of the vehicle body surface. Moreover, the inclination of the windshield is such that the surface could be nearly perpendicular to the sun radiation flux, thereby enabling a maximum impact of the radiation energy. Hence, glazing largely contributes to the energy exchange between the vehicle and its outdoor environment.

Heat transfer through the glazing can be described in a similar way (and thus with the same set of equations) than that through the cabin body.

As a major difference, glazing is not opaque and allows for the direct transfer of a fraction of solar radiation. Hence, the optical properties of the glazing are of tremendous importance in energy transfer. Also, the glazing thickness may be considered uniform, without the distinction of different layers.

3.3.4.1 Optical Properties of Glazing

As illustrated in Figure 3.12, when solar radiation (both direct and diffuse) reaches the glazing, a fraction is transmitted through the glazing, a fraction is reflected, and the rest is absorbed by the glazing material. The transmitted fraction is the transmissivity τ, the reflected fraction is the reflectivity ρ, and the absorbed fraction is the absorptivity α. Expressing the conservation of energy on the glazing gives the following relationship between these glazing optical properties

$$\tau + \rho + \alpha = 1 \tag{3.58}$$

where

τ is the transmissivity (or transmittance) of the glazing, [−]
ρ is the reflectivity (or reflectance) of the glazing, [−]
α is the absorptivity (or absorptance) of the glazing, [−]

The transmissivity, reflectivity, and absorptivity vary with the wavelength and with the incident angle. These optical properties can be wavelength-integrated, leading to the definition of total transmissivity, reflectivity, and absorptivity by opposition to spectral properties (Table 3.4).

3.3.4.1.1 Visible Transmissivity The transmissivity of the visible portion of the solar spectrum is the visible transmissivity or "visible light transmissivity." It is the capacity of the glazing to transmit the daylight inside the vehicle cabin and to offer to the driver a good visibility on outside. The visible transmissivity of standard glass is typically larger than 70%. In contrary, dark tinted glass

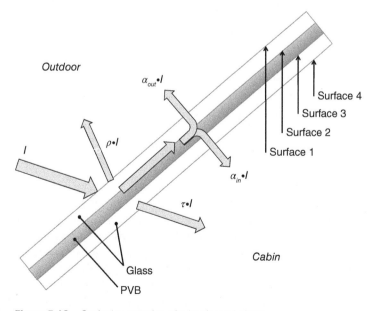

Figure 3.12 Optical properties of a laminated glass.

Table 3.4 Optical properties of conventional glazing and advanced glazing.

Glass type	Visible transmissivity [−]	Solar transmissivity [−]	Solar reflectivity [−]	Solar absorptivity [−]
Conventional	0.80	0.76	0.08	0.16
IR reflective	0.73	0.52	0.37	0.11
IR absorptive	0.76	0.53	0.06	0.41

Source: Adapted from Sullivan and Selkowitz (1990).

used for ensuring privacy show lower values of visible transmissivity, typically between 10% and 45% (Saint Gobain, 2013). For the windshield and front door glasses, regulation commonly imposes a minimum visible transmissivity of 70–75%.

3.3.4.1.2 Solar Heat Gain Coefficient The transmitted radiation is distributed over the surfaces inside the cabin and finally contributes to heat the cabin. The impact on the cabin energy balance of the absorbed fraction of the solar radiation by the glazing is more complex to describe. As depicted in Figure 3.12, a fraction α_{in} of the absorbed energy is transferred inside the cabin by conduction (through the glazing), convection (between the glazing indoor side and the cabin air), and long-wave radiation (from the glazing indoor surface to the surfaces inside the cabin). The remaining fraction α_{out} is released to the outdoor. That is,

$$\tau + \rho + \alpha_{in} + \alpha_{out} = 1 \tag{3.59}$$

where

α_{in} is the inward-flowing fraction of the absorbed solar radiation, [−]
α_{out} is the outward-flowing fraction of the absorbed solar radiation, [−]

The sharing between these two parts is in particular dependent on the vehicle speed. Actually, the convection associated with the motion of the vehicle tends to release the absorbed energy to the outdoor. When the vehicle speed increases, α_{in} decreases and α_{out} increases.

As a consequence, the solar gain through the glazing has two contributions: the fraction of the incident irradiance that is transmitted through the glazing and the inward-flowing fraction of the absorbed radiation (Chapter 31 of the ASHRAE Fundamentals Handbook [ASHRAE, 2005c]). In practice, to simply take into account these two contributions, the solar heat gain coefficient (SHGC) is often used. The latter is defined as the fraction of the incident solar irradiance that enters the cabin under the form of thermal energy. That is,

$$SHGC = \tau + \alpha_{in} \tag{3.60}$$

3.3.4.2 Advanced Glazing Technologies

Ideally, a glass should reduce the transmissivity of the ultraviolet portion of the spectrum (to protect the skin), maximize the transmissivity of the visible portion of the spectrum (to ensure the visibility and benefit for daylight), and reduce the transmissivity τ_{IR} and the inward-flowing absorptivity $\alpha_{in,IR}$ of the infrared portion of the spectrum (to decrease the solar heat gain inside the cabin in hot environment condition). Decreasing the solar heat gain allows for a reduction in the time needed to reach comfortable conditions inside the cabin and a reduction in the cooling load.

Advanced glazing technologies or "spectrally selective glazing" can be used to decrease the transmission of the infrared portion of the solar radiation (the actual fraction that is transmitted plus the inward-flowing fraction of the absorbed infrared radiation). Among these technologies, one can mention the absorbing glazing and the reflective glazing. Another advanced glazing technology is the low-emissivity glazing, which is also described hereunder. Reduction of the infrared rays should not impact the transmission of visible light.

Other aspects of the glazing have also to be considered to improve the performance during the demisting or defrosting phase. This can be achieved with an electrically heated windshield and an electrically heated rear glazing.

The windshield is always constituted of laminated glazing with two glass layers and a polyvinyl butyral (PVB) layer (see Figure 3.12).

The other parts of the glazing are generally constituted of tempered glazing with a simple glass layer.

3.3.4.2.1 Absorbing Glazing
This type of glazing decreases the infrared transmissivity by increasing the infrared absorptivity. The reflectivity remains similar to a conventional glass (Saint Gobain, 2013).

The absorption of the infrared radiation can be achieved by the glass itself in a tinted tempered glazing or by the interlayer in the laminated glass (see Figure 3.12) or both. In the case of absorption by the interlayer, the PVB interlayer is doped with metallic absorbing particles.

An over-tinted tempered glazing is also proposed and gives a larger infrared absorptivity but the visible transmissivity becomes lower than the accepted minimum value of 70% in the front zone of the cabin. This technology is very efficient for the glazing in the rear zone of the cabin.

3.3.4.2.2 Reflective Glazing
Another way to decrease the infrared transmissivity is to increase the reflectivity, keeping the absorptivity at a similar level than in conventional tinted glass (Saint Gobain, 2013). In such a glass which is a laminated glass with a reflective layer, the coating can be placed on surfaces 2 or 3 (between the PVB and the outer glass or between the PVB and the inner glass layer as shown in Figure 3.12). The reflective layer is made of widespread metal particles.

Since this solution is dedicated to the laminated glazing, it is well adapted to the windshield to limit the sun radiation.

Another kind of reflective glazing is the low-emissivity glazing. In such a glass, a coating is placed on surface 4. It has a low emissivity and thus a low absorptivity and high reflectivity, with respect to long-wave infrared radiations. As a consequence, the low-emissivity glazing reflects the radiations emitted by surfaces inside the cabin and thus traps the heat inside the cabin, which is a benefit in winter. In summer, the low emissivity glass reduces both the transmitted fraction of the solar infrared radiation (τ) and the part of the absorbed fraction that is released inside the cabin (α_{in}), which reduces the solar heat gain.

Therefore, the low-emissivity glazing limits the cold wall sensation in winter or hot wall sensation in summer. This layer constitutes a thermal barrier and allows for increasing (decreasing, respectively) the temperature of the pulsed air in a hot environment (cold environment, respectively).

As explained before, an efficient glazing system is composed of a laminated glass with a reflective layer for the windshield, tinted tempered glasses with visible transmissivity higher than 70% for the front lateral zones and over-tinted tempered glasses for the rear zones. A low-emissivity layer can be added on the lateral glasses.

Figure 3.13 shows an example of typical gain on the dashboard temperature and on the head air temperature obtained with a reflective glazing compared with a tinted glazing (for a vehicle at a standstill).

After 2 hours for a vehicle parked under the sun radiation, the maximum dashboard temperature decreases of around 10 K and the head air temperature decreases of around 5 K. Note that for a radiation of 850 W/m^2 and an outdoor temperature of 27°C, with a tinted glazing, the dashboard

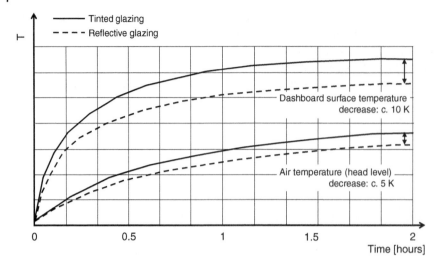

Figure 3.13 Time evolution of the measured dashboard and head air temperature obtained with two types of glazing for a vehicle at a standstill.

temperature can reach 90°C after soaking. The main gain is on the inertia of the dashboard during the starting phase of the trip.

3.3.4.2.3 *Electrically Heated Glazing* For demisting and defrosting purpose, there are two different technologies for an electrically heated windshield. One solution consists of very thin electric wires integrated between the two glass layers in the laminated glazing. The other solution can be used in the case of a windshield with a reflective layer and proposes to take benefit of the deposited conductive metallic layer and supply it with electric voltage.

These solutions bring very fast performance for demisting and defrost and do not disturb the thermal comfort. The electric power can be important and around 500 W.

This function also allows for consolidating the strategies of partial air recycling while limiting mist formation. It also shows the advantage of treating the whole windshield surface and also beneficiating from the heat reflecting function.

However, with 14 V supply, this electric power is difficult to obtain with a reflective layer. In an electrified vehicle equipped with a high voltage, this solution could give a fast performance and could be applied in the future.

Introducing thin electric wires inside the windshield has been investigated, but the wires imply a problem of visibility through the glazing particularly during night situations, which reduces the application on the vehicle.

Concerning the rear glazing, the electric resistance is realized with the introduction of a conductive ink. The electrical power is around 200–300 W.

3.3.5 Ventilation

The purpose of ventilation is to bring a given quantity of fresh outdoor air inside the cabin. Ventilation air also supplies the heating and air conditioning system to control the cabin temperature. The ventilation function is ensured by the HVAC unit, which will be described later. Bringing fresh air is necessary to remove in-cabin pollutants such as carbon dioxide, cigarettes fumes, or emissions from materials in new vehicles. Bringing outdoor air is also a mean to decrease the humidity content of the cabin. The outdoor air entering the cabin should also be as free as possible of outdoor pollutants (carbon dioxide and monoxide, volatile organic compounds, particles, etc.). Outdoor air is thus filtered before being pulsed into the cabin. Filters used to filter outdoor and indoor air are integrated inside the HVAC unit, which will be described later in this chapter.

The cabin air change rate can be expressed in *air change per hour* (ACH), which is the number of times per hour that the whole cabin air is replaced by fresh outdoor air.

The fresh air pulsed into the cabin can be cooled down, heated up, and dehumidified to cover sensible and latent heat loads of the cabin. Fresh air can be mixed with recirculated air from the cabin to decrease the energy necessary to prepare the pulsed air to set-point conditions. The HVAC unit ensures all these functions.

As a consequence, the specific enthalpy $h_{a,su,cab}$ of the air pulsed into the cabin may differ from the specific enthalpy $h_{a,out}$ of the outdoor air.

The enthalpy flow rate of the air entering the cabin through the ventilation system is given by

$$\dot{H}_{a,su,cab} = \dot{m}_{a,vent} \cdot h_{a,su,cab} \qquad (3.61)$$

where

$\dot{m}_{a,vent}$ is the dry air mass flow rate pulsed into the cabin, [kg s^{-1}]
$h_{a,su,cab}$ is the specific enthalpy of the air pulsed into the cabin, [J kg^{-1}]

Under the assumption that the air inside the cabin is fully mixed, we can write

$$\dot{H}_{a,ex,cab} = \dot{m}_{a,vent} \cdot h_{a,cab} \qquad (3.62)$$

where

$\dot{H}_{a,ex,cab}$ is the enthalpy flow rate of the air at the exhaust of the cabin, [W]
$h_{a,cab}$ is the specific enthalpy of the air inside the cabin, [J kg^{-1}]

Hence, the net enthalpy flow rate brought by ventilation is given by:

$$\Delta \dot{H}_{a,vent} = \dot{m}_{a,vent} \cdot (h_{a,su,cab} - h_{a,cab}) \qquad (3.63)$$

The sensible fraction of the enthalpy flow rate is given by

$$\Delta \dot{H}_{a,vent,sens} = \dot{m}_{a,vent} \cdot c_{p,ha} \cdot (T_{a,su,cab} - T_{a,cab}) \qquad (3.64)$$

In the previous equation, $c_{p,ha}$ is the specific heat at constant pressure of humid air and is assumed to be computed at the cabin supply. Hence,

$$c_{p,ha} = c_{p,a} + \omega_{su,cab}\, c_{p,w} \qquad (3.65)$$

where

$c_{p,a}$ is the constant pressure specific heat of pure air, ~1005 [J kg^{-1}K^{-1}]
$\omega_{su,cab}$ is specific humidity of air entering the cabin, [kg kg^{-1}]
$c_{p,w}$ is the constant pressure specific heat of water vapor, ~1820 [J kg^{-1}K^{-1}]

The latent contribution of the enthalpy flow rate is given by

$$\Delta \dot{H}_{a,vent,lat} = \Delta \dot{H}_{a,vent} - \Delta \dot{H}_{a,vent,sens} \qquad (3.66)$$

The latter contribution corresponds to the net water mass flow rate brought by the ventilation. This mass flow rate is given by

$$\Delta \dot{m}_{w,vent} = \dot{m}_{a,vent} (\omega_{su,cab} - \omega_{cab}) \qquad (3.67)$$

where

ω_{cab} is specific humidity of air inside the cabin, [kg kg^{-1}]

3.3.6 Infiltration

The vehicle body surrounding the cabin in not fully tight: outdoor air enters the cabin even when the ventilation is off, the air supply vents are closed, and the windows are closed. According to Knibbs et al. (2009), the major pathway of outdoor air infiltration is the recirculation flap inside the HVAC unit, and windows and doors seals are other potential pathways. The tightness of door seals has been largely increased during the later decade. The infiltration rate depends on the tightness of the vehicle (and consequently to some extent on its age), but also on the vehicle speed. Even in stationary conditions, air infiltration takes place. Knibbs et al. (2009) measured the infiltration rates for a set of vehicles showing different tightness levels. They observed that the air change rate by infiltration moderately increases with the wind speed. However, no effect of temperature gradient between cabin indoor/outdoor on the infiltration rate was observed.

Infiltration (but also exfiltration) can be treated in the same way than ventilation. The net enthalpy flow rate brought by infiltration is given by:

$$\Delta \dot{H}_{a,inf} = \dot{m}_{a,inf}.(h_{a,out} - h_{a,cab}) \tag{3.68}$$

where

$\dot{m}_{a,inf}$ is the infiltration mass flow rate, [kg s^{-1}]
$h_{a,out}$ is the specific enthalpy of outdoor air, [J kg^{-1}]
$h_{a,cab}$ is the specific enthalpy of cabin air, [J kg^{-1}]

In the previous equation, it is assumed that the infiltration air has the same thermodynamic state as outdoor air. Also, it is assumed that the air inside the cabin is fully mixed, so that the air leaving the cabin is at the same thermodynamic state as the cabin air.

Here also, a sensible contribution $\Delta \dot{H}_{a,inf,sens}$ and a net water vapor supply $\Delta \dot{m}_{w,inf}$ can be defined.

3.3.7 Internal Gains

3.3.7.1 Occupants

The driver and the passengers reject both sensible and latent heats inside the cabin.

$$\dot{H}_{occ} = \dot{Q}_{occ,sens} + \dot{H}_{occ,lat} \tag{3.69}$$

According to Chapter 8 of ASHRAE Fundamentals Handbook (ASHRAE, 2005a), a human being rejects approximately 70 W of sensible heat ($\dot{Q}_{occ,sens}$) and 35 W of latent heat ($\dot{H}_{occ,lat}$) when seated. The latent contribution corresponds to an injected water vapor flow rate $\dot{m}_{w,occ}$. As explained in Section 3.2.2, this water production is associated with respiration, sweat, and water diffusion through the skin. It can be considered that about 1.2 L of water is transferred to the air per day and per person. This corresponds to 50 g of water per hour and per person. In a nonventilated cabin with a volume of 3 m^3, this would yield an increase of the specific humidity of 13.9 g kg^{-1} of dry air per hour and per person.

When the cabin is ventilated, the introduction of an external air flow and/or the use of the air conditioning evaporator allow for the limitation of the humidity inside the cabin.

3.3.7.2 Other Internal Gains

The cabin could comprise other internal sources of sensible heat, such as heated seats and steering wheels or any other decentralized heating systems. They are denoted $\dot{Q}_{int,other}$.

3.3.8 Other Energy Transfer Mechanisms

For conventional and hybrid vehicles, Fayazbakhsh and Bahrami (2013) also suggested to account for the heat transfer rates from the high temperature exhaust gas pipe ($\dot{Q}_{gas}$) and engine ($\dot{Q}_{eng}$). Part of the heat of the exhaust gas is transferred into the cabin through the cabin floor.

3.3.9 Lumped Modeling Approach

There exist several techniques to model the thermal behavior of vehicle cabins. In this section, only the lumped-parameter modeling technique is described. Lumped-parameter models are more appropriate for investigating the coupling between the cabin model and the A/C system than 3D CFD models (Torregrosa-Jaime et al., 2015). Actually, lumped-parameter models offer a good trade-off between model accuracy and computational time effort. 3D CFD models are more appropriate to investigate, for instance, the occupant thermal comfort by better characterizing the field of temperatures and air velocities within the cabin.

The lumped modeling approach consists of defining a limited number of temperature nodes representing the major elements constituting the cabin. Energy and mass balances are associated with each of these nodes.

The simplest model only considers one node representing the control volume delimited by the cabin body and the glazing. This control volume contains air and the different components comprised inside the cabin, such as the seats and dashboard. However, the model does not make the distinction between the air and the masses inside the cabin.

To better represent the transient regimes (pull-down and heating-up), whose impact on thermal comfort is of high importance for short-duration trips, the thermal inertia of the cabin body can be taken into account. Hence, other thermal nodes associated with the cabin body must be introduced into the modeling.

A more sophisticated model introduces an additional node for the internal masses inside the cabin. This is particularly useful to describe the radiative heat transfers between the different surfaces delimiting the cabin and comprised inside the latter. Also, distinguishing the cabin indoor air and internal masses allows for a better prediction of the cabin air temperature during the transient evolution (Breque, 2017).

Finally, the cabin can be divided into several zones, each one being described by a temperature node. The multizone approach allows for a better representation of the nonuniform temperature inside the cabin. This approach is also more suitable to describe dual zone, 3-zone, and 4-zone HVAC climate control systems.

The model presented hereunder, inspired from Breque (2017), considers temperature nodes for the cabin body, cabin glazing, internal air, and internal masses. The cabin body and glazing can possibly be split into several elements to account for their different orientations with respect to solar radiation. The proposed model considers only one single zone. For a more detailed description considering two zones (for instance, the driver's region and the passengers' region), the interested reader should refer to the model proposed by Torregrosa-Jaime et al. (2015).

The lumped-parameter modeling technique is based on the expression of the equations of conservation of energy and mass associated with the different nodes. These equations are described hereunder.

3.3.9.1 Energy Balance on the Cabin Body
In the lumped parameter approach, the cabin body thermal inertia could be described by one single capacity as illustrated in Figure 3.14.

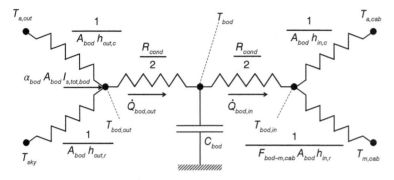

Figure 3.14 Modeling of the cabin body by a R-C approach.

Introducing the overall heat capacity of the cabin body C_{bod} [J K^{-1}] and the cabin body temperature T_{bod}[°C], the energy balance at the body temperature node gives

$$C_{bod}\frac{dT_{bod}}{dt} = \dot{Q}_{bod,out} - \dot{Q}_{bod,in} \tag{3.70}$$

In the previous equation, $\dot{Q}_{bod,out}$ represents the heat transfer rate coming from the outdoor environment and $\dot{Q}_{bod,in}$ the heat transfer rate going toward the indoor environment. To simplify the representation of the cabin body, only some major portions of it could be described. For instance, a simple model would at least represent the roof (Marcos et al., 2014).

3.3.9.2 Energy Balance on the Cabin Glazing
The energy balance on the cabin glazing is similar to that on the cabin body. The energy balance on the glazing temperature node is given by

$$C_{glaz}\frac{dT_{glaz}}{dt} = \dot{Q}_{glaz,out} - \dot{Q}_{glaz,in} \tag{3.71}$$

As proposed by Marcos et al. (2014), four different glazed surfaces are considered: the windshield (ws), the rear window (rw), the left side window (lsw), and the right side window (rsw).

3.3.9.3 Energy Balance on the Cabin Internal Masses
All internal masses comprised inside the vehicle (dashboard, seats, etc.) are lumped into one single temperature node $T_{m,cab}$. This temperature node receives the transmitted fraction of the total solar radiation $\dot{Q}_{glaz,sol,\tau}$, exchanges heat by convection with the cabin air and exchanges heat by radiation with the cabin body and glazing indoor surfaces. Hence, the energy balance on this temperature node is given by

$$C_{m,cab}\frac{dT_{m,cab}}{dt} = \dot{Q}_{glaz,sol,\tau} + \dot{Q}_{bod,in,r} + \dot{Q}_{glaz,in,r} - \dot{Q}_{m,in,c} \tag{3.72}$$

The heat capacity $C_{m,cab}$ [J K^{-1}] of the internal masses is the product of the internal masses $m_{m,cab}$ [kg] and their specific heat capacity $c_{m,cab}$ [J kg^{-1}K^{-1}].

The transmitted fraction of the total solar radiation is given by

$$\dot{Q}_{glaz,sol,\tau} = \sum_{j=1}^{4} A_j\, \tau_{j,s}\, I_{s,tot,j} \tag{3.73}$$

where

j is the glazed surface ($1 = ws$; $2 = rw$; $3 = lsw$; $4 = rsw$)
A_j is the area of glazed surface j, [m^2]
$\tau_{j,s}$ is the solar transmissivity of glazed surface j, [$-$]
$I_{s,tot,j}$ is the total solar irradiance on glazed surface j (Eq. (3.37)), [Wm^{-2}]

The radiative heat transfer rate from the indoor surfaces of the different glazed surfaces is computed as

$$\dot{Q}_{glaz,in,r} = \sum_{j=1}^{4} F_{j-m,cab} \, A_j \, \varepsilon_{j,r} \, \sigma \, \left(T_{j,in}^4 - T_{m,cab}^4 \right) \tag{3.74}$$

where

$F_{j-m,cab}$ is the view factor between glazed surface j and internal masses, [$-$]
$\varepsilon_{j,r}$ is the longwave emissivity of glazed surface j, [$-$]
$T_{j,in}$ is the temperature of indoor glazed surface j, [K]

The convective heat transfer rate $\dot{Q}_{m,in,c}$ between the internal masses and the internal air is given by

$$\dot{Q}_{m,in,c} = A_m \, h_{in,c} \, (T_{m,cab} - T_{a,cab}) \tag{3.75}$$

where

A_m is the external heat transfer area of the internal masses, [m^2]
$h_{in,c}$ is the convective heat transfer coefficient between the internal mass surface and the cabin air, [Wm^{-2}K^{-1}]

3.3.9.4 Mass and Energy Balances on the Cabin Air, Water, and CO$_2$

Air inside the cabin is atmospheric air, which comprises water vapor ("moisture"). It also contains CO$_2$, mainly generated by the occupants. Hence, the distinction is done between the dry air mass balance, the moisture balance (or "humidity balance") and the CO$_2$ mass balance.

3.3.9.4.1 Air Mass Balance Since there is no dry air mass accumulation in a zone, the sum of the entering air mass flow rates equals the sum of the leaving air mass flow rates. Air mass balance must account for the ventilation air mass flows pulsed through the supply vents, the air mass flow extracted from the cabin through the air exhauster at the rear of the vehicle, the infiltration air mass flows, the exfiltration air mass flows, the air mass flow that is recirculated and the air mass transfer between the different zones. If the model comprises one single zone, as depicted in Figure 3.10, the dry air mass balance on the zone becomes

$$\dot{m}_{a,vent} + \dot{m}_{a,inf} = \dot{m}_{a,rec} + \dot{m}_{a,ex} + \dot{m}_{a,exf} \tag{3.76}$$

where

$\dot{m}_{a,vent}$ is the ventilation air mass flow rate pulsed through the supply vents, [kg s^{-1}]
$\dot{m}_{a,rec}$ is the air mass flow rate that is recirculated, [kg s^{-1}]
$\dot{m}_{a,ex}$ is the air mass flow rate that leaves the cabin through the air exhauster, [kg s^{-1}]
$\dot{m}_{a,exf}$ is the exfiltration air mass flow rate, [kg s^{-1}]
$\dot{m}_{a,inf}$ is the infiltration air mass flow rate, [kg s^{-1}]

3.3.9.4.2 Moisture Mass Balance

In contrast, the moisture content of the cabin can vary with time. This could be described by the capacitance model (Klein, 2007). Considering a single-zone model (Figure 3.15), this approach gives

$$\frac{dm_{w,cab}}{dt} = C_w \frac{d\omega_{cab}}{dt} = \dot{m}_{w,vent} + \dot{m}_{w,inf} - \dot{m}_{w,rec} - \dot{m}_{w,ex} - \dot{m}_{w,exf} + \dot{m}_{w,occ} + \dot{m}_{w,other}$$

$$\Longleftrightarrow C_w \frac{d\omega_{cab}}{dt} = \dot{m}_{a,vent} \cdot \omega_{su,cab} + \dot{m}_{a,inf} \cdot \omega_{out} - \dot{m}_{a,rec} \cdot \omega_{cab} - \dot{m}_{a,ex} \cdot \omega_{cab} - \dot{m}_{a,exf} \cdot \omega_{cab}$$

$$+ \dot{m}_{w,occ} + \dot{m}_{w,other}$$

$$\Longleftrightarrow C_w \frac{d\omega_{cab}}{dt} = \Delta \dot{m}_{w,vent} + \Delta \dot{m}_{w,inf} + \dot{m}_{w,occ} + \dot{m}_{w,other} \qquad (3.77)$$

where

C_w is the moisture capacitance of the cabin, [kg]
$\dot{m}_{w,vent}$ is the water flow rate entering through ventilation, [kg s^{-1}]
$\dot{m}_{w,inf}$ is the water flow rate entering through infiltration, [kg s^{-1}]
$\dot{m}_{w,rec}$ is the water flow rate recirculated, [kg s^{-1}]
$\dot{m}_{w,ex}$ is the water flow rate leaving through exhausters, [kg s^{-1}]
$\dot{m}_{w,exf}$ is the water flow rate leaving through exfiltration, [kg s^{-1}]
$\omega_{su,cab}$ is the specific humidity of air supplied to the cabin by the ventilation, [kg kg^{-1}]
ω_{out} is the specific humidity of outdoor air, [kg kg^{-1}]
ω_{cab} is the specific humidity of cabin air, [kg kg^{-1}]
$\dot{m}_{w,occ}$ is the rate of moisture production of occupants, [kg s^{-1}]
$\dot{m}_{w,other}$ is the rate of moisture production of other internal sources, [kg s^{-1}]

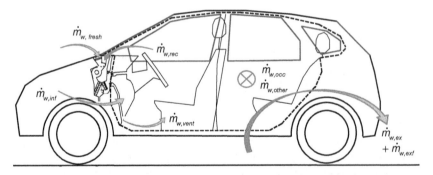

Figure 3.15 Moisture transfer mechanisms between the cabin and its environment.

Similar to that done for building energy simulation, the moisture capacitance of the cabin is the air mass in the zone multiplied by a factor between 1 and 10 (Klein, 2007).

3.3.9.4.3 CO$_2$ Mass Balance

As explained previously, ventilation is also important for limiting the CO$_2$ concentration inside the cabin. Similar to the moisture mass balance, the CO$_2$ mass balance across the cabin can be expressed as follows

$$\frac{dm_{CO_2,cab}}{dt} = \dot{m}_{CO_2,vent} + \dot{m}_{CO_2,inf} - \dot{m}_{CO_2,rec} - \dot{m}_{CO_2,ex} - \dot{m}_{CO_2,exf} + \dot{m}_{CO_2,occ}$$

$$\Longleftrightarrow \frac{dm_{CO_2,cab}}{dt} = \Delta \dot{m}_{CO_2,vent} + \Delta \dot{m}_{CO_2,inf} + \dot{m}_{CO_2,occ} \qquad (3.78)$$

Pulsed ventilation air brings CO_2 inside the cabin, while extracted air removes another quantity of CO_2. That is, the net CO_2 mass flow rate entering the cabin through the ventilation system is given by

$$\Delta \dot{m}_{CO_2,vent} = \frac{MM_{CO_2}}{MM_a} \frac{1}{1 \times 10^6} \dot{m}_{a,vent} \left(X_{CO_2,su,cab} - X_{CO_2,cab} \right) \tag{3.79}$$

where

MM_{CO_2} is the molar mass of CO_2, [g mol^{-1}]
MM_a is the molar mass of dry air, [g mol^{-1}]
$X_{CO_2,su,cab}$ is the concentration of CO_2 in the air supplied to the cabin, [ppmv]
$X_{CO_2,cab}$ is the concentration of CO_2 in the cabin air, [ppmv]

The concentration of CO_2 in the air supplied to the cabin by means of the ventilation system differs from the concentration of CO_2 in the outdoor air when fresh air is mixed with recirculated air in the HVAC unit.

A similar approach can be used to compute the net CO_2 gain due to infiltrations. The last term on the right-hand side of Eq. (3.78) is the CO_2 mass production by the occupants and has been introduced previously (Eq. (3.10)).

The variation of the concentration of CO_2 in the cabin can be related to the variation of mass of CO_2 by the following equation

$$\frac{dm_{CO_2,cab}}{dt} = \frac{V_{cab}}{v_{a,cab}} \frac{MM_{CO_2}}{MM_a} \frac{1}{1 \times 10^6} \frac{dX_{CO_2,cab}}{dt} \tag{3.80}$$

where

V_{cab} is the cabin volume, [m^3]
$v_{a,cab}$ is the specific volume of cabin air, [m^{3}kg^{-1}]

The concentration of CO_2 in outdoor air is approximately ranging from 300 to 450 ppmv. Standards generally recommend a maximum concentration of CO_2 in indoor environment of 1000 ppm A concentration ranging from 1000 to 2500 ppm leads to general drowsiness (Seo and Choi, 2013).

Example 3.1 *CO$_2$ Mass Balance in a Cabin*

A car cabin of 3.1 m^3 is occupied by a person of 75 kg and 1.75 m tall driving on a surfaced road. The outdoor air CO_2 concentration is 412 ppm and atmospheric pressure is 101325 Pa. The car indoor temperature is 25°C. Compute the time evolution of the CO_2 concentration in the cabin during 1 hour of driving assuming that the initial CO_2 concentration is the outdoor concentration. The ventilation flow rate is 40 m^3 h^{-1}. Compute the ideal ventilation volume flow rate to maintain a CO_2 concentration of 1000 ppm.

Analysis

The first question is answered by expressing the mass balance of CO_2 across the cabin in the transient regime. The second question is simply answered by expressing the steady-state mass balance (assuming that a steady-state regime has been reached for the CO_2 concentration)

Solution

```
"1. INPUTS"
"1.1. Outdoor data"
P_atm=101325 [Pa] "atmospheric pressure"
T_a_cab=25 [C] "cabin temperature"
X_CO2_out=412 [ppm] "outdoor CO2 concentration"
m_dot_a_vent=40/3600 "[m^3/h]" "ventilation flow rate"

"1.2. Occupants"
E_dot_m/A_sk=70 [W/m^2] "metabolism per unit of skin surface area
    (driving on a surfaced road)"
N_occ=1 [-] "number of occupants in the cabin"
M=75 [kg] "mass of each occupant"
H=1,75 [m] "height of each occupant"

"2. OUTPUTS"
"X_CO2_cab=?? [ppm]" "Evolution of cabin CO2 concentration with
    time"

"3. PARAMETERS"
V_cab=3,1 [m^3] "cabin volume"
tau_1=0 [s] "initial integration time"
tau_2=3600 [s] "final integration time"

"4. MODEL"
"4.1. CO2 emission by occupant"
A_sk=0,203*M^0,425*H^0,725 "[m^2]" "skin surface area"
m_dot_CO2_occ=N_occ*1,05e-7*E_dot_m "[kg/s]" "mass flow rate of CO2
    emitted by occupant"
V_dot_CO2_occ_lh=m_dot_CO2_occ/MM_CO2*V_m*1000*3600 "volume flow
    rate of CO2 emitted by occupant in l/h"
V_m=R#*(T_a_cab+273,15)/P_atm "[L/Mol]""molar volume"
V_dot_CO2_occ_m3h=V_dot_CO2_occ_lh/1000 "[m^3/h]" "volume flow rate
    of CO2 emitted by occupant in m^3/h"
MM_CO2=molarmass('CO2') "[g/mol]" "molar mass of CO2"
MM_air=molarmass('air') "[g/mol]" "molar mass of air"

"4.2. Transient evolution of CO2 concentration"
V_cab/v_a_cab*MM_CO2/MM_air/1e6*dX_CO2_cabdtau=Delta_m_dot_CO2_vent+
    m_dot_CO2_occ
v_a_cab=volume(Air;P=P_atm;T=T_a_cab)
Delta_m_dot_CO2_vent=MM_CO2/MM_air/1e6*m_dot_a_vent*(X_CO2_su_cab-
    X_CO2_cab)
X_CO2_su_cab=X_CO2_out "ventilation air is pulsed with outdoor CO2
    concentration"
X_CO2_cab_i=X_CO2_out "initial CO2 concentration in cabin"
DELTAX_CO2_cab=integral(dX_CO2_cabdtau;tau;tau_1;tau_2)
```

```
DELTAX_CO2_cab=X_CO2_cab-X_CO2_cab_i

$integraltable tau:60 X_CO2_cab

"4.3. Calculation of ideal ventilation flow rate "
X_CO2_out/1e6+V_dot_CO2_occ_m3h/V_dot_vent_ideal_m3h=1000/1e6
```

Results

The time evolution of CO_2 concentration is shown in Figure 3.16.

```
V_dot_CO2_occ_lh=28,07 [L/h]
V_dot_vent_ideal_m3h=47,73 [m^3/h]
```

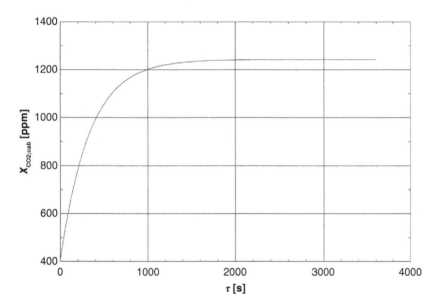

Figure 3.16 Time evolution of CO_2 concentration in cabin.

3.3.9.4.4 Energy Balance In the case of a single-zone model, the energy balance across the cabin air in the transient regime expresses that the rate of variation of the internal energy $\dot{U}_{a,cab}$ is equal to the sum of the previously defined energy transfer mechanisms:

$$\dot{U}_{a,cab} = \dot{Q}_{glaz,in,c} + \dot{Q}_{bod,in,c} + \dot{Q}_{m,in,c} + \Delta\dot{H}_{a,vent} + \Delta\dot{H}_{a,inf} + \Delta\dot{H}_{occ} \tag{3.81}$$

For short-duration trips, the fraction of time when the cabin is in transient regime is important. As mentioned previously, this period is usually called "convergence" period. When the temperature inside the cabin stabilizes, the regime can be considered as steady-state ("comfort" regime). In the latter case, the rate of variation of the internal energy is null.

The equation of conservation of energy is usually split into the sensible energy balance and the latent energy balance. The latent energy balance is related to the humidity balance across the cabin (Eq. (3.77)). The sensible energy balance can be written as

$$C_{a,cab}\frac{dT_{a,cab}}{dt} = \dot{Q}_{glaz,in,c} + \dot{Q}_{bod,in,c} + \dot{Q}_{m,in,c} + \Delta\dot{H}_{a,vent,sens} + \Delta\dot{H}_{a,inf,sens} + \dot{Q}_{occ,sens} \tag{3.82}$$

where

$C_{a,cab}$ is the heat capacity of the cabin air, [J K^{-1}]
$\dot{Q}_{occ,sens}$ is the rate of sensible heat generated by the occupants [W]

The heat capacity $C_{a,cab}$ of the cabin depends on the cabin volume (Eq. [3.83]). In the steady-state regime, the time derivative of the cabin indoor air temperature is null.

$$C_{a,cab} = \frac{V_{cab}}{v_{a,cab}} c_{p,ha,cab} \tag{3.83}$$

Example 3.2   *Passive Cooling of the Vehicle Cabin at Night*

A vehicle is parked at night after a road trip. When the occupants leave the vehicle, the indoor temperature was 22°C. The outdoor sol–air temperature is 10°C. Compute the cabin indoor temperature after 60 minutes if the heat capacity of the cabin is 72 000 [J K^{-1}] and the conductance of the cabin body and glazing is 48 [W K^{-1}]. The thermal capacity of the vehicle body and glazing are neglected. The heat capacity of the cabin includes internal masses.

Analysis

The cabin indoor temperature evolution is a transient evolution. Since we are at night, there is no solar gain. Since the vehicle is not occupied, there is no internal heat gain, and the ventilation system is switched off. AU_{cab} represents here the combined conductance of the cabin and glazing. The model developed previously has been largely simplified since we have here only one thermal capacity.

Solution

In the EES code given hereunder, the integral function is used to numerically solve the differential equation.

```
"1. INPUTS"
T_a_cab_i=22 [C]
T_out_sa=10 [C]

"2. OUTPUTS"
"T_a_cab=?? [C]" "temperature of the cabin air with time"

"3. PARAMETERS"
C_cab=72000 [J/K]
AU_cab=48 [W/K]

"4. MODEL"
C_cab*dT_a_cabdt=AU_cab*(T_out_sa-T_a_cab)
DeltaT_a_cab=integral(dT_a_cabdt;t;t_1;t_2)
t_1=0
t_2=3600 [s]
DeltaT_a_cab=T_a_cab-T_a_cab_i

$integraltable t:10; T_a_cab
```

Results

The time evolution of the cabin temperature is represented graphically. It could be observed that after 1 hour (3600 seconds), the cabin indoor temperature is close to $11[°C]$ (Figure 3.17).

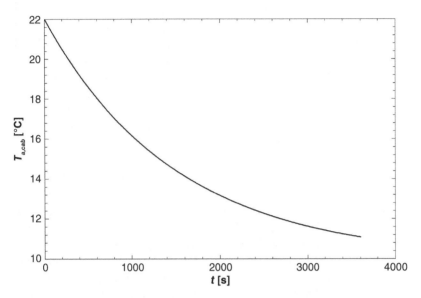

Figure 3.17   Time evolution of the cabin indoor temperature (1-hour simulation).

3.4   Distribution of Thermal Energy Through the Cabin

In most of the vehicles, thermal energy (heat and cold) is supplied to the cabin by means of the ventilation system. The cabin cooling (sensible and latent) and heating loads are covered by supplying air at a controlled temperature and humidity. The pulsed air is conditioned in the HVAC unit (Figure 3.18), the components and working principle of which will be described in Section 3.4.1.

Figure 3.18   Different HVAC units. Source: Courtesy of Valeo.

3.4.1 HVAC Unit Components and Working Principle

The HVAC unit controls the air pulsed into the cabin in terms of temperature, humidity, and flow rate. The HVAC unit also filters the air.

In most of the vehicles, the HVAC unit is located underneath the control panel. However, other positions are possible. Larger vehicle can comprise two HVAC units to provide a better comfort to the front and rear seats.

A typical schematic of a HVAC unit is given in Figure 3.19. Fresh outdoor air enters the HVAC unit through the air inlet (1). The latter is located between the bonnet and the windshield, in a zone that is slightly over-pressurized when the vehicle is in motion. A grille that prevents large bodies, such as insects or leaves, from entering the unit covers the air intake. Air passes then through a filter (2) that retains small particles, allergens, soot, etc. The filter can contain activated carbon to adsorb gaseous pollutants. Recirculated air from the cabin enters the unit through a specific inlet (3) possibly before the filter so that the filter treats both the air from the outside and from the cabin. The position of the recirculation flap (4) is adjusted to control the amount of recirculated air mixed with fresh outdoor air. Generally, the position of the recirculation flap is limited to fully open position and fully closed position and possibly one or several intermediate positions. However, it procures advantages to adjust gradually the position of the recirculation flap and of the air recirculation flow rate to minimize the energy consumption associated with the cabin climate control and to reduce the time to reach the thermal comfort while controlling the cabin specific humidity and the cabin air quality. The purpose of recirculation will be explained in Section 3.4.2. A centrifugal fan (5) is used to circulate the air through the unit. The fan (also called blower) is equipped with its own electric motor. The fan speed can be adjusted to control the air mass flow rate flowing through the HVAC unit. Alternatively, the filter (6) could be placed after the fan. Such a position is better suited to treat the air from the cabin (for instance, to capture Volatile Organic Compounds [VOCs] released by the cabin materials). The entire airflow then passes through the evaporator (7). The purpose of the evaporator is to cool down and possibly dehumidify the air. The evaporator is active only if the compressor of the refrigerant circuit is engaged. The working

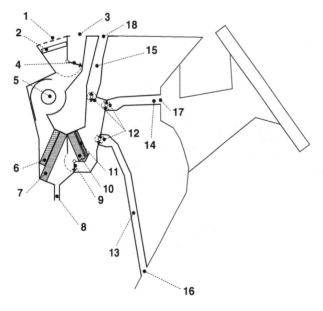

Figure 3.19 Sectional drawing indicating components of the HVAC unit.

principle of the refrigeration loop (also called air-conditioning loop) will be explained in Section 3.5. Dehumidification occurs when the air temperature is decreased underneath its dew point. In such a case, a quantity of water vapor in the air condenses, which decreases the air specific humidity. Condensates are evacuated from the HVAC unit through a drain (8). The unit is equipped with a mixing flap (9) whose position is adjusted to control the amount of air flowing through the heater core (10). The positions of the mixing flap can be gradually adjusted between fully open position and fully closed position. The purpose of the heater core is to heat the air. It is kind of a radiator that is fed by a hot coolant from the engine cooling loop. This component will be described more in detail in Section 3.6.1. An additional electric heater (11) can be used if the heating capacity of the heater core is not large enough to cover the heating demand. This could occur with highly efficient engines in particular during the engine warm-up phase at the starting of the engine. Auxiliary heaters are typically positive temperature coefficient (PTC) electrical heaters.

Downstream is a mixing zone where the amount of air heated by the heater core is mixed with the part of the air which by-passes the heater core duct. The temperature of the mixed air corresponds to the target.

Conditioned air is then directed to different zones of the cabin through channels (13, 14, 15) ending to pulsing vents. Many vent locations are possible, but at least floor vents (16), upper vents (17), and demisting/defrosting vents (18) are available. Conditioned air can also be directed and pulsed in the rear passenger zone by means of piping. Central and lateral vents can also be considered. The amount of air pulsed through the different vents is controlled by means of flaps (12) whose positions can be adjusted. The flaps inside the HVAC unit can be controlled by Bowden cables, pneumatic control actuators, or electric control motors (Daly, 2006).

3.4.2 Cabin Air Recirculation

The airflow pulsed into the cabin can be entirely fresh outdoor air, a mix of outdoor air and recirculated air, or entirely recirculated air. Air recirculation consists of adjusting the position of the recirculation flap in such a way to recycle in the HVAC unit part of the air flow rate blown into the cabin. The other part is rejected outside the vehicle through the cabin air extraction and is compensated with an equivalent fresh air flow rate. An example of partial recirculation is shown in Figure 3.20.

Figure 3.20 Partial recirculation of cabin indoor air.

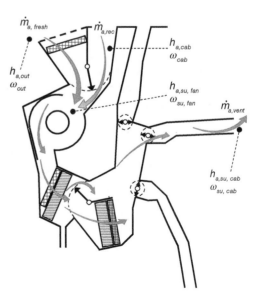

Recirculation of stale air is a mean to more quickly achieve the desired temperature set point in the cabin and defrost the glazing after the start of the vehicle; it also aims to decrease the heating or cooling load associated with the pulsed air (and hence to achieve, with a limited cost, energy saving of the air conditioning/heating system) and to increase the performance and lifespan of the gas and particle filtration system integrated in the HVAC.

However, the fraction of recirculated air can be limited by the cabin humidity content that could yield mist formation on the glazing or cabin CO_2 concentration that could limit the oxygen quantity to the passengers. The fresh air flow rate must be determined based on water vapor and CO_2 quantity generated by the occupants inside the cabin (roughly 35 to 50 g h^{-1} of water per occupant). Since humidity and CO_2 are mainly generated by the cabin occupants (see Eqs. (3.77) and (3.78)), their production rates depend on the number of occupants. Hence, it is necessary to renew the air with a minimum amount of fresh air function of the number of occupants to prevent any risk of mist formation and an exposure time under a too large concentration of CO_2 leading to uncomfortable levels or to dangerous situation (drowsiness).

The main constraints of air recirculation are the following ones:

- The risk of mist formation on glazing, mainly at low outdoor temperature, not only during driving but also during parking (after driving with too high air recirculation) because of condensation generated by the temperature drop in the cabin.
- The increased concentration of CO_2 and the decreased concentration of oxygen in the air.
- The increase of the aeraulic noise coming from the opening of the recirculation flap through the recirculation air duct. The design of the recirculation flap must be meticulous to prevent any risk of whistling during opening and closing of the flap.
- The decrease of thermal comfort in the rear seats because of the reduction of the mass amount of air sent to this zone when the recirculation intake is located in the front zone leading to direct re-aspiration after the air supply vents. This effect is mitigated if dedicated air vents and recirculation intake are located in the rear zone.
- The design and position of the recirculation flap must be meticulous in such a way to prevent, according to the speed of the vehicle, the direct passage of outdoor air to the cabin through the recirculation flaps by-passing the fan. The control can hence adapt the speed of the fan as a function of the vehicle speed in such a way to control the recirculation rate and the fresh air flow rate.
- The proximity of the floor vents (for feet heating) and of the recirculation intake under the dashboard can cause an undesirable extraction of air pulsed from the floor vents to the cabin (short-circuit), which can deteriorate the thermal comfort because of a large recirculation rate under a low pulsed air flow rate. The air flow rate pulsed through the windshield vents allows for limiting this risk.

In practice, the recirculation fraction, defined as the ratio of the recirculated mass flow rate to the mass flow rate pulsed into the cabin (Eq. [3.84]), is often used. The recirculation fraction depends on the position of the mixing flap. In the full recirculation mode, $X_{rec} = 1$.

$$X_{rec} = \frac{\dot{m}_{a,rec}}{\dot{m}_{a,vent}} \tag{3.84}$$

where

$\dot{m}_{a,rec}$ is the recirculated air mass flow rate, [kg s^{-1}]
$\dot{m}_{a,vent}$ is the air mass flow rate pulsed into the cabin, [kg s^{-1}]

Since humid air comprises both dry air and water vapor, two equations of conservation of mass describe the mixing between the fresh and recirculated air. These equations can be expressed as

$$\dot{m}_{a,vent} = \dot{m}_{a,fresh} + \dot{m}_{a,rec} \tag{3.85}$$

$$\dot{m}_{a,vent} \cdot \omega_{su,fan} = \dot{m}_{a,fresh} \cdot \omega_{out} + \dot{m}_{a,rec} \cdot \omega_{cab} \tag{3.86}$$

where $\omega_{su,fan}$ is the specific humidity of air at the fan supply, [kg kg^{-1}]

Assuming that the mixing between the two airstreams is adiabatic, the conservation of energy can be written as

$$\dot{m}_{a,vent} \cdot h_{a,su,fan} = \dot{m}_{a,fresh} \cdot h_{a,out} + \dot{m}_{a,rec} \cdot h_{a,cab} \tag{3.87}$$

where $h_{a,su,fan}$ is the specific enthalpy of air at the fan supply, [J kg^{-1}]
Under the assumption that the specific heat of humid air is constant in the range of operating conditions associated with the mixing process, combining previous equation with the equation of conservation of mass of moisture yields, as a first approximation

$$\dot{m}_{a,vent} \cdot T_{a,su,fan} \approx \dot{m}_{a,fresh} \cdot T_{a,out} + \dot{m}_{a,rec} \cdot T_{a,cab} \tag{3.88}$$

For a given mass flow rate $\dot{m}_{a,vent}$ pulsed into the cabin, recirculation of a fraction of cabin air allows for a reduction of the energy consumption associated with heating or cooling the air. Actually, in the cooling mode, recirculation decreases the temperature at the inlet of the evaporator as soon as the cabin temperature is under the outdoor temperature after engine starting. In the heating mode, recirculation increases the temperature at the supply of the heater core.

Considering physiological criteria, existing standards in buildings and the moisture and CO_2 production per occupant (see Example 3.1), an optimal outdoor fresh air flow rate of around 30 to 40 m^3 h^{-1} per occupant, in a cabin of approximately 3 m^3, can be enough. The recirculation fraction can thus be large and the resulting gain important, which is the case when one single passenger is in the vehicle cabin. The increase in the number of passengers yields a decrease of the optimal recirculation rate, especially in winter conditions when the indoor hygrometry is the limiting factor that determines the required minimal outdoor air flow rate to prevent at low outdoor temperature any risk of mist formation in the cabin.

Full recirculation ($X_{rec} = 1$) serves two functions:

- Decreasing the time needed to reach comfortable conditions inside the cabin, when the cooling or heating system is switched on.
- Isolate the cabin from the vehicle outside, preventing outdoor pollutants to enter into the cabin.

Full recirculation should be limited in time to limit the quantity of CO_2 and humidity (rejected by occupants) in the cabin. As mentioned by Colinet (1993), during full recirculation, the refrigeration system (air-conditioning loop) should be activated if humidity ratio is too high to dehumidify the air by means of the evaporator. This would decrease the mist formation on the windshield and ensure the visibility.

Note that it is necessary after some time to renew the air in the cabin with a minimum amount of fresh air in function of the number of occupants.

In winter, 100% outdoor air, which contains less moisture than indoor air, is usually supplied to the cabin to limit the mist formation on the windshield. Inversely, in summer, a large fraction of indoor air is recirculated. As it will be explained in Chapter 4, for electrified vehicles, recirculating indoor air in winter is a mean to limit the impact of cabin heating on the driving range in the electric mode. However, care should be taken to ensure the demisting of the windshield.

In hot climate, the gain is interesting. The air-conditioning system, if operating, causes the air to be partially dried, which limits the risk of mist formation, especially when the outdoor temperature is high. Consequently, the recirculation rate can reach 100% without any risk of mist formation for temperature over 25–30°C. However, at vehicle start, the indoor air temperature is larger than the outdoor air temperature and can be very high. As soon as this temperature reaches a low enough level, partial recirculation allows the temperature drop to be accelerated to reach the thermal comfort and limit the consumption associated to air-conditioning. If the demisting mode is necessary because it is imposed by the occupant or by the automatism, air recirculation can be reduced or even deactivated, and the fresh air flow rate is increased in such a way to guarantee a good visibility while not using the air-conditioning system to dry the air. Also, as explained before, stratification of the air supply, with 100% of outdoor air toward the windshield and recycled air towards the feet, allows for a given recirculation rate to reduce the risks of mist formation on glazing.

Partial air recirculation can be achieved with a recirculation flap having several positions, for instance 3 positions (100% fresh air, 100% recycled air, and intermediate position), or showing a modulating position.

The design of the air inlet and of the motorized flap must allow, through a smart management, for an accurate and progressive control of the fresh air supplied to the cabin and for a good linearity of the recirculation rate as a function of the opening rate of the flap, of the desired fresh air flow rate, of the speed of the fan and of the vehicle speed.

Advanced control systems coupled to the use of sensors can use measurements of humidity or mist formation to control the fraction of renewed air and to allow for the adjustment of the kinematics and position of the air inlet flap (the recirculation flap):

- Pollutants sensors that detect an excessive pollution in the intake of outdoor air or in the cabin. Such sensors allow for considering a full air recirculation in such a way to prevent polluted air to enter the cabin or considering an increase of fresh air in such a way to reduce the pollution of the cabin air. In the case of risk of mist formation, it is advised to prioritize the demisting action and not to satisfy the demand of pollutants sensors. The acoustic criterion can also be taken into consideration.
- Presence sensors through IR camera or sensors embedded in seats and detecting the number of car occupants or CO_2 sensors that evaluate the quantity of CO_2 produced by occupants.
- Indoor temperature-humidity sensors, outdoor temperature sensors and glazing temperature sensors that determine preventively the risk of mist formation and allow for the comfort to be optimized.

Moreover, coupled with a heated windshield with layers, the partial recirculation rate could be larger.

Finally, the optimization of partial recirculation with a reduced fresh air flow rate eases the integration of an efficient filter on the outdoor fresh air.

3.4.3 HVAC Unit Operating Modes

In function of the climatic conditions, the HVAC unit can be operated according to different modes.

3.4.3.1 Ventilation

The purpose of this mode is to only bring fresh outdoor air inside the cabin. The air is neither heated nor cooled down. This means that the refrigeration system is not activated and the heater core is by-passed or not fed with high temperature coolant. Air is usually pulsed through the upper vents. An electric centrifugal fan is used, with a rotational speed that can either have different values or be adjusted continuously. The generated air flow rate depends on the vehicle. The maximum flow rate is of the order of 400–800 m³ h⁻¹.

3.4.3.2 Cooling
Cabin cooling is achieved by pulsing cold air inside the cabin, usually through the upper and/or floor vents. If the outdoor air is cold enough, the refrigeration system (A/C loop) is not used and the evaporator does not further cool the air. This mode is called passive cooling. The air mass flow rate pulsed into the cabin is controlled by adjusting the fan rotational speed. If it is necessary to cool the outdoor air, the refrigerant compressor is engaged, which circulates refrigerant through the evaporator (Figure 3.21). In this active cooling mode, the evaporator cools the air and possibly dehumidifies it. Dehumidification occurs if the temperature of the air is brought underneath the dew point temperature at the evaporator supply.

3.4.3.3 Heating
Cabin heating is ensured by pulsing hot air inside the cabin. There are two main mechanisms to control the heating power supplied to the cabin: the air mix type and the water flow type (Daly, 2006).

The *air mix type*, which is depicted in Figure 3.22, consist of mixing air flowing through the heater core and air bypassing the heater core to achieve the desired air temperature. The air mass flow rate flowing through the heater core is controlled by means of the mixing flap. Full heating capacity is achieved by directing all the airstream through the heater core.

In a *water flow type* system, the whole airstream flows through the heater core, and the heating power is controlled by adjusting the hot coolant flow rate through the heater core. This is achieved by means of a valve on the coolant side. Daly (2006) lists as follows the advantages and drawback of both types of control. The air mix type is more reactive than the water flow type. However, it is bulkier because of the presence of the mixing chamber. When the heater core is fully bypassed, air is still heated by radiation from the heater core (and potential leaks). This could, however, be prevented by using a shut-off valve on the coolant loop. One of the difficulties of the water flow type system is that the coolant flow rate and temperature depend on the engine regime (Daly, 2006).

Figure 3.21 Active cooling mode.

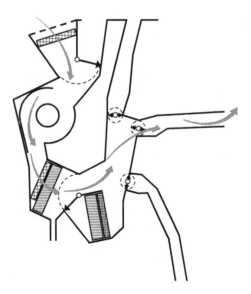

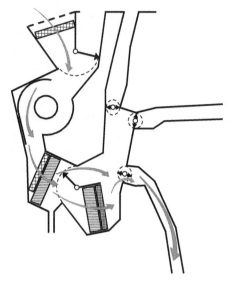

Figure 3.22 Control of the heating capacity with an air mix-type configuration.

In the heating mode, the hot air is pulsed generally through the floor vents (Colinet, 1993) in the front zone and possibly in the rear zone.

3.4.3.4 Demisting and Defrosting

The air humidity can cause an important security issue when mist or frost degrade the visibility of the driver.

Mist formation on the windshield and other glazed surfaces appears when the air in the boundary layer gets saturated in water vapor, i.e. when the indoor surface temperature is lower than the cabin air dew point temperature. Water vapor then condenses on the glazed surface. One solution would be to decrease the absolute humidity at the vicinity of the glazing by pulsing dehumidified air (therefore decreasing the dew point). Another solution consists of increasing the indoor surface temperature by pulsing hot air.

Frost appears on the glazing indoor and outdoor surfaces when the temperature of these surfaces is lower than 0°C.

Demisting and defrosting are achieved by pulsing air at the appropriate temperature and speed on the glazed surfaces (Colinet, 1993). The HVAC unit is operated as depicted in Figure 3.23.

Demisting of the indoor surface of the windshield and side windows is ensured by pulsing high speed, hot and dehumidified air on these surfaces (Colinet, 1993). High air speeds are reached by the shape of the air outlets. When this mode is activated after starting the engine, all the airstream should be directed toward the demisting vents.

Other actions facilitate the demisting: use of the refrigerant loop to dehumidify the air, adjustment of the position of the mixing flap to increase the air temperature, increase the fan rotational speed to increase the air flow rate, and close the recirculation flap to pulse only fresh air.

To defrost the windshield and side windows, it is not necessary to dehumidify the air pulsed in the cabin since frost is formed on outdoor surfaces. Hence, the whole capacity of the heater core is used to heat as much as possible the pulsed air. Once again, after starting the engine, only the defrost vents should be opened to quickly ensure the visibility (Colinet, 1993).

Demisting and defrosting could also be achieved by using an electrically heated glazing. It should also be noted that the demisting-defrosting mode could also be coupled to the heating mode.

Figure 3.23 Demisting and defrosting modes.

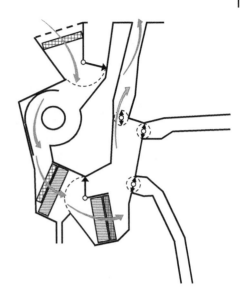

3.4.3.5 Ventilation and Heating

In this mode, air is pulsed through both the upper vents and floor vents. As shown in Figure 3.24, the position of the mixing flap and the connections of piping at the outlet of the mixing chamber that mixes air coming from the heater core and from the bypass duct ensures that air at a higher temperature is pulsed through the floor vents than the upper vents (Colinet, 1993). A difference of approximately 7 K between the air impinging the face and the feet is achieved (Daly, 2006), which improves the thermal comfort of occupants. Ensuring a temperature difference between the feet and the head is called the bi-level mode.

Other intermediate modes of air distribution can be achieved, such as with air supply through floor and demisting vents or through floor, upper, and demisting vents.

Figure 3.24 Ventilation and heating mode.

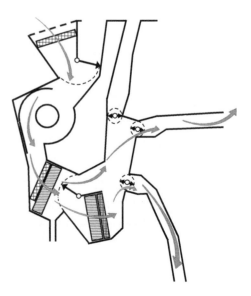

3.4.3.6 Temperature and Flow Rate of the Air Flow Pulsed by the HVAC Unit

As illustrated in Figure 3.25, the supply air temperature can be a function of the outdoor climate conditions (for instance, temperature and solar radiation). Usually, colder air is supplied to the upper level of the body than to the feet (bi-level operation). In winter, lower and upper blown temperatures largely differ from each other. Such a large temperature gradient is required to obtain an appropriate degree of comfort and take into account the natural convection (hot air moves from the bottom to the top and cold air from the top to the bottom).

The air mass flow rate supplied to the cabin is also typically a function of the outdoor air temperature. As shown in Figure 3.26, larger air flow rates are pulsed into the cabin when extreme outdoor temperatures are reached. This figure also indicates that larger flow rates are pulsed during the transient ("convergence") period than in the steady-state regime. This allows increasing the heat

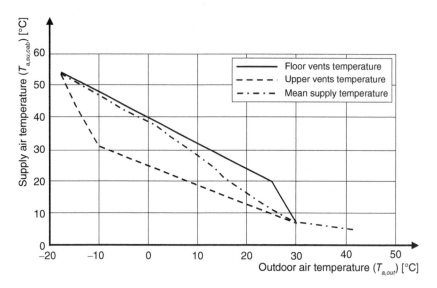

Figure 3.25 Example of evolution of supply air temperature with the outdoor air temperature.

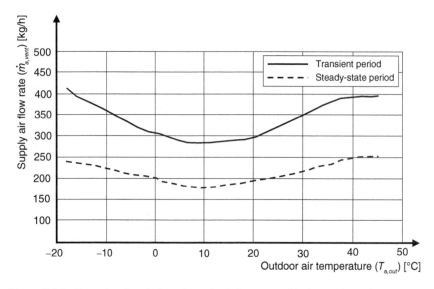

Figure 3.26 Example of evolution of supply air flow rate with the outdoor air temperature.

transfer rate during transient evolutions. As explained through the equivalent temperature, the air velocity increases the cold sensation. This is a positive effect in hot climates. In contrary, in cold climates, attention should be paid not to create discomfort when the temperature of pulsed air is still not high enough.

The cabin climate can also be adjusted by varying the fraction of the total mass flow rate pulsed through the vents at the upper body position, at the feet position, and underneath the windshield (used mainly for demisting or defrosting).

3.4.4 Cabin Air Quality

Car manufacturers and customers pay an increasing attention to the air quality inside the cabin. The air quality treatment system is integrated inside the HVAC unit. Sometimes, additional components are located in the cabin to treat the cabin air.

The objectives consist of filtering gases and particles from the outdoor air and indoor air supplied to the cabin, eliminating the odors, and protecting the evaporator.

Two types of filters are used. The *particle filter* is able to filter particles with large diameters, such as pollen and soot. The *gas filter* is able to filter VOCs, SO_2, NO_2, and ozone. The particle filter and the gas filter are integrated in a common combined filter.

Such a filter is folded and composed of very large developed surface, which decreases the air speed and pressure drop through the filter.

The particle filtration section of the filter is generally composed of fiber media. Pressure drop and efficiency evolve with the clogging of the filter. The filter lifespan depends on the driving conditions.

The gas filtration section of the filter is generally composed of activated carbon that adsorbs the gaseous molecules.

An outdoor environment gas sensor sometimes is added near the air entrance and in case of over pollution close the recirculation flap preventing the cabin against outdoor pollution.

3.5 Production of Cooling Capacity

Cooling capacity is typically provided by means of a vapor-compression refrigerator, which is also named air-conditioning loop (A/C loop). The vapor-compression refrigerator can also provide the cooling effect necessary for cooling the battery (see Chapter 4). As it will be explained latter (still in Chapter 4), the same machine can work in heating production and is then called a heat pump.

3.5.1 Working Principle of a Vapor-Compression Refrigerator

As depicted in Figure 3.27, the vapor-compression refrigerator comprises four main components: the evaporator, the compressor, the condenser, and a throttling device (typically an expansion valve or an orifice tube). Different pipes connect all these components. The working fluid, called refrigerant, is flowing successively and cyclically through these components. In practical realization, the vapor-compression refrigerator comprises other components that aim at improving its performance and reliability.

The evolution of the refrigerant describes the thermodynamic cycle represented in Figure 3.28. In this idealized cycle, there is no pressure variation outside the expansion valve and compressor. This means that there is pressure drop neither in the heat exchangers nor in the piping. As a consequence, two levels of pressure characterize the cycle: the evaporating pressure P_{ev} and the condensing pressure P_{cd}. It is also assumed that there is no temperature glide during the liquid–vapor phase changes (in the case of a pure fluid or zeotropic mixture).

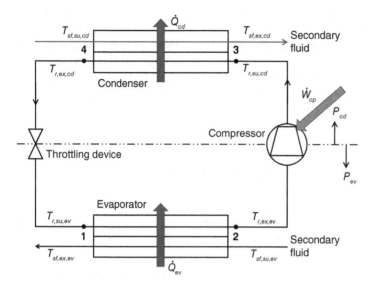

Figure 3.27 Schematic representation of a vapor-compression refrigerator.

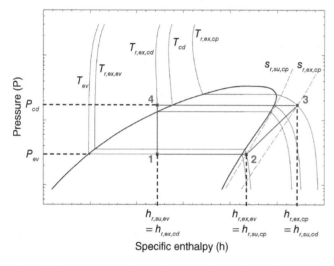

Figure 3.28 Representation of the subcritical refrigeration cycle in a P-h diagram.

3.5.1.1 Evaporator

The evaporator is a heat exchanger that allows for the heat transfer from the secondary fluid (typically humid air or glycol water) to the refrigerant.

The refrigerant enters the evaporator at state *1* and leaves it at state *2*. At the evaporator supply, the refrigerant is in a two-phase state (a mixture of a saturated liquid and a saturated vapor) at the evaporating pressure P_{ev} and evaporating temperature T_{ev}. From state *1* to state *2*, the refrigerant is vaporized at constant pressure and eventually slightly superheated. The heat necessary to bring the fluid from state *1* to state *2* is absorbed from the secondary fluid, which creates the cooling effect.

3.5.1.2 Compressor

At state *2*, the superheated vapor enters the compressor. The latter compresses the refrigerant, thereby reaching state *3* and increasing its pressure until P_{cd}.

3.5.1.3 Condenser

The condenser is a heat exchanger that allows for the heat transfer between a refrigerant and a secondary fluid (typically humid air or glycol water).

The refrigerant enters the condenser at state *3* and leaves it at state *4*. In the condenser, the superheated vapor is first cooled down at constant pressure until reaching the saturation temperature T_{cd}. The saturated vapor is then condensed at constant pressure until saturated liquid state. The liquid is finally subcooled at constant pressure until state *4* at temperature $T_{ex,cd}$. From state 3 to state 4, heat is rejected from the refrigerant into the secondary fluid. According to the cooling fluid, technical literature usually distinguishes air-cooled condensers and water-cooled condensers.

3.5.1.4 Throttling Device

The high-pressure subcooled liquid enters the throttling device at state *4* and is throttled to state 1, closing the cycle.

The slight superheat at the compressor supply (state 2) is necessary to protect the compressor against liquid ingestion. The liquid subcooling at the supply of the throttling device is also necessary to ensure the proper operation of the latter. It will be shown later in this chapter that an additional heat exchanger, named subcooler, could be used to create this subcooling.

The useful effect of the refrigerator is to produce a cooling effect at the evaporator, by absorbing heat from a secondary fluid. To compress the refrigerant, the compressor consumes electrical or mechanical work. A heat pump operates according to the same vapor compression cycle than the refrigerator. However, in the case of a heat pump, the useful effect is to produce a heating effect at the condenser, by rejecting heat in a secondary fluid.

The performance of the refrigerator is typically assessed with the coefficient of performance (COP), defined in the cooling mode as the ratio of the cooling capacity to the compressor power.

$$COP_c = \frac{\dot{Q}_{ev}}{\dot{W}_{cp}} \tag{3.89}$$

3.5.2 Integration of the Air-Conditioning Loop into the Vehicle

As illustrated in Figure 3.30, the A/C loop is integrated inside the engine compartment. In this figure, the visible front heat exchanger is the condenser. The compressor and its pulley can also be easily recognized. Most often, the condenser is cooled by air and is located in the front-end module. It should be mentioned that water-cooled condensers can also be used. The air-heated evaporator is integrated into the HVAC unit and is fed by ventilation air, which is pulsed into the cabin. The HVAC unit is most of the time integrated inside the dashboard, but other locations could also be considered. As indicated in Figure 3.29, the HVAC unit also comprises the heater core (HC) and

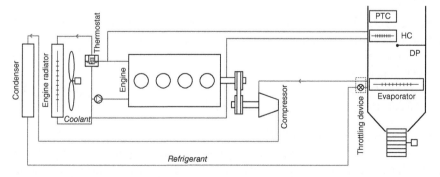

Figure 3.29 Schematic of the integration of the A/C system in the vehicle thermal management system.

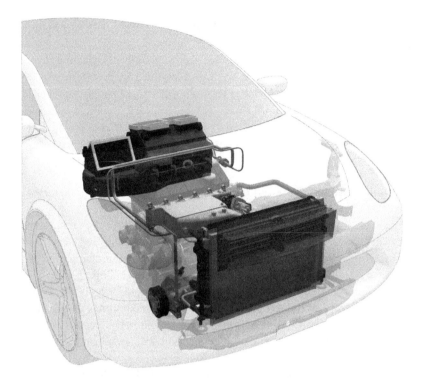

Figure 3.30 Integration of the A/C loop in the engine compartment. Source: Courtesy of Vaelo.

possibly electric resistances (PTC) used for heating purpose. The air flow rate through the heater core can be adjusted by means of the mixing flap or damper (DP).

Different compressor configurations are possible, which will be described later. The throttling device is located at the evaporator supply, but it is regulated with information from the evaporator exhaust, which explains its position in Figure 3.29.

Besides these four major components (compressor, condenser, evaporator, and throttling device), the A/C loop comprises secondary components the functions of which will be described in the subsequent sections: receiver, accumulator, drier, filter, internal heat exchangers, etc. These components are connected by tubes and hoses. A practical realization of a vapor compression A/C loop is shown in Figure 3.31.

3.5.3 Compressor

The compressor is a critical component of the A/C loop. It allows for the circulation of the refrigerant inside the loop, creates the pressure increase from the evaporator to the condenser, and can possibly be used to modify the cooling capacity of the A/C loop by adjusting the refrigerant mass flow rate through the evaporator.

Compressors used in A/C loops are positive displacement (or "volumetric") compressors. The most often used compressors are the piston, the scroll, and the vane compressors.

3.5.3.1 Mechanical Versus Electrical Compressors

As mentioned in Chapter 1, a shaft work is necessary to move the pistons or the rotor and communicate energy to the fluid. This shaft work is either provided by the vehicle engine ("mechanical compressor") or by an electric motor ("electric compressor").

Figure 3.31 Components of an A/C loop including an air-cooled condenser and an open-drive compressor. Source: Courtesy of Valeo.

Figure 3.32 Practical realization of a mechanical compressor. Source: Courtesy of Valeo.

For *mechanical compressors* (Figure 3.32), the compressor shaft is connected to the vehicle engine shaft through a belt and pulley coupling. The speed of the compressor is generally slightly larger than that of the engine, with a speed ratio of 1.3 for instance. This ratio is selected in such a way to satisfy the compressor mechanical constraints at high engine speed, and the compressor displacement is selected to have enough capacity at engine idling. Vibration and noise, which are more severe at high compressor rotational speed, are other criteria to take into consideration.

Mechanical compressors are often equipped with an electromagnetic clutch that is energized in order to engage the compressor. However, clutchless architectures also exist.

Electric compressors (Figure 3.33) are made up of the compressor element, an electric motor and an inverter embedded in a semi-hermetic envelope. Because of the good tightness of their envelopes, such compressors show less refrigerant leakages than mechanical compressors.

Also, with such compressors, the A/C cooling capacity is independent of the vehicle engine regime. Actually, the speed of the compressor is adjusted by the inverter. This means that the

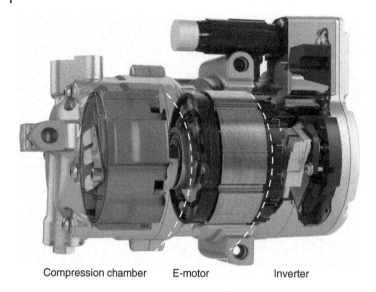

Compression chamber   E-motor   Inverter

Figure 3.33 Internal structure of an electric scroll compressor. Source: Courtesy of Valeo.

A/C loop can be activated even if the vehicle is at standstill, allowing for instance for cabin pre-conditioning. Today, electric compressors are of scroll type. Such technology will be described later.

3.5.3.2 Compressor Capacity

For given evaporating and condensing pressures, subcooling and superheating degrees, the A/C cooling capacity $\dot{Q}_{ev}$ is proportional to the refrigerant mass flow rate flowing through the evaporator.

$$\dot{Q}_{ev} = \dot{m}_r(h_{r,ex,ev} - h_{r,su,ev}) \tag{3.90}$$

where

   $\dot{m}_r$ is the refrigerant mass flow rate through the evaporator, [kg s^{-1}]
   $h_{r,ex,ev}$ is the refrigerant specific enthalpy at the evaporator exhaust, [J kg^{-1}]
   $h_{r,su,ev}$ is the refrigerant specific enthalpy at the evaporator supply, [J kg^{-1}]

Hence, to vary the cooling capacity $\dot{Q}_{ev}$, the refrigerant mass flow rate through the evaporator can be adjusted.

   In the steady-flow regime, the refrigerant mass flow rate traveling through the different components of the loop is related to the displaced volume flow rate $\dot{V}_s$ of the compressor. Assuming, as a first approximation, that the volumetric effectiveness of the compressor is 100%, the refrigerant mass flow rate can be related to the compressor displacement and rotational speed by

$$\dot{m}_r = \dot{V}_s \, \rho_{r,su,cp} = N V_s \rho_{r,su,cp} \tag{3.91}$$

where
   $\dot{V}_s$ is the compressor displaced volume flow rate, [m^3 s^{-1}]
   $\rho_{r,su,cp} = 1/v_{r,su,cp}$ is the refrigerant density at the compressor supply, [kg m^{-3}]
   N is the compressor rotational speed, [Hz]
   V_s is the compressor displacement, [m^3]

The density of the refrigerant at the compressor supply is function of the fluid pressure and temperature. The displaced volume flow rate is the product of the rotational speed N times the displacement (or "swept volume") V_s. Sizing the compressor means identifying the appropriate value of the displacement. The displacement is selected to cool down the vehicle cabin and reach a comfortable temperature within a limited time ("cooldown period"), ranging approximately from 10 to 20 minutes, in the worst-case scenario. This scenario corresponds to a low vehicle speed or idling (yielding a higher cooling load), low vehicle engine speed (for a mechanical compressor), and high outdoor temperature (for instance, 45°C).

Eq. (3.91) indicates that during operation of the A/C loop, its cooling capacity can be varied by either adjusting the rotational speed N or the displacement V_s of the compressor.

Since mechanical compressors are mechanically coupled to the engine shaft, their rotational speed cannot be controlled. Some mechanical compressors have a mechanism to control the displacement V_s. In the absence of such a mechanism, there is no other means to control the cooling capacity than cycling on and off of the compressor. Cycling is achieved by means of the compressor electromagnetic clutch. As a drawback, cycling of the compressor can yield discomfort inside the cabin.

3.5.3.3 Piston Compressors
A piston compressor comprises one of several cylinders inside which the pistons are moving in a reciprocating fashion. The evolution of the refrigerant through the compressor can be described by means of the theoretical indicator diagram shown in Figure 3.34.

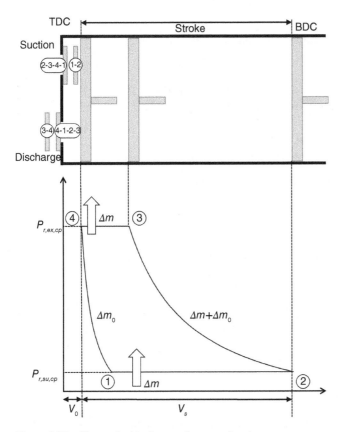

Figure 3.34 Theoretical indicator diagram of a piston compressor.

From position *1* to position *2*, the piston moves towards the *bottom dead center* (BDC), which causes a mass of refrigerant Δm at the supply pressure P_{su} to enter inside the cylinder. This is the suction phase. During this phase, the suction valve is maintained open by pressure difference between the supply line and the inside of the cylinder. As the piston reaches the BDC (position *2*), it starts to travel again toward the *top dead center* (TDC). As the volume occupied by the refrigerant is decreased, the refrigerant pressure is increased. This is the compression phase. The suction valve is now maintained closed by the pressure difference between the inside and outside of the cylinder. At position *3*, the in-cylinder pressure reaches the discharge pressure P_{ex} and the discharge valve opens. From position *3* to position *4*, the piston continues to move toward to the TDC, which causes a mass Δm of high-pressure vapor to leave the cylinder. This is the discharge phase. At position *4*, the piston reaches the TDC and starts to move back toward the BDC. In this position, the volume occupied by the refrigerant is minimal and is equal to the clearance volume V_0. From position *4* to position *1*, the cylinder volume increases and both valves are closed. The pressure of the mass of refrigerant Δm_0 that was trapped inside the clearance volume is decreased until reaching, at position *1*, the supply pressure P_{su}. This "re-expansion of the clearance volume" consumes part of the swept volume V_s. As a consequence, the actual displaced volume during the suction phase is V_{12}. It could be shown that this volume can be expressed at a function of the swept volume V_s by the following approximation:

$$V_{12} = \varepsilon_{v,0} \, V_s = \left(1 - C \left(\frac{v_{r,su,cp}}{v_{r,ex,cp}} - 1 \right) \right) V_s \approx \left(1 - C \left(\left(\frac{P_{r,ex,cp}}{P_{r,su,cp}} \right)^{1/\gamma} - 1 \right) \right) V_s \qquad (3.92)$$

where

$\varepsilon_{v,0}$ is the clearance volume contribution to the volumetric effectiveness, [−]
V_s is the swept volume, [m^3]
C is the clearance factor, [−]
$v_{r,su,cp}$ is the compressor specific volume at compressor supply, [m^3 kg^{-1}]
$v_{r,ex,cp}$ is the compressor specific volume at compressor exhaust, [m^3 kg^{-1}]
$P_{r,ex,cp}$ is the compressor discharge pressure, [Pa]
$P_{r,su,cp}$ is the compressor supply pressure, [Pa]
γ is the specific heat ratio, assuming the refrigerant to be an ideal gas, [−]

The compressor clearance factor C [−] is defined as the ratio between the clearance volume and the swept volume.

$$C = \frac{V_0}{V_s} \qquad (3.93)$$

The swept volume is related to the cylinder bore B and piston stroke S by

$$V_s = \frac{\pi B^2}{4} S \qquad (3.94)$$

In practice, the clearance factor is around a few percent. As illustrated in Figure 3.35, the penalty associated with the re-expansion of the clearance volume increases with both the pressure ratio $r_{p,cp}$ and the clearance factor C.

The pressure ratio imposed to the compressor is defined by

$$r_{p,cp} = \frac{P_{r,ex,cp}}{P_{r,su,cp}} \qquad (3.95)$$

The piston compressor is the oldest technology of compressors used in mobile A/C. In the 1950s, crank-type piston compressors were in use. Today, piston compressors show an axial configuration,

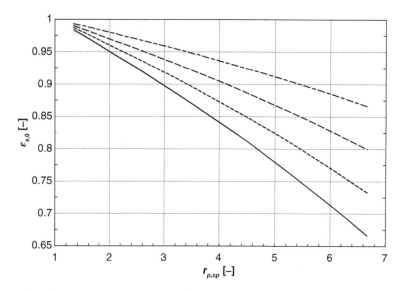

Figure 3.35 Evolution of the clearance volume contribution to the volumetric effectiveness with the pressure ratio for different clearance factors.

which means that the pistons (usually 3 to 7) are parallel to the drive shaft. Such compressors are equipped with a plate tilted with respect to the drive shaft. Pistons are in contact with the plate. The movement of the plate, when the shaft rotates, forces the piston to reciprocate inside cylinders. Each cylinder has supply and discharge orifices covered by reed valves. One can distinguish two types of axial piston compressors: the swash plate and the wobble plate.

In a swash plate compressor (invented in the 1960s), the extremities of the pistons slide, by means of shoes, on both faces of a plate (named "swash plate") that rotates with the drive shaft (Figure 3.36). Such compressors can be equipped with single-acting (Figure 3.37) or double-acting pistons (Figure 3.36). In the latter case, cylinders are present on both sides of the swash plate and extremities of the piston reciprocate in opposite cylinders. Double-acting piston compressors, also called "dual swash plate" (versus "single swash plate"), have 3 to 5 pairs of pistons and deliver a flow with less pulsations.

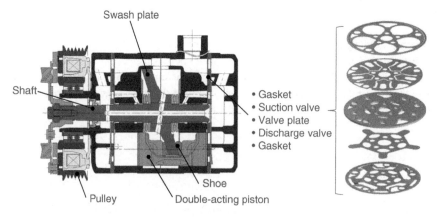

Figure 3.36 Section view of a fixed-displacement swash plate compressor with a double-acting piston. Source: Courtesy of Valeo.

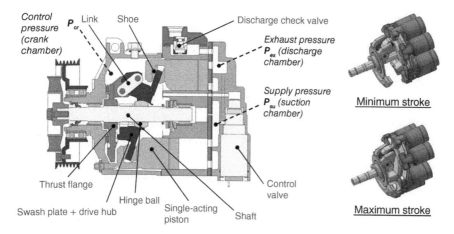

Figure 3.37 Section view of a variable-displacement swash plate compressor with a single-acting piston. Source: Courtesy of Valeo.

In a wobble plate compressor (invented in the 1950s), ball-ended piston rods are connected to a plate (named "wobble plate") that shows a wobbling motion. The wobbling motion is created by the drive plate, which rotates with the drive shaft, bearings located between both plates, and an anti-rotation mechanism that constrains the rotation of the wobble plate (Ishii et al., 1990).

Indicator diagrams of piston compressors (Figure 3.34) make appear two major differences with respect to sliding vane and scroll compressors described hereunder (see Figure 3.40). First, there is no fixed built-in volume ratio. Consequently, the compressor internal pressure ratio is automatically adapted to the external pressure ratio (the pressure at the end of the compression process is equal to the pressure in the discharge line augmented by the discharge pressure losses). Second, piston compressors are characterized by a clearance volume, the penalty of which has been described previously.

The axial piston compressor offers a mean of controlling continuously its displacement and hence the A/C loop cooling capacity. Such a control is achieved by adjusting the swash plate or wobble plate tilt angle, which results in a variation of the piston stroke. Modulation of cooling capacity allows for the improvement of the energy performance of the A/C loop and for a better thermal comfort of passengers.

Modulation of the displacement of piston compressors will be described more in detail in Section 3.5.10.2.2.

Example 3.3 *Volumetric Performance of a Piston Compressor*

A fixed-displacement piston compressor operates with R1234yf between a supply pressure of 3 bar and an exhaust pressure of 20 bar. The superheat at the compressor supply is 5 K. If the compressor displacement is 162.5 cm^3 and its clearance factor is 5.2%, estimate the mass flow rate displaced by the compressor when it rotates at 2000 rpm (data for this example are inspired from Cuevas et al. [2008]).

Analysis

It is assumed here that only the re-expansion of the clearance volume affects the volumetric performance of the machine. The clearance volume contribution to the volumetric effectiveness $\varepsilon_{v,0}$ is computed and then the actual displaced volume flow rate. Finally, the mass flow rate is computed.

Solution

```
"1. INPUTS"
fluid$='R1234yf'
P_r_su_cp=3e5 [Pa] "compressor supply pressure"
P_r_ex_cp=20e5 [Pa] "compressor exhaust pressure"
DELTAT_r_su_cp=5 [K] "superheat at the compressor supply"
N=2000/60 "[Hz]" "compressor rotational speed"


"2. OUTPUTS"
"m_dot_r_cp=?? [kg/s]" "mass flow rate displaced by the compressor"


"3. PARAMETERS"
V_s_cp=162,5e-6 [m^3] "compressor displacement"
C=0,052 [-] "clearance factor"


 "4. MODEL"
"4.1. Compressor supply state"
T_r_su_cp=T_r_sat_su_cp+DELTAT_r_su_cp "compressor supply
   temperature"
h_r_su_cp=enthalpy(fluid$;P=P_r_su_cp;T=T_r_su_cp)
T_r_sat_su_cp=temperature(fluid$;P=P_r_su_cp;x=1)
s_r_su_cp=entropy(fluid$;P=P_r_su_cp;T=T_r_su_cp)
v_r_su_cp=volume(fluid$;P=P_r_su_cp;T=T_r_su_cp)


"4.2. Compressor exhaust state"
v_r_ex_cp=volume(fluid$;P=P_r_ex_cp;s=s_r_su_cp) "assuming
   isentropic compression"


"4.3. Displaced mass flow rate"
V_dot_s_cp=N*V_s_cp "[m^3/s] displaced volume flow rate"
epsilon_v_0=1-C*(r_v-1) "volumetric effectiveness"
epsilon_v_0=V_dot_12_cp/V_dot_s_cp "to compute the actual
   displaced volume flow rate"
r_v=v_r_su_cp/v_r_ex_cp

m_dot_r_cp=V_dot_12_cp/v_r_su_cp
```

Results

```
epsilon_v_0=0,653 [-]
m_dot_r_cp=0,05787 [kg/s]
```

3.5.3.4 Sliding Vane Compressors

The sliding vane compressor is a specific type of rotary vane compressor. A sliding vane compressor is made up of a cylindrical rotor eccentrically or centrally located inside a stator. The stator has typically a cylindrical or elliptical shape. The rotor is equipped with slots inside which vanes are reciprocating. Vanes can be either positioned radially or canted. In the former case, frictional forces are reduced (because of lower centrifugal forces) and longer vane can be used. In the cylindrical configuration, the stator has a supply port and a discharge port. The inner wall of the stator, the outer wall of the rotor, and the vanes define a series of chambers. As the rotor rotates, these chambers move along the stator. The chambers are kept separated by the vanes, which slide inside the slots, due to centrifugal forces, ensuring a good contact with the stator. As long as a working chamber is in communication with the supply port, it is filled with refrigerant. The suction process ends when the chamber is isolated from the supply line. The chamber volume is then reduced and the entrapped refrigerant pressure is increased. The discharge process starts when the leading vane reaches the discharge port. The discharge port can be covered by a check valve to prevent rotation in the opposite direction during shutdown.

Sliding vane compressors are used in A/C loops since the late 1970s. Their low mass-to-displacement ratio and their compactness make vane compressors a suitable technology for mobile A/C (Chapter 37 of the 2008 ASHRAE Handbook – HVAC Systems and Equipment [ASHRAE, 2008]). Moreover, such compressors are robust, characterized by a high volumetric effectiveness and make little noise. As a major drawback, vane compressors show lower efficiency than scroll compressors (ASHRAE, 2008). It is difficult to adapt the displacement of such compressors. Hence, they are equipped with electromagnetic clutches for cooling capacity control (on/off control).

As shown in Figure 3.38 and Figure 3.39, the elliptical configuration of compressor shows two pairs of compression chambers, as well as two supply and discharge ports. Such configuration

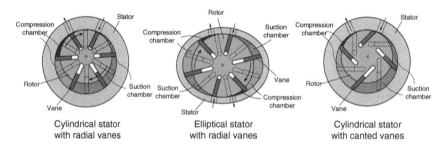

Figure 3.38 Schematics of different designs of sliding vane compressors. Source: Reproduced from Aw and Ooi (2021) (licensed under CC BY 4.0).

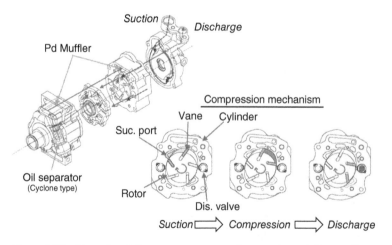

Figure 3.39 View of a sliding vane compressor (centered rotor). Source: Courtesy of Valeo.

allows for a better balance of forces on bearings, extending their lifespan. However, the challenge is to machine an accurate geometry of the stator to ensure a good contact at the vane tips and limit the leakages (Choo, 2021). The presence of 2 supply and 2 discharge ports helps to reduce pulsation and associated noise and vibration.

Sliding vane compressors are characterized by a fixed internal built-in volume ratio $r_{v,in}$. The latter is defined as the ratio between the volume of the suction chamber at the end of the suction process (V_s) and the volume of the compression chamber at the end of the compression process. This ratio, typically comprised between 1.6 and 3, is a characteristic of the machine imposed by its geometry. As a result of the fixed built-in volume ratio, the pressure $P_{r,in,cp}$ at the end of the compression process may differ from the pressure $P_{r,ex,cp}$ in the discharge line. If the pressure $P_{r,in,cp}$ in the compression chamber is lower than the pressure $P_{r,ex,cp}$ in the discharge line (Figure 3.40b), some fluid flows back into the compression chamber until pressures equalize (under-compression phenomenon). If the pressure $P_{r,in,cp}$ in the compression chamber is higher than the pressure $P_{r,ex,cp}$ in the discharge line (Figure 3.40c), some fluid has to leave the chamber for pressures to equalize (over-compression phenomenon). Neither under- nor over-compression losses occur if the pressures in the discharge line and in the compression chamber are equal (Figure 3.40a). In the theoretical indicator diagram shown in Figure 3.40, both pressure equalization irreversible processes happen instantaneously before the discharge process begins. The mismatch between the internal pressure ratio (ratio of the pressure at the end of the compression process to the pressure at the beginning of the compression process) and the system pressure ratio (ratio of the pressure in the discharge line to the pressure in the suction line) is a major source of decrease in performance.

As a consequence of under- and over-compression losses, the evolution of the isentropic effectiveness of the vane compressor with the system pressure ratio shows a maximum. This will be shown in Section 3.5.3.7.

3.5.3.5 Scroll Compressors

The scroll compressor is the most recently used compressor in automotive applications. Scroll compressors are composed of a fixed scroll and of an orbiting scroll (Figure 3.41). The latter is the central symmetry of the former. The relative position of both scrolls defines a series of pairs of crescent-shaped chambers. As illustrated in Figure 3.42, the low-pressure fluid enters the machine by filling two peripheral chambers. The suction process extends over one full crankshaft revolution, until the chambers cease to be in communication with the supply line. Suction chambers are then closed off and become compression chambers. With the movement of the orbiting scroll, the two compression chambers evolve towards the center of the machine, and their volume is decreased leading to the increase in the fluid pressure. When the crank angle θ reaches the discharge angle θ_d, the two chambers open to the discharge region, which is in communication with the discharge line. This is the beginning of the discharge process, which extends over one full crankshaft revolution.

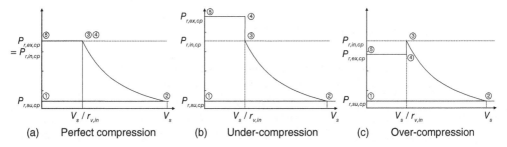

Figure 3.40 Theoretical indicator diagram of compressors with a fixed built-in volume ratio.

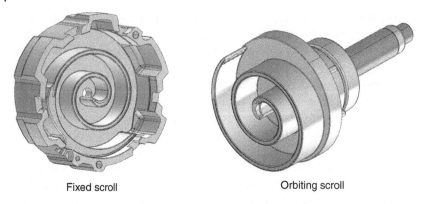

Fixed scroll   Orbiting scroll

Figure 3.41 View of the fixed and orbiting scrolls of an electrically driven scroll compressor. Source: Adapted from Valeo.

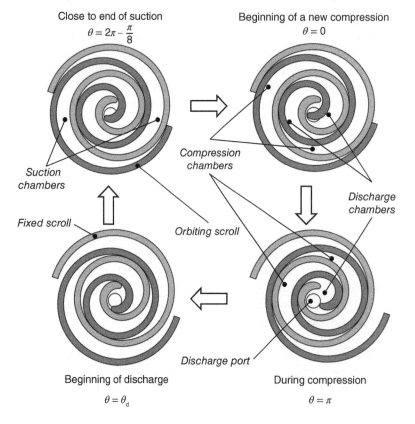

Figure 3.42 Operating principle of a scroll compressor.

Similar to rotary vane compressors, scroll compressors are characterized by a fixed built-in volume ratio, yielding under- and over-compression losses. Under-compression losses can be diminished by introducing a reed valve that covers the discharge port of the scrolls set. Over-compression losses can also be diminished by means of intermediate discharge valves.

Due to their simple design, scroll compressors are a cheaper technology of compressors than piston compressors. Also, they do not allow for displacement control. Hence, mechanical scroll

Table 3.5 Values of working cycle frequency for different compressor technologies.

Compressor technology	Working cycle frequency i [−]
single-acting piston	Number of cylinders
double-acting piston	2 × Number of cylinders
rotary vane	Number of vanes
rotary vane with a centered rotor	2 × Number of vanes
scroll	1

compressors are equipped with clutches (for on/off cycling). On the contrary, the capacity of electric scroll compressors is controlled by varying their speed.

3.5.3.6 Expression of the Compressor Displaced Mass Flow Rate

In the most general case, the volumetric effectiveness of the compressor can be disaggregated into several factors as

$$\varepsilon_v = \frac{\dot{m}_r}{\dot{m}_{r,th}} = \frac{\dot{m}_r}{\dot{m}_{r,in}} \frac{\dot{m}_{r,in}}{\dot{m}_{r,in,th}} \frac{\dot{m}_{r,in,th}}{\dot{m}_{r,th}} = \varepsilon_{v,l} \cdot \varepsilon_{v,PT} \cdot \varepsilon_{v,0} \tag{3.96}$$

That is, the mass flow rate displaced by the compressor is equal to

$$\dot{m}_r = \varepsilon_{v,l} \cdot \varepsilon_{v,PT} \cdot \varepsilon_{v,0} \cdot \dot{m}_{r,th} \tag{3.97}$$

The theoretical mass flow rate $\dot{m}_{r,th}$ displaced by the compressor depends on the compressor displacement V_s, rotational speed N, cycle frequency i (already introduced for ICE), and thermodynamic state of the fluid at the compressor suction. That is,

$$\dot{m}_{r,th} = i\, N \frac{V_s}{v_{r,su,cp}} \tag{3.98}$$

where

i is the working cycle frequency, [−]
N is the rotational speed, [Hz]
V_s is the compressor displacement, [m^3]
$v_{r,su,cp}$ is the specific volume of the refrigerant at the compressor supply, [m^{3}kg^{-1}]

The working cycle frequency i is the number of pressure–volume diagrams described by the fluid per full rotation of the compressor shaft. The working cycle frequency for different compressor technologies is given in Table 3.5. For piston compressor, V_s is defined as the volume of one cylinder.

The factor $\varepsilon_{v,l}$ quantifies the impact of the internal leakages inside the compressor. Assuming that the internal leakage flows connect the compressor discharge to its suction, the mass flow rate $\dot{m}_r$ entering (and leaving) the compressor is related to the internal flow rate $\dot{m}_{r,in}$ by

$$\dot{m}_r = \dot{m}_{r,in} - \dot{m}_{r,l} \tag{3.99}$$

In the latter expression, $\dot{m}_{r,l}$ [kg s^{-1}] is the lumped leakage mass flow rate representing the net mass flow rate from the discharge to the suction.

The factor $\varepsilon_{v,PT}$ accounts for the impact of the pressure drops and heat transfers encountered by the refrigerant during suction and discharge processes. These losses are not represented on the theoretical indicator diagram of Figure 3.34. The factor $\varepsilon_{v,0}$ has been introduced previously for the piston compressor (see Eq. (3.92)).

3.5.3.7 Expression of the Compressor Power

The area of the theoretical indicator diagram of Figure 3.34 and Figure 3.40 is the theoretical internal (or "indicated") work $W_{in,th}$. The actual internal work W_{in} differs from the theoretical one due to pressure losses and heat transfers that are not taken into account in Figure 3.40. The diagram factor (or "internal effectiveness") ε_{in} is defined as the ratio of the theoretical internal work to the actual one.

$$\varepsilon_{in} = \frac{W_{in,th}}{W_{in}} \tag{3.100}$$

The internal power is related to the internal work by

$$\dot{W}_{in} = i \cdot N \cdot W_{in} \tag{3.101}$$

In the case of a piston compressor, the theoretical power can be computed by

$$\dot{W}_{in,th} = \dot{m}_{r,in,th}(h_{r,ex,cp,s} - h_{r,su,cp}) \tag{3.102}$$

In this expression, $h_{r,ex,cp,s}$ is the enthalpy of the refrigerant at the exhaust pressure $P_{r,ex,cp}$ and the supply entropy $s_{r,su,cp}$. The subscript s denotes an isentropic process.

In the case of a scroll and a rotary vane compressor, the previous equation must be modified to account for under- and over-compression losses.

$$\dot{W}_{in,th} = \dot{m}_{r,in,th} \left[(h_{r,in,cp} - h_{r,su,cp}) + \frac{v_{r,su,cp}}{r_{v,in}}(P_{r,ex,cp} - P_{r,in,cp}) \right] \tag{3.103}$$

In this expression, $h_{r,in,cp}$ is the enthalpy of the refrigerant at the intermediate pressure $P_{r,in,cp}$ and the supply entropy $s_{r,su,cp}$. The theoretical isentropic effectiveness can be defined as

$$\varepsilon_{s,th} = \frac{\dot{m}_{r,in,th}(h_{r,ex,cp,s} - h_{r,su,cp})}{\dot{W}_{in,th}} \tag{3.104}$$

Figure 3.43 shows the evolution of the theoretical isentropic effectiveness with the pressure ratio in the case of a compressor characterized by a built-in volume ratio (vane or scroll compressor). This indicator only accounts for under- and over-compression losses.

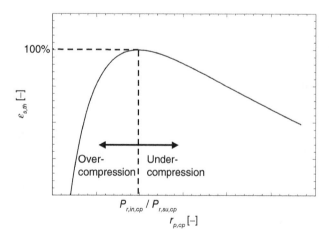

Figure 3.43 Evolution of the theoretical isentropic effectiveness with the pressure ratio for a compressor characterized by a built-in volume ratio.

It can be observed that the theoretical isentropic effectiveness goes through 100% when the discharge pressure $P_{r,ex,cp}$ is equal to the internal pressure $P_{r,in,cp}$.

The overall isentropic effectiveness ε_s of a compressor can be used to compute the actual electrical consumption $\dot{W}_{el,cp}$ of the compressor.

$$\varepsilon_s = \frac{\dot{m}_r \left(h_{r,ex,cp,s} - h_{r,su,cp} \right)}{\dot{W}_{el,cp}} \tag{3.105}$$

This overall isentropic effectiveness is defined as

$$\varepsilon_s = \varepsilon_{v,l} \cdot \varepsilon_{v,PT} \cdot \frac{\dot{W}_{sh,cp}}{\dot{W}_{el,cp}} \cdot \frac{\dot{W}_{in}}{\dot{W}_{sh,cp}} \cdot \frac{\dot{W}_{in,th}}{\dot{W}_{in}} \cdot \frac{\dot{m}_{r,in,th} \left(h_{r,ex,cp,s} - h_{r,su,cp} \right)}{\dot{W}_{in,th}}$$

$$\Leftrightarrow \varepsilon_s = \varepsilon_{v,l} \cdot \varepsilon_{v,PT} \cdot \eta_{el} \cdot \eta_m \cdot \varepsilon_{in} \cdot \varepsilon_{s,th} \tag{3.106}$$

where

η_{el} is the electrical efficiency of the electric motor, [−]
η_m is the mechanical efficiency of the compressor, [−]

The previous equation makes appear the different sources of losses inside a compressor. In the case of a mechanical compressor, the electrical efficiency is equal to 1.

The mechanical losses in the compressor and the electrical losses in the electric motor actually contribute to increase the consumption of the compressor.

$$\dot{W}_{el,cp} = \dot{W}_{sh,cp} + \dot{W}_{loss,el} = \dot{W}_{in} + \dot{W}_{loss,m} + \dot{W}_{loss,el} \tag{3.107}$$

In the case of electric compressors, assuming no heat losses to the ambient, the discharge enthalpy of the refrigerant can be expressed by means of the energy balance on the compressor. That is,

$$\dot{W}_{el,cp} = \dot{m}_r \left(h_{r,ex,cp} - h_{r,su,cp} \right) \tag{3.108}$$

Example 3.4   *Isentropic Effectiveness of a Scroll Compressor*
A scroll compressor operating with R1234yf between 3 bar and 19 bar shows an isentropic effectiveness of 60%. Its built-in volume ratio is 2.0. Estimate its isentropic effectiveness if the discharge pressure is increased up to 21 bar. Assume that the superheat at the compressor supply is 5 K in both cases.

Analysis

There are 2 different operating points. With the first operating point, the theoretical isentropic effectiveness $\varepsilon_{s,th,1}$ can be calculated. Knowing the overall isentropic effectiveness $\varepsilon_{s,1}$, it is possible to identifiy the product of factors $\varepsilon_{v,l} \cdot \varepsilon_{v,PT} \cdot \eta_{el} \cdot \eta_m \cdot \varepsilon_{in} = K$. If it is assumed that K is maintained identical for the second operating point, $\varepsilon_{s,2}$ can be estimated from $\varepsilon_{s,th,2}$.

Solution

```
"1. INPUTS"
fluid$='R1234yf'
P_r_su_cp_1=3e5 [Pa] "compressor supply pressure on operating
   point 1"
P_r_ex_cp_1=19e5 [Pa] "compressor exhaust pressure on operating
   point 1"
```

```
epsilon_s_1=0,6  [-] "compressor isentropic effectiveness on
    operating point 1"
DELTAT_r_su_cp=5 [K] "superheat at the compressor supply"
P_r_su_cp_2=3e5 [Pa] "compressor supply pressure on operating
    point 2"
P_r_ex_cp_2=21e5 [Pa] "compressor exhaust pressure on operating
    point 2"

"2. OUTPUTS"
"epsilon_s_2=?? [-]" "compressor isentropic effectiveness on
    operating point 2"

"3. PARAMETERS"
r_v_in=2 [-] "built-in volume ratio"

"4. MODEL"
"4.1. Operating point 1"
T_r_su_cp_1=T_r_sat_su_cp_1+DELTAT_r_su_cp "compressor supply
    temperature"
h_r_su_cp_1=enthalpy(fluid$;P=P_r_su_cp_1;T=T_r_su_cp_1)
T_r_sat_su_cp_1=temperature(fluid$;P=P_r_su_cp_1;x=1)
s_r_su_cp_1=entropy(fluid$;P=P_r_su_cp_1;T=T_r_su_cp_1)
v_r_su_cp_1=volume(fluid$;P=P_r_su_cp_1;T=T_r_su_cp_1)
v_r_in_cp_1=v_r_su_cp_1/r_v_in "specific volume at the end of
    isentropic compression"
P_r_in_cp_1=pressure(fluid$;v=v_r_in_cp_1;s=s_r_su_cp_1)
h_r_in_cp_1=enthalpy(fluid$;v=v_r_in_cp_1;s=s_r_su_cp_1)

h_r_ex_cp_1_s=enthalpy(fluid$;P=P_r_ex_cp_1;s=s_r_su_cp_1) "exhaust
    enthalpy if isentropic compression"
w_s_1=h_r_ex_cp_1_s-h_r_su_cp_1
w_1=h_r_in_cp_1-h_r_su_cp_1+v_r_in_cp_1*(P_r_ex_cp_1-P_r_in_cp_1)
epsilon_s_th_1=w_s_1/w_1 "theoretical isentropic effectiveness"
K=epsilon_s_1/epsilon_s_th_1

"4.2. Operating point 2"
T_r_su_cp_2=T_r_sat_su_cp_2+DELTAT_r_su_cp "compressor supply
    temperature"
h_r_su_cp_2=enthalpy(fluid$;P=P_r_su_cp_2;T=T_r_su_cp_2)
T_r_sat_su_cp_2=temperature(fluid$;P=P_r_su_cp_2;x=1)
s_r_su_cp_2=entropy(fluid$;P=P_r_su_cp_2;T=T_r_su_cp_2)
v_r_su_cp_2=volume(fluid$;P=P_r_su_cp_2;T=T_r_su_cp_2)
v_r_in_cp_2=v_r_su_cp_2/r_v_in "specific volume at the end of
    isentropic compression"
P_r_in_cp_2=pressure(fluid$;v=v_r_in_cp_2;s=s_r_su_cp_2)
h_r_in_cp_2=enthalpy(fluid$;v=v_r_in_cp_2;s=s_r_su_cp_2)
```

```
h_r_ex_cp_2_s=enthalpy(fluid$;P=P_r_ex_cp_2;s=s_r_su_cp_2) "exhaust
   enthalpy if isentropic compression"
w_s_2=h_r_ex_cp_2_s-h_r_su_cp_2
w_2=h_r_in_cp_2-h_r_su_cp_2+v_r_in_cp_2*(P_r_ex_cp_2-P_r_in_cp_2)
epsilon_s_th_2=w_s_2/w_2 "theoretical isentropic effectiveness"
epsilon_s_2=K*epsilon_s_th_2 "K is supposed constant"
```

Results

```
epsilon_s_2=0,5633 [-]
epsilon_s_th_1=0,6304 [-]
epsilon_s_th_2=0,5918 [-]
K=0,9518 [-]
```

3.5.3.8 Oil Circulation Ratio

Mobile A/C compressors are lubricated by oil. To limit the amount of oil circulating from the compressor through the rest of the refrigerant circuit, an oil separator is embedded inside the compressor. The mass fraction of oil in the mixture at the outlet of the compressor is the oil circulating ratio (OCR). It is defined as

$$OCR = \frac{\dot{m}_{oil}}{\dot{m}_r + \dot{m}_{oil}} \tag{3.109}$$

OCR of a few percent are typically achieved.

3.5.4 Evaporator

In an A/C loop, the purpose of the evaporator is to produce a cooling effect consisting of decreasing the temperature of a secondary fluid, typically air. However, in some applications (such as battery cooling), the evaporator can cool down a liquid coolant (glycol water). As it will be shown hereunder, when humid air is cooled down, water vapor contained in the air is likely to condense. In that specific case, the evaporator decreases not only the temperature but also the water content in the air. As explained more in detail in Chapter 4, in heat pump operation, the purpose of the evaporator is to extract heat typically from the outdoor air for cabin heating purpose.

3.5.4.1 Air-Heated Evaporators

In most cases, the secondary fluid is humid air. As mentioned hereunder, humid air is cooled down and possibly dehumidified in the evaporator. Air dehumidification is necessary for windshield demisting and inner surface defrosting. Dehumidification is achieved by decreasing the air temperature underneath its dew point. The energy balance on the air-side can be written as (neglecting the enthalpy flow rate of the condensates)

$$\dot{Q}_{ev} = \dot{m}_{a,ev}(h_{a,su,ev} - h_{a,ex,ev}) \tag{3.110}$$

The cooling capacity is the sum of the sensible cooling capacity and latent cooling capacity. The sensible cooling capacity is a function of only the temperature drop:

$$\dot{Q}_{ev,sens} = \dot{m}_{a,ev} \cdot c_{p,ha} \cdot (T_{a,su,ev} - T_{a,ex,ev}) \tag{3.111}$$

As indicated previously, the specific heat at constant pressure of humid air $c_{p,ha}$ can be computed as

$$c_{p,ha} = c_{p,a} + \omega_{su,ev}\, c_{p,w} \qquad (3.112)$$

where

$c_{p,a}$ is the constant pressure specific heat of pure air, ~1005 [J kg^{-1}K^{-1}]
$\omega_{su,ev}$ is specific humidity of air at the evaporator supply, [kg kg^{-1}]
$c_{p,w}$ is the constant pressure specific heat of water vapor, ~1820 [J kg^{-1}K^{-1}]

The latent capacity is given by

$$\dot{Q}_{ev,lat} = \dot{Q}_{ev} - \dot{Q}_{ev,sens} \qquad (3.113)$$

The mass flow rate of condensates leaving the evaporator is given by:

$$\dot{m}_{cond} = \dot{m}_{a,ev}\,(\omega_{su,ev} - \omega_{ex,ev}) \qquad (3.114)$$

As illustrated in Figure 3.19, the air-heated evaporator is integrated inside the HVAC unit. It is slightly tilted to allow for condensate drainage.

The following constraints must be considered when designing the air-heated evaporator:

– Limited pressure drops on the air-side to limit the blower consumption and on the refrigerant side.
– Resistance to corrosion (presence of humidity and pollutants, such as chloride and sulfur).
– Absence of odors to ensure cabin comfort.
– Appropriate draining of condensates.
– Limited noise.
– No projection of liquid inside the cabin.

First evaporators were of round tube-fin design. Today, two major technologies of air-heated evaporator are considered: flat tube-fin and plate-fin. In the latter two designs, inner fins are installed on the refrigerant side to improve the performance. For a passenger car, the thermal capacity of this heat exchanger is of the order of 5–6 kW. Figure 3.44 shows a practical realization of a plate-fin evaporator. In this example, circulation of the refrigerant in the heat exchanger occurs in six passes. Other designs are characterized by four passes.

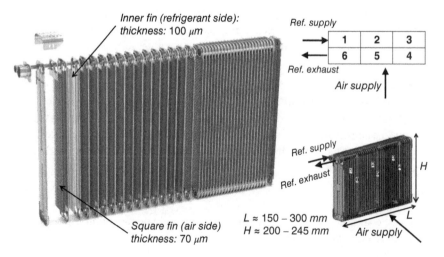

Figure 3.44 Practical realization of a plate-fin evaporator. Source: Courtesy of Valeo.

Evaporators are made up of aluminum with sometimes special hydrophilic coatings to ease the drainage of condensate and limit sources of odors.

The operation of the A/C loop must be adapted in such a way to prevent frosting of the evaporator. This will be explained more in detail in Section 3.5.9. Frosting appears when the evaporator surface temperature is lower than 0°C. Penalty associated with frost formation is twofold. First, it increases the thermal resistance between the air and the refrigerant. Second, it increases the air pressure drop through the heat exchanger.

It will be explained in Chapter 4 that electric vehicles can be equipped with heat pumps for cabin heating. In such a case, the heat pump evaporator can extract heat from outdoor air and is generally located in the front-end module.

3.5.4.2 Water-Heated Evaporators ("Chillers")

In an electric vehicle, the battery pack can be cooled down by a liquid coolant (see Chapter 4). In that specific case, the liquid coolant (chilled glycol water) can be itself cooled down in a water-heated evaporator called chiller. Cabin cooling could also be ensured by a cooling coil (cooler core) embedded inside the HVAC unit and fed with chilled glycol water produced by a chiller.

3.5.5 Condenser

In an A/C loop, the purpose of the condenser is to cool down, condense, and possibly subcool the high-pressure refrigerant. As explained previously, subcooling is necessary for the proper operation of the expansion valve. As shown in Figure 3.45, three states can exist in the condenser: superheated vapor, two-phase (a mixture of saturated liquid and vapor), and subcooled liquid. The refrigerant enters in the condenser at a temperature typically 10–30 K higher than the saturation temperature. Also, the refrigerant leaves the condenser with a subcooling degree up to 12 K. Condensers can be either air-cooled or water-cooled. As it will be shown later, in a heat pump, the desired effect is the heating production at the condenser.

3.5.5.1 Air-Cooled Condensers

Air-cooled condensers are located in the front-end module. A practical realization is depicted in Figure 3.46. As explained in Chapter 2, the air flow through the condenser is ensured by the ram

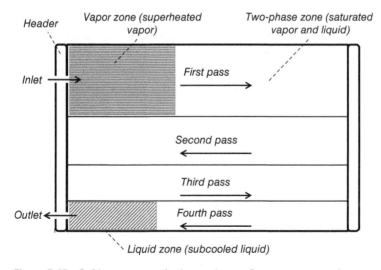

Figure 3.45 Refrigerant states in the condenser: 3-zone representation.

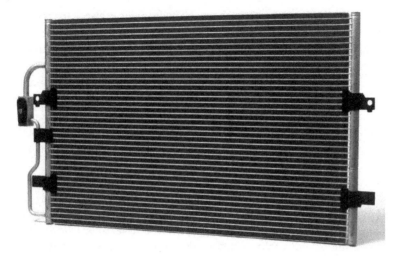

Figure 3.46 Air-cooled tube-fin condenser. Source: Courtesy of Valeo.

effect created by the motion of the vehicle. In idling conditions and also for limited speed of the vehicle, a fan can be activated to ensure a large enough flow rate.

The energy balance on the air-side is expressed as

$$\dot{Q}_{cd} = \dot{m}_{a,cd} \cdot c_{p,a} \cdot (T_{a,ex,cd} - T_{a,su,cd}) \tag{3.115}$$

For passenger car applications, the maximum heating capacity of the condenser is around 15 kW. In idling regime when vehicle is at standstill, the performance of the condenser can be impacted by hot air return. Internal circulation of hot air from the back to the front of the condenser results in an increase in air supply temperature $T_{a,su,cd}$ up to 20 K over ambient temperature. Recirculated air is indeed heated by the radiator and the condenser and other front-end heat exchangers.

Today, most of the air-cooled condensers are flat tube-fin heat exchangers with multi-passes. They show high thermal performance and a reduced internal volume, which allows for decreasing the refrigerant charge. Performance on the air-side is improved by employing louvered fins brazed on the tubes. The pressure drop on the air-side should not be too large to avoid impacting the air flow rate or fan consumption. Also, the air flow through the condenser should be homogeneous.

There are two types of tubes: extruded and folded tubes (Figure 3.47). As illustrated in Figure 3.45, the different passes do not show the same number of tubes: there are more tubes in the first pass than in the last one. This is related to decrease in the specific volume of the refrigerant between the supply and the exhaust of the condenser. The feeding of the different tubes of a same pass is ensured by manifolds. On the refrigerant side, the pressure drop $(P_{r,su,cd} - P_{r,ex,cd})$ should also be as small as possible to limit the pressure ratio imposed to the compressor and the consumption of the latter one.

As explained in Section 3.5.7, the air-cooled condenser can integrate a liquid receiver before the last refrigerant pass. The latter is used to subcool the liquid refrigerant.

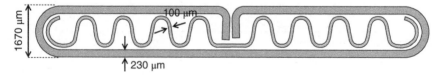

Figure 3.47 Cross-section of a folded tube used in a condenser with common dimensions. Source: Courtesy of Valeo.

3.5.5.2 Water-Cooled Condensers

Condensers can be cooled down by a liquid coolant, typically an aqueous solution of glycol ("glycol water"). In such a configuration, the condenser is connected to a low-temperature radiator by means of a low-temperature water loop. This radiator can be shared among different components cooled by the same coolant loop. Water-cooled condensers are more compact than air-cooled condensers. Since they are not located in the front-end module, more room is available for air-cooled heat exchangers, as long as the low-temperature radiator of the water-cooled condenser ensures other cooling functionalities.

Water-cooled condensers are also used in heat pumps to produce hot water. This will be explained in Chapter 4.

Most of water-cooled condensers are corrugated plate heat exchangers. As illustrated in Figure 3.48 the condenser can integrate a receiver–drier. The functionality of this component will be explained in Section 3.5.7. Such a configuration of the condenser allows for subcooling the refrigerant. In the configuration of Figure 3.48, some plates of the heat exchanger are dedicated to subcooling ("integrated design"). Another configuration ("split configuration") consists of having the subcooler separated from the condensation section.

Figure 3.48 Water-cooled condenser with an integrated receiver-drier. Source: Courtesy of Valeo.

3.5.6 Throttling Device

The purpose of the throttling device is to create the pressure drop between the system high and low pressures. Three major types of expansion valves are employed: the thermostatic expansion valve (TXV), the electronic expansion valve (EXV), and the orifice tube (OT). Research and development works also indicated the potential use of work-producing expansion machines such as volumetric expanders.

Besides creating the pressure drop, TXV and EXV fulfill a second function, which is the control of the superheat degree at the evaporator exhaust. This superheat is defined as the difference between the temperature at the evaporator exhaust and the saturation temperature corresponding to the pressure at the evaporator exhaust.

$$\Delta T_{r,ex,ev} = T_{r,ex,ev} - T_{r,sat,ex,ev} \left(P_{r,ex,ev} \right) \tag{3.116}$$

The superheat control is achieved by adjusting the refrigerant mass flow rate supplied to the evaporator. In both the TXV and EXV, the pressure drop is created by means of a flow restriction. Adjusting the cross-sectional area of this orifice ("valve port") controls the refrigerant mass flow rate flowing through the valve and supplied to the evaporator. The difference between the TXV and the EXV lays in the mechanism used to adapt the valve port cross-sectional area in response to the superheat degree at the evaporator exhaust.

3.5.6.1 Thermostatic Expansion Valve (TXV)

A sectional representation of a block valve TXV is shown in Figure 3.49. The valve has two inlets and two outlets. The inlets are connected to the condenser and evaporator outlets, while the outlets are connected to the evaporator and compressor inlets. The TXV is made up of the main valve body inside which a vertical rod can slide up and down. When this rod slides down, it pushes a ball valve, which creates a passageway ("valve port"). The high-pressure liquid from the condenser is throttled through this port, resulting in a low-pressure two-phase mixture flowing to the evaporator.

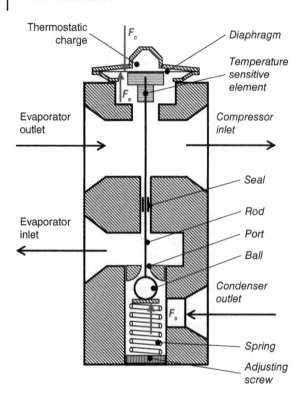

Thermostatic charge — Diaphragm — Temperature sensitive element — Evaporator outlet — Compressor inlet — Evaporator inlet — Seal — Rod — Port — Ball — Condenser outlet — Spring — Adjusting screw

Figure 3.49 Sectional drawing indicating the components of a thermostatic expansion valve (TXV).

The size of the valve port, which is the opening rate of the valve, depends on the position of the rod. The top extremity of the rod is connected to a diaphragm on which several forces are applied:

- An opening force F_c resulting from the pressure of the thermostatic charge in the diaphragm chamber on top of the diaphragm. The thermostatic charge senses the temperature at the outlet of the evaporator by means of the temperature sensitive element. As the evaporator exhaust temperature increases, the thermostatic charge temperature increases, and its pressure increases accordingly. The pressure in the diaphragm chamber is thus a function of the superheat of the vapor. The nature of the thermostatic charge is discussed later.
- A closing force F_e resulting from the pressure underneath the diaphragm. The cavity underneath the diaphragm communicates with the exhaust of the evaporator, so that the pressure at the outlet of the evaporator applies underneath the diaphragm. Such a TXV allows for an external equalization, in contrast to TXVs that sense the evaporator inlet pressure ("internal equalization"). External equalization allows for a more accurate control of the superheat when pressure losses through the evaporator are not negligible.
- A closing force F_s applied on the rod by a spring of constant k [N m^{-1}] ("superheat spring force"), which depends on the valve lift L [mm]. This force, and hence the resulting superheat, can be controlled by adjusting a screw (during the settings of the TXV).

The force balance (Eq. [3.117]) on the diaphragm determines the position of the rod and hence the opening rate of the valve. L_0 indicates the initial compression of the spring.

$$F_c(T_{r,ex,ev}) = F_e(P_{r,ex,ev}) + F_s = F_e(P_{r,ex,ev}) + k(L + L_0) \tag{3.117}$$

The operating principle of a TXV can be explained while it is used along with an evaporator. Let's imagine that the combined system operate in a stable regime. The refrigerant leaves the TXV in two-phase state and flows through the evaporator where it is successively vaporized and slightly superheated. The mass flow rate through the evaporator is such that it allows, for the current evaporator thermal load, to fully vaporize and slightly superheat the refrigerant. For a given volume flow rate displaced by the compressor, this stable operating regime also results in a stable evaporating pressure. If the thermal load applied to the evaporator increases, for an identical refrigerant mass flow rate, the superheat increases. This temperature increase is seen by the thermostatic charge and is turned into an increase of the pressure applied on top of the diaphragm. Consequently, the rod is pushed down, and the size of the valve orifice increases yielding a larger mass flow rate supplied to the evaporator and a reduction of the superheat. The system will find another equilibrium point with a larger evaporating pressure and a slightly larger superheat (due to the stiffness of the membrane and spring). Starting from the same stable operating condition, if the heat load applied to the evaporator decreases, the superheat decreases and the valve orifice closes to supply a lower mass flow rate through the evaporator, which results in a lower evaporating pressure and slightly lower superheat.

The characteristic curve of a TXV is depicted in Figure 3.50. It shows the evolution of the valve capacity (actually, the cooling capacity of the associated evaporator) with the superheat for given evaporating and condensing temperatures (Huelle, 1972). Instead of the cooling capacity, the y-axis can indicate the opening rate (or valve lift). The characteristic curve makes appear the static superheat ΔT_{ss}. For superheats lower than the static superheat, the pressure in the diaphragm head is not large enough to counterbalance the pressure at the evaporator outlet and the spring pressure and the valve remains closed. For superheats larger than the static superheat, the valve opens. As explained previously, the superheat increases with the cooling capacity. The operating superheat ΔT_{oper} is the actual superheat including the static superheat. The opening superheat ΔT_{open} corresponds to the additional superheat with respect to the static superheat. The maximum cooling capacity $\dot{Q}_{ev,max}$ corresponds to the full opening of the valve. This maximum cooling capacity is generally 10–40% larger than the rated cooling capacity $\dot{Q}_{ev,rated}$, hence providing a reserve of capacity (Chapter 45 of the 2002 ASHRAE Refrigeration handbook [ASHRAE, 2002]). The valve characteristic depends on the adjustment of the spring. From factory settings, compressing the spring shifts the valve characteristics to the right. The capacity of the TXV (also called tonnage) must be adapted to the cooling capacity of the A/C system.

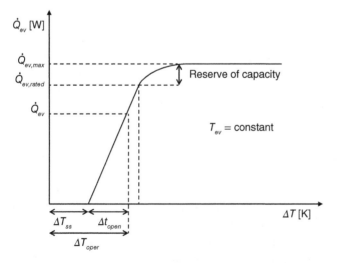

Figure 3.50 Characteristic curve of a thermostatic expansion valve.

Figure 3.51 Coupling between the evaporator and thermostatic expansion valve. Source: Courtesy of Valeo.

Since the block valve is equipped with an external equalization system, which requires a measurement of the evaporator outlet pressure, it comes along with the evaporator and is located in the HVAC module (Figure 3.51).

For a proper operation of the TXV, the latter must be supplied with subcooled liquid. Supplying the valve with refrigerant in the two-phase state could yield a blockage of the valve port by a gas bubble.

In contrast with the expansion valve having a remote temperature-sensing bulb connected to the valve by means of a capillary tube, block valves are bulbless and integrate the temperature-sensing function (ASHRAE, 2002). As explained above, the diaphragm head is filled with a thermostatic charge that applies a pressure on the diaphragm. This pressure varies with the temperature "read" at the outlet of the evaporator. Under normal operating conditions, the diaphragm chamber contains a mixture of saturated liquid and vapor. The saturating pressure of the mixture increases with its temperature. The block valve is gas-charged, meaning that the diaphragm head contains a limited amount of refrigerant liquid that totally vaporizes for temperatures higher than a given threshold. Over this threshold, an increase of the temperature yields almost no increase of the pressure. Gas-charged valves allow operating suction pressure to be limited (MOP: Maximum Operating Suction Pressure) and contribute to limit the ingestion of liquid by the compressor during starting (Chapter 45 of the 2002 ASHRAE Refrigeration Handbook (ASHRAE, 2002)). Limiting the maximum operating suction pressure prevents motor overloading (Wang, 2001).

The valve characteristics can also be represented as a curve showing the pressure at the outlet of the evaporator as a function of the temperature at the outlet of the evaporator (pressure and temperature seen by the valve) for a given valve opening rate. The TXV suppliers give the maximum pressures at the evaporator outlet for temperatures at the evaporator outlet of 0 and 10°C, respectively (Monforte, 2001). These pressures correspond to the minimum opening rate of the valve ("closing position"). The curve corresponding to the minimum opening can be determined by solving Eq. (3.117) when imposing the valve lift L to be minimum (Gillet, 2018). Above this curve, the valve is totally closed. Under this curve, the valve opening rate increases. The valve characteristic makes also appear the MOP temperature threshold: if the temperature "seen" by the bulb continues increasing, the pressure at the compressor suction does not increase anymore.

The fluid inside the diaphragm chamber could be the same as the refrigerant (parallel charge, as shown in Figure 3.52) or different (cross charge). Using a cross charge allows to get a more stable superheat with a variable capacity compressor.

Figure 3.52 Parallel charge-type valve characteristic. Source: Adapted from Monforte (2001).

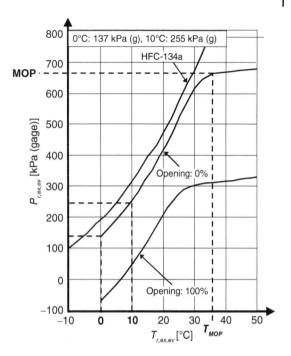

The mass flow rate through the valve depends on the refrigerant state at the valve supply (density and pressure) as well as the valve exhaust pressure. The mass flow rate can be approximated by

$$\dot{m}_{r,TXV} = C_d \, A_{port} \, \sqrt{2 \, \rho_{r,su,TXV} (P_{r,su,TXV} - P_{r,ex,TXV})} \tag{3.118}$$

where

C_d is the valve discharge coefficient, $[-]$

A_{port} is the valve port cross-sectional area, $[\mathrm{m}^2]$

$\rho_{r,su,TXV}$ is the refrigerant density at the valve supply, $[\mathrm{kg \, m}^{-3}]$

$P_{r,su,TXV}$ is the refrigerant pressure at the valve supply, $[\mathrm{Pa}]$

$P_{r,ex,TXV}$ is the refrigerant pressure at the valve exhaust, $[\mathrm{Pa}]$

The discharge coefficient varies with the operating conditions. The valve port cross-sectional area depends on the valve lift L [mm], which is a function of $P_{r,ex,ev}$ and $T_{r,ex,ev}$.

Figure 3.53 shows the 4 quadrants of a typical 4-quadrant representation of a cross-charge TXV. This information can be used to establish a thermostatic valve model, as explained by Gillet (2018). The curves of quadrant Q2 allow to establish the relationship between the valve lift L [m] and $P_{r,ex,ev}$ and $T_{r,ex,ev}$. Quadrant Q3 illustrates Eq. (3.118). From the curve of Q3, the relation between the valve port cross-sectional area A_{port} [m^2] and the lift L [m] can be identified.

For a position of the TXV screw, the superheat at the evaporator exhaust results from the coupling between the evaporator and the TXV. More globally, the superheat, the evaporating pressure, and the refrigerant mass flow rate result from the coupling between the evaporator, the TXV and the compressor. This is schematically represented in Figure 3.65, which shows the different interactions between the components.

3.5.6.2 Electronic Expansion Valve (EXV)

Electronic expansion valve use an apparatus that measures both pressure and temperature at the evaporator exhaust to compute the superheat. A motor connected to a controller adapts the orifice

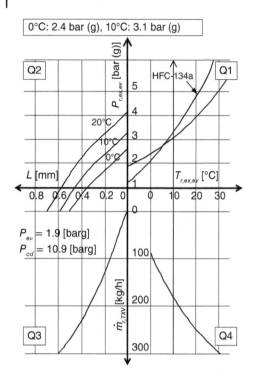

0°C: 2.4 bar (g), 10°C: 3.1 bar (g)

Figure 3.53 Quadrant Q1: Evaporator exhaust temperature as function of the exhaust temperature for the minimum opening rate of the valve. Quadrant Q2: Valve lift as function of the refrigerant pressure and temperature at the exhaust of the evaporator. Quadrant Q3: Refrigerant mass flow rate as function of the valve lift (for given valve supply and exhaust pressures). Quadrant Q4: Refrigerant mass flow rate as function of the evaporator exhaust temperature.

port size to bring the superheat close to a given set point. EXVs allow for a more accurate control of the superheat on a larger operating range of the A/C unit. In CO_2 transcritical cycles, EXVs are also used to control the high pressure inside the gas cooler.

3.5.6.3 Orifice Tube (OT)

An orifice tube (OT) is located inside the line between the condenser exhaust and evaporator supply. Inside the OT, refrigerant enters into an internal tube of smaller diameter. This restriction creates a pressure drop. Typically, the internal tube diameter is of fixed size between 1 and 2 mm, and the length over diameter ratio ranges from 21 to 35 (Chapter 45 of the 2002 ASHRAE handbook of refrigeration, 2002 (ASHRAE, 2002)). The OT also comprises filters at its inlet and outlet.

Because of the fixed internal tube diameter, orifice tubes do not offer a control of the superheat at the evaporator exhaust. Hence, they are commonly used in conjunction with an accumulator, placed between the evaporator and the compressor. The functionalities of the accumulator are described latter. Actually, the OT is sized on the highest cooling capacity. In part load, the evaporator is flooded.

The OT must be correctly sized. Too small OT could yield excessive condensing pressures, while too large OT could yield too small subcooling.

Variable orifice tubes (VOTs) have an orifice size that varies with either the inlet temperature or the inlet pressure.

3.5.7 Receiver, Accumulator, Drier, and Filter

Besides the major components that have been described in the previous sections (compressor, evaporator, throttling device, and condenser), the A/C loop comprises secondary components that are necessary to ensure its functionality and improve its performance.

Among these components are the receiver, the accumulator, the drier, and the filter. The A/C loop also comprises sensors and switches used for the control (see section 3.5.10 about cabin climate control). Receivers and accumulators are metal cylinders filled with both liquid and vapor refrigerant. The A/C loop is equipped with either a receiver or an accumulator. They share some functions:

– The receiver/accumulator stores the mass of refrigerant that is not circulating through the A/C loop when the operating conditions vary (part load operation).
– The receiver/accumulator builds a reserve of refrigerant to compensate for leakages.
– The receiver/accumulator removes impurities and debris from the refrigerant by means of filters. These particles may damage the compressor and clog the expansion valve.
– The receiver/accumulator removes moisture from the refrigerant by means of desiccant material. Moisture can be trapped inside the A/C loop during its construction or servicing. Moisture inside the A/C loop can contribute to internal corrosion (Daly, 2006). Actually, in presence of water, oil degrades at high temperature, leading to the formation of acids and risks of corrosion. Water freezing can also cause a blockage of the expansion valve.

However, the receiver and the accumulator serve other different specific functions and are characterized by different constructions and locations inside the A/C loop.

3.5.7.1 In-line Receiver

A receiver is used in association with a thermostatic expansion valve. It is either located on the piping between the outlet of the condenser and the inlet of the expansion valve (*in-line configuration*) or it is directly attached to the condenser (*cartridge or integrated configuration*). The position of the receiver inside the A/C unit is shown in Figure 3.54.

The main function of the in-line receiver is to ensure that the expansion valve is fed with liquid and not gas.

Figure 3.55 details the internal structure of an in-line receiver drier. The latter comprises an inlet port connected to the condenser outlet and an outlet port connected to the expansion valve inlet.

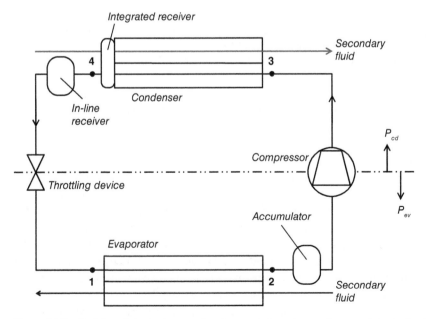

Figure 3.54 Position of the in-line receiver, integrated receiver and accumulator in the A/C loop.

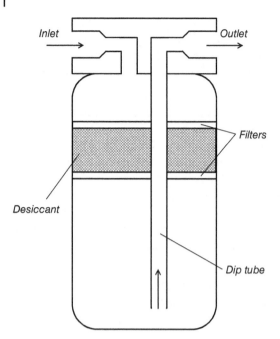

Figure 3.55 Internal structure of a receiver.

The refrigerant (in liquid state or in two-phase state during transient operation) enters into the receiver by the inlet port. By gravity, the liquid phase falls in the bottom of the receiver. It passes through filters and desiccant material, which remove particles and moisture from the refrigerant. A straight dip tube, connected to the outlet port, picks up the liquid refrigerant from the bottom of the receiver to feed the expansion valve. Receivers can be equipped with a sight glass and pressure switch (to limit the high pressure inside the system).

Provided that the receiver contains a mixture of vapor and liquid (which means that the receiver is not overflowed), the refrigerant leaves the receiver in the saturated liquid state, with nearly zero subcooling (Yamanaka et al., 1997). A given level of liquid (corresponding to a mass of refrigerant around 150 g) is necessary inside the receiver to prevent gas to leave the receiver (Yamanaka et al., 1997). In the steady-state regime, the refrigerant entering the receiver is also in the saturated liquid state. However, in transient conditions, the refrigerant could enter the receiver in the two-phase state.

To create subcooling at the outlet of the receiver, the latter must be overflowed. In that particular case, a portion of the condenser downstream the receiver (corresponding to the last pass) is flooded with liquid, which is therefore subcooled.

3.5.7.2 Integrated Receiver

An integrated receiver (sometimes called *modulator*) is a receiver that is integrated with the condenser. As shown in the example of configuration given in Figure 3.56, the receiver is associated to one of the headers of a condenser having four passes. On the refrigerant circuit, the receiver is located between the third and fourth pass. The refrigerant enters the condenser in the superheated vapor phase. The three first passes allow for the cooling of the vapor and its condensation. In normal operation, the receiver is partially filled with liquid, and the fourth pass is flooded with liquid. In the steady-state regime, the refrigerant enters in the saturated liquid state into the receiver. The saturated liquid leaves the receiver and is subcooled in the fourth pass. Refrigerant in the subcooled liquid state leaves the condenser and is routed toward the expansion valve. Such configurations allow for forced subcooling to be achieved.

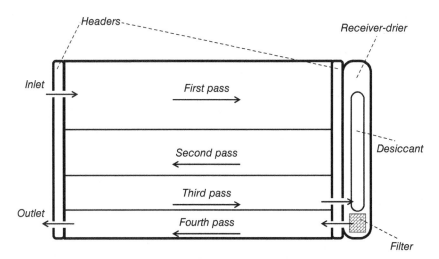

Figure 3.56 Condenser with an integrated receiver–drier.

Increasing the subcooling increases the cooling capacity of the system. Actually increasing the subcooling decreases the enthalpy at the outlet of the condenser, which is approximately the enthalpy $h_{r,su,ev}$ at the evaporator inlet. Consequently, the enthalpy difference $\Delta h_{r,ev} = (h_{r,ex,ev} - h_{r,su,ev})$ across the evaporator increases, leading to a cooling effect increase.

In the meantime, the condensing pressure increases, since a smaller area of the heat exchanger is dedicated to the condensation. The increase in the pressure tends to increase the compressor specific work. This pressure increase also tends to increase the enthalpy at the condenser outlet and hence decrease the enthalpy difference $\Delta h_{r,ev}$ across the evaporator (Pomme, 1999). An increase in the condensing pressure also yields a lower refrigerant mass flow rate. This is especially the case with piston compressors, the volumetric performance of which is highly sensitive to the pressure ratio (see Figure 3.35). Hence, the subcooling value that maximizes the enthalpy difference $\Delta h_{r,ev}$ is not the value that maximizes the cooling capacity $\dot{Q}_{ev}$ (Pomme, 1999).

As a consequence of these different antagonistic effects, there exist optimal subcooling values that maximize the coefficient of performance and/or the cooling capacity of the A/C system. The antagonistic effects of the subcooling on the cooling effect and compressor specific work, which yield an optimal value of the subcooling that maximizes the COP of the A/C system, have been described by Pottker and Hrnjak (2012a). These authors show that the optimal subcooling value and the improvement of the A/C system COP increase with the difference between the air and the refrigerant temperatures in the condenser (i.e. the A/C cooling load for a given condenser size) (Pottker and Hrnjak, 2012b).

As mentioned previously, in normal operation, the last pass is entirely filled with liquid, and the receiver is partially filled with liquid. However, other situations may occur in the condenser and receiver. The evolution of the liquid zone and subcooling at the outlet of the condenser have been described in detail by Cola et al. (2016). This explanation is summarized in the following paragraph related to Figure 3.57 and Figure 3.58. Figure 3.57 shows the position of the liquid zones inside the condenser and receiver. Figure 3.58 shows the evolution of the subcooling at the inlet of the expansion valve with the charge of refrigerant inside the A/C loop. The behaviors observed with integrated receiver and in-line receiver are compared. In the case of an integrated receiver, the inlet of the expansion valve is the outlet of the condenser, while in the case of an in-line receiver, it is the outlet of the receiver.

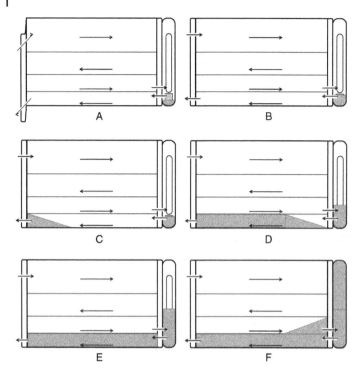

Figure 3.57 Liquid zones inside the condenser-receiver assembly for different charges of refrigerant inside the A/C loop. Source: Adapted from Cola et al. (2016).

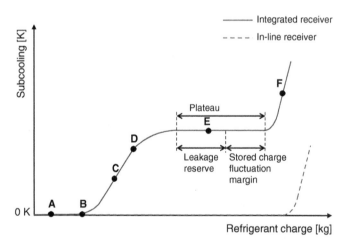

Figure 3.58 Evolution of the subcooling with the charge of refrigerant inside the A/C loop. Source: Adapted from Cola et al. (2016) and Yamanaka et al. (1997)).

In Figure 3.58, moving from the left to the right (situation A) on the horizontal line corresponding to no-subcooling at the outlet of the condenser, the receiver is progressively filled with liquid until the liquid level reaches the upper edge of the outlet port of the receiver (situation B). Along the first part of the ramp, an additional mass of refrigerant introduced inside the system contributes to flood the last pass of the condenser leading to an increase of the subcooling (situation C). The level of liquid inside the receiver is not increased. At the end of the ramp, the last pass is almost completely flooded, and part of additional refrigerant contributes to increase the level of liquid inside the receiver (situation D). The slope of the ramp is decreasing until reaching a "plateau".

In the plateau region (situation E), the fourth pass is completely flooded, and all the additional charge contributes to increase the liquid level inside the receiver. At the right-hand extremity of the plateau, the receiver is entirely filled with liquid, and the third pass starts to be flooded. As a consequence, the subcooling starts to increase again and the condensing pressure as well (situation F). Under optimal operating conditions, the A/C system should be located on the plateau. The sizing of the receiver must allow to compensate for refrigerant leakages and fluctuations of the stored quantity of refrigerant when the operating conditions of the A/C system vary (Abraham et al., 2006).

As proposed and explained by Yamanaka et al. (1997), Figure 3.58 also compares the evolution of the subcooling with the charge of refrigerant inside the system for an in-line receiver and an integrated receiver. In the case of the in-line configuration, the system is stable in the no-subcooling region. With an integrated receiver, the system is stable in the region of the plateau.

Because of the plateau (of approximately 90 g), A/C systems with integrated receivers are less sensitive to refrigerant leakages than those with in-line receivers. They maintain good cooling performance and correct operation of the thermostatic expansion valve for a longer time.

Integrated receivers show some other advantages over in-line receivers than increasing the cooling capacity and/or COP. First, they allow for a reduction of the refrigerant charge inside the system even if there is a small increase in the refrigerant charge inside the condenser to the sub-cooling section (Yamanaka et al., 1997). Second, the cost and weight of the system are reduced. Actually, the system is more compact because of the reduced number of connecting pipes.

As mentioned previously, forced subcooling configuration can be used with a water-cooled condenser. In that case, standard receiver–driers are used and assembled to a condenser and a subcooler. Such an arrangement is shown in Figure 3.48.

3.5.7.3 Accumulator

An accumulator is typically used in conjunction with an orifice tube. It is mounted on the line between the evaporator outlet and the compressor inlet. Its main function is to protect the compressor by ensuring that the latter is fed with vapor refrigerant and not liquid refrigerant.

The internal structure of an accumulator is illustrated in Figure 3.59. The accumulator comprises an inlet port connected to the evaporator exhaust and an outlet port connected to the compressor

Figure 3.59 Internal structure of an accumulator.

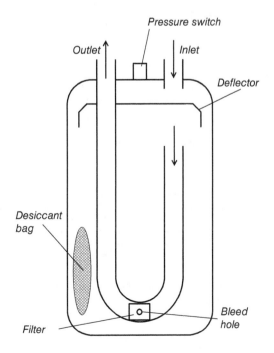

supply. The refrigerant in the two-phase state enters the accumulator through the inlet port. It flows around a deflector. Liquid refrigerant falls in the bottom of the accumulator by gravity. The accumulator comprises a U tube, of which one extremity is connected to the accumulator outlet port and the other extremity is open under the deflector. The latter extremity picks up vapor refrigerant, which flows through the U tube and leaves the accumulator to feed the compressor. At the bottom of the U tube, a bleed hole allows for a small quantity of oil (and liquid refrigerant) to be mixed with vapor refrigerant to ensure the oil return to the compressor. The accumulator also comprises a desiccant bag.

3.5.8 Internal Heat Exchanger

Some A/C systems comprise an additional heat exchanger, which is located between the condenser exhaust and the evaporator exhaust (Figure 3.60). This heat exchanger, called *internal heat exchanger*, allows for the heat transfer from the high-pressure liquid to the low-pressure vapor. Consequently, both the subcooling at the condenser exhaust and the superheating at the evaporator exhaust are increased, respectively, from the points 4 to 4′ and from the points 2 to 2′ in Figure 3.60. The cooling effect is increased because of the larger enthalpy increase in the evaporator. This is visible in Figure 3.61, which compares the P-h diagrams for cycles with and without internal heat exchanger. With the internal heat exchanger, the enthalpy increase in the evaporator is $(h_{r,ex,ev} - h'_{r,su,ev})$ instead of $(h_{r,ex,ev} - h_{r,su,ev})$. Consequently, the performance of the A/C loop is also increased. The major drawback is the increase in the compressor exhaust temperature.

The two major technologies of internal heat exchanger are the plate and the coaxial ones. Coaxial heat exchangers are more often used because they are cheaper, but they are less efficient than plate heat exchangers. Internal heat exchangers are typically used to compensate for COP and cooling capacity loss when switching from R134a to R1234yf.

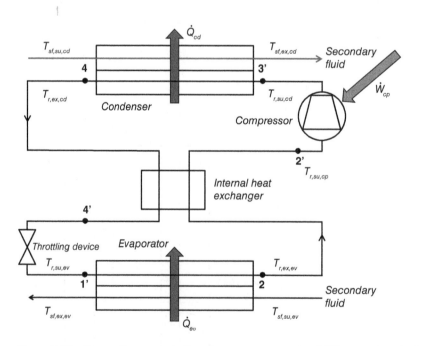

Figure 3.60 Schematic representation of a vapor-compression A/C loop equipped with an internal heat exchanger (IHX).

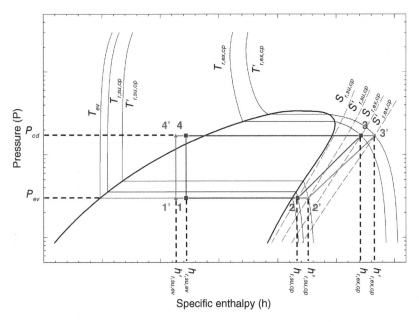

Figure 3.61 Representation of the vapor-compression A/C cycle with an internal heat exchanger in a P-h diagram.

3.5.9 R744 (CO$_2$) as Working Fluid

Carbon dioxide (R744) constitutes an alternative to HFO-1234yf and was already investigated around 2000 (Bullard et al. 2000). R744 is a natural fluid characterized with a GWP of 1. It is a nontoxic, nonflammable fluid that is easily available. Its major drawback is the high operating pressure of the A/C loop. While components of A/C loops designed for R134a are compatible with R1234yf, this is not the case with R744.

Because of its low critical temperature (30.98°C), the R744 cycle is likely to operate in the transcritical regime, i.e. with compressor discharge pressure larger than the critical pressure (73.77 bar).

3.5.9.1 Internal Heat Exchanger with R744

The use of an internal heat exchanger is particularly interesting for cycles operating with R744. Indeed, it not only increases the COP and cooling capacity but also reduces COP-optimal and capacity-optimal compressor discharge pressures and brings them closer to each other, in comparison with a system without an internal heat exchanger, thus making easier the control (Boewe et al., 2001). The use of an internal heat exchanger was less interesting with HFC-134a, because the additional cost and weight outweigh the performance increase. The particular interest for CO$_2$ cycles is related to the higher irreversibility during expansion in the throttling device (Boewe et al., 2001).

3.5.9.2 Gas Cooler

In R744 (CO$_2$) transcritical cycles, there is no condenser but a gas cooler. Actually, the supercritical fluid at the compressor exhaust does not condense but is cooled down to the throttling device supply. Gas can be cooled either by air or by water. Example 3.5 illustrates the operating conditions of a gas cooler.

3.5.9.3 R744 Versus R1234yf

In the cooling mode, R744 yields a lower cooling COP than R1234yf for outdoor temperatures higher than 35°C, but higher cooling COP for outdoor temperatures lower than 25°C. Potentially, the refrigeration cycle architecture could be improved, for instance, by using ejectors, to improve the COP at high outdoor temperatures.

In the heating mode, R1234yf shows low COP and heating capacity for outdoor temperatures lower than −10°C (COP typically lower than 1.5 for temperatures lower than −10°C). In contrast, CO_2 allows maintaining good COP and capacity at temperatures down to −20°C (a COP larger than 2.0 could be achieved for a temperature of −20°C). It also provides high heating capacity, because of high evaporating pressure. The higher evaporating pressure remove the risk of subatmospherical operation, which could weaken the tightness and constitutes a limitation of the operating map of the heat pump. The higher evaporating pressure also contributes to limit the pressure drop in the compressor suction line.

Example 3.5 *Heat Balance Across an Air Gas Cooler*

R744 (CO2) is flowing through an air gas cooler with a mass flow rate of $160 \, kg \, h^{-1}$. The supply pressure of R744 is 125 bar, and the supply and exhaust temperatures are 150 and 50°C, respectively. Refrigerant undergoes a pressure drop of 0.5 bar across the heat exchanger. Air is entering the gas cooler at 20°C and is leaving it at 55°C. Determine (i) the heat transfer rate at the air gas cooler and (ii) the air mass flow rate.

Analysis

Knowing the pressure drop across the gas cooler, we can deduce the exhaust pressure. Using the pressure and the temperature, the enthalpy of the exiting refrigerant can be calculated. The enthalpy of the refrigerant at the supply of the heat exchanger is also calculated based on the pressure and temperature. The heat transfer rate at the air gas cooler is calculated by expressing the heat balance on the R744 side. The air mass flow rate is found by expressing the heat balance on the air-side.

Solution

```
"1. Inputs"
fluid$='R744'
m_dot_r_gc=160/3600 "[kg/s]" "mass flow rate of refrigerant in the
   gas cooler"
P_r_su_gc=125e5 [Pa] "refrigerant pressure at the gas cooler supply"
DELTAP_r_gc=0,5e5 [Pa] "refrigerant pressure drop in the gas cooler"
T_r_su_gc=150 [C] "refrigerant temperature at the gas cooler supply"
T_r_ex_gc=50 [C] "refrigerant temperature at the gas cooler exhaust"
T_a_su_gc=20 [C] "air temperature at the gas cooler supply"
T_a_ex_gc=55 [C] "air temperature at the gas cooler exhaust"

"2. Outputs"
"Q_dot_gc=?? [W]" "heat transfer rate through the gas cooler"
"m_dot_a_gc=?? [kg/s]" "air mass flow rate"

"3. Parameters"
"_"
```

```
"4. Model"
"4.1. Energy balance on refrigerant side"
P_r_ex_gc=P_r_su_gc-DELTAP_r_gc "refrigerant pressure at the gas
    cooler exhaust"
h_r_ex_gc=enthalpy(fluid$;P=P_r_ex_gc;T=T_r_ex_gc)
h_r_su_gc=enthalpy(fluid$;P=P_r_su_gc;T=T_r_su_gc)
Q_dot_gc=m_dot_r_gc*(h_r_su_gc-h_r_ex_gc)

"4.2. Energy balance on air side"
Q_dot_gc=m_dot_a*c_p_a*(T_a_ex_gc-T_a_su_gc)
c_p_a=CP(air;T=T_a_su_gc)
```

Results

```
c_p_a=1004 [J/kg-K]
m_dot_a=0,2885 [kg/s]
Q_dot_gc=10143 [W]
```

3.5.10 Cabin Climate Control

The primary purpose of the cabin climate control is to meet the cooling/heating demands of the cabin. This is achieved by supplying the cabin with cold air or hot air at the desired temperature. The A/C loop can also be used to control the humidity.

The control of the climate in the cabin can be

- *Manual*. The occupant adjusts the pulsed air flow rate and temperature as well as the air distribution through the different vents. The air flow rate can be adjusted manually by means of a multi-speed blower motor. The occupants can switch on and off the A/C system, can adjust air heating through the mixing flap, and can select the recirculation mode. Doors and flaps inside the HVAC unit are controlled via Bowden cables or step motors.
- *Automatic*: An electronic climate control (ECC) or automatic climate control (ACC) system regulates the air flow rate, temperature, distribution and recirculation as function of comfort settings imposed by the occupants and constraints imposed by the vehicle. The control system utilizes information provided by sensors.

As indicated in the example of control configuration of Figure 3.62, in an automatic climate control system, the A/C loop control module interacts with the cabin climate control module, the engine control module and the coolant loop management. For an electrified vehicle with a heat pump, an additional specific heat pump control module can be added. For such vehicles, the control system must allow for the selection of cabin cooling and heating modes and battery cooling (by controlling 2-way and 3.way solenoid valves). These valves will be presented in Chapter 4.

The cabin climate control module determines the cabin supply air temperature set point based on:

- Settings imposed by the cabin occupants (e) through the control panel (a). A second control panel can be used at the rear of the car.

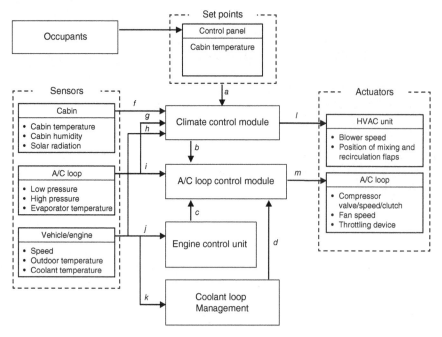

Figure 3.62 Automatic climate control.

- Cabin sensors (f): temperature, humidity (in the indoor rear-view mirror), solar radiation (on the dashboard). The measured temperature is corrected to better represent the temperature felt by the occupants.
- Car data provided by the engine control unit (outdoor temperature [underneath one of the outdoor rear-view mirrors], vehicle speed) (h).

The cabin climate control module also adapts the HVAC unit blower speed and the positions of the mixing and recirculation flaps (l).

The A/C loop control module utilizes information provided by different sensors and switches and operates actuators. These sensors and switches will be described latter. The major functions fulfilled by the A/C loop control module are as follows:

- To provide the desired cooling capacity. The climate control module can impose an evaporating temperature to be reached by the A/C loop (b). This is, for instance, the case with piston compressors equipped with externally controlled valves.
- To control the vapor superheat at the compressor supply. If a thermostatic expansion valve is used, it achieves itself the superheat control.
- To protect the evaporator from frosting.
- To limit the high pressure inside the system.
- To optimize the condenser fan rotational speed.

The coolant loop management system engages the front-end module fan if the coolant temperature reaches a first threshold and switches off the A/C loop if the coolant temperature reaches a second maximum threshold, for instance 126°C (Daly, 2006) (d).

The engine control unit can also switch off the A/C loop if the engine requests the whole power (during acceleration for instance) (c).

In a manual climate control configuration, HVAC unit actuators are directly operated through the control panel (for instance, through Bowden cables).

3.5.10.1 A/C Loop Pressure and Temperature Switches/Sensors

As mentioned hereunder, the climate control module and the A/C loop control module utilize information provided by switches and sensors integrated inside the A/C loop. The following switches and sensors are usually integrated inside the A/C loop:

3.5.10.1.1 Thermostat The thermostat senses the temperature of either the air or the refrigerant at the outlet of the evaporator (Schnubel, 2016). The thermostat cycles the compressor on and off according to a temperature setting. This mechanism allows for the cabin temperature control. For a heat pump (Chapter 4), the sensor is located below the air-heating heat exchanger in the HVAC module to control cabin heating.

3.5.10.1.2 Evaporator Temperature Sensors The evaporator temperature sensor is mounted on the evaporator surface (attached to the fins on the coldest point) or in the air at the outlet of the evaporator. The evaporator temperature sensor disengages the compressor if the surface temperature goes below a given limit $T_{ev,min}$. The compressor is reactivated when the temperature reaches an upper limit $T_{ev,min} + \Delta T$. Typical lower and upper limits are 1 and 2.5°C, respectively (Daly, 2006). The evaporator temperature sensor prevents the evaporator from frosting. In an electronic climate control system, the climate control module uses the information to

- turn the compressor on and off (fixed displacement compressor)
- adapt the compressor displacement (variable displacement compressor)
- adapt the compressor speed (electric compressor)

For heat pumps systems introduced in Chapter 4, a temperature sensor can be located on the pipe surface at the outlet of the outdoor heat exchanger to control frosting/defrosting phases and to prevent the pressure to go underneath a given limit.

3.5.10.1.3 Temperature at Compressor Discharge A surface temperature at the compressor discharge can not only offer an additional security but also be used for controlling heat production in heating mode. It could be located at the supply of the inner condenser in the case of a heat pump (Chapter 4).

3.5.10.1.4 Pressure Switches Different types of pressure switches can be used.

The *low-pressure switch* disengages the compressor if the low pressure goes under a lower limit, which is typically around 1.5 bar. The low-pressure switch is often mounted on the accumulator. Abnormal low pressures can be encountered when refrigerant leaks from the A/C loop.

The *cycling switch* prevents the evaporator from frosting by imposing a lower limit $P_{ev,min}$ on the low-pressure of the system. This limit is generally close to 1.5 bar. When this limit is reached, the compressor is disengaged. It re-engages again when the low pressure reaches $P_{ev,min} + \Delta P$, for instance 2.9 bar (Daly, 2006). The low-pressure and cycle switches can be the same devices.

Similarly, the *high-pressure switch* disengages the compressor if the high pressure exceeds an upper limit, which is around 30–35 bar. This switch is usually mounted on the receiver–drier or at the compressor discharge.

The dual pressure switch associates 2 switches in the same device. The first one, a high-pressure switch, limits the system high pressure. The second one, a condenser fan switch, increases the fan rotational speed if the high pressure exceeds a given limit (for instance, around 18 bar).

The trinary switch, located on the receiver–drier, contains a high-pressure switch, a condenser fan switch, and a low-pressure switch (Daly, 2006). Examples of switch-off and switch-on pressures are given by Daly (2006): $2.0 \pm 0.25/2.15 \pm 0.35$ (low-pressure switch), $12.5 \pm 1.5/16.5 \pm 1.2$ (condenser fan switch), and $30.0 \pm 2.0/24.0 \pm 2.0$ (high-pressure switch).

3.5.10.1.5 *Pressure Sensors* A single pressure sensor is used instead of pressure switches in more modern engines (Daly, 2006).

3.5.10.2 Control of the A/C Loop Cooling Capacity

The cooling capacity provided by the A/C loop has to be adjusted, because the cabin cooling load varies with time. This is due to the variation of outdoor climate, vehicle speed, and transition between pull-down and convergence periods.

Adapting the A/C loop cooling capacity is achieved by adjusting the refrigerant flow rate through the A/C loop. As explained previously, the refrigerant mass flow rate can be related to the compressor displacement V_s, speed N and refrigerant density $\rho_{r,su,cp}$ at the compressor supply by Eq. (3.91). From this equation, considering that the refrigerant density is constant, it can be observed that the refrigerant flow rate can be adjusted by controlling the compressor speed N or displacement V_s. The cooling capacity can also be adjusted by cycling on and off the compressor to achieve an average refrigerant flow rate. These mechanisms are discussed in more detail hereunder.

It should also be mentioned that in the case of a mechanical compressor, for a given cooling capacity, the displacement V_s or the cycling rate has to be adapted if the engine rotational speed, and hence the compressor speed, varies.

3.5.10.2.1 *Compressor Clutch Cycling* The on–off cycling of the compressor, by means of the clutch, is controlled by a measurement of either the system low pressure (low pressure switch) or the system evaporator temperature.

3.5.10.2.2 *Variable Displacement Compressors* As mentioned previously, piston compressors offer a mean of variation of their displacement. The plate angle is adjusted by means of a compressor control valve. This valve is used to adjust the pressure in the crankcase (P_{cr} in Figure 3.37) and consequently the force balance on the plate, resulting in a variation of the tilt angle. The compressor control valve can be internally controlled or externally controlled.

Internal control valves (also named "mechanical" valve, in use since the beginning of the 90s) react to a variation in the compressor suction pressure (P_{su} in Figure 3.37), which tends to vary with the cooling load. If the cabin temperature (and consequently the cooling load) increases, the evaporating pressure tends to increase. In response, the internal valve allows for the compressor displacement to be increased in order to maintain the suction pressure. The larger displacement yields a larger cooling capacity. The opposite behavior happens if the cabin temperature decreases.

External control valves (also named "electronic" valve, in use since the end of the 90s) are solenoid operated electromagnetic valves that are controlled by an external signal from a control unit. Such a valve allows for an accurate control of the suction pressure (evaporating pressure). External control valves allow for an adjustment of the displacement from 0% to 100% (Benouali et al. 2002), thus making the clutch unnecessary (Figure 3.63). They can also cope with fast variation of the engine rotational speed (Benouali et al. 2002). External control allows to operate the compressor at optimal displacement, yielding energy savings (Toyota, 2022).

As shown in Figure 3.64, externally controlled piston compressors allow for a better control of the air temperature at the evaporator outlet in part load conditions. Hence, the reheating of the air in the heater core is unnecessary. Consequently, the energy consumption of the A/C loop is decreased. In an automatic climate control system, the optimal evaporating temperature set point is determined by the climate control module and communicated to the A/C loop control module (arrow (b) in Figure 3.62).

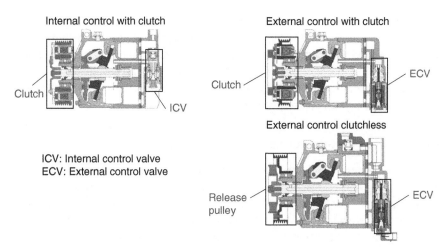

Figure 3.63 Internally and externally controlled displacement valve Source: Courtesy of Valeo.

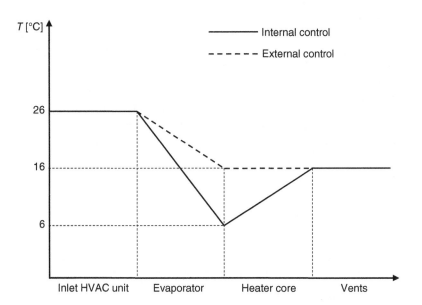

Figure 3.64 Evolution of the air temperature in the HVAC unit: comparison between piston compressors equipped with internal and external control valves.

3.5.10.2.3 Variable Speed Compressors In the case of variable speed compressors, the compressor speed is independent of the engine speed. This offers a convenient means of cooling capacity control. Especially, large enough cooling capacities can be ensured in the case of engine idling conditions.

3.5.10.3 Optimization of the Condenser Fan Speed

The total power consumption of the A/C loop is the sum of the compressor consumption and condenser fan consumption. In the case of an electrical compressor, it gives

$$\dot{W}_{AC} = \dot{W}_{el,cp} + \dot{W}_{el,fan} \tag{3.119}$$

An increase in the fan speed allows for a decrease in the condensing pressure. This could be shown by a simulation model relying on the block diagram presented in Figure 3.66. Simultaneously, the compressor consumption decreases because the system operates under a smaller pressure ratio. However, the fan consumption increases. Hence, there exists an optimal fan speed to minimize the A/C loop electrical consumption. This minimum can be approached or reached with a multiple speed fan or a variable speed fan, respectively.

3.5.11 Interaction Between the Major Components of the A/C Loop

The block diagram representation of Figure 3.65 represents the interaction between the 4 main components of the A/C loop. This representation serves as the basis for the modeling of the A/C loop system. Each component is represented by a box. Left arrows represent model inputs and right arrow models outputs. Model parameters are not represented here. The representation of Figure 3.65 uses the thermostatic expansion valve model described previously.

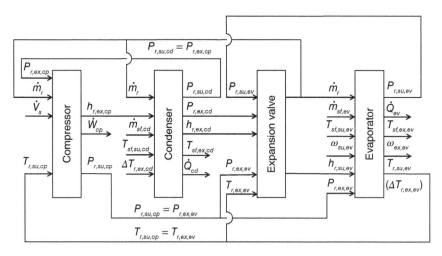

Figure 3.65 Simple model of an A/C loop equipped with a TXV.

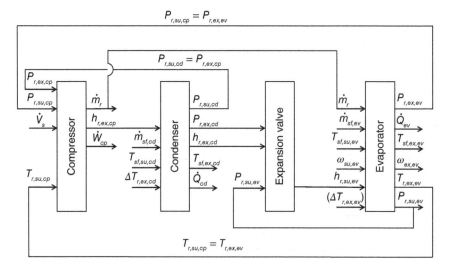

Figure 3.66 Simple model of an A/C loop equipped with a TXV (assumption of constant evaporator superheat).

To simplify the model, it can be assumed that the thermostatic expansion valve is able to impose the superheat set point at the evaporator exhaust, this block diagram can be modified as shown in Figure 3.66. The superheat $\Delta T_{r,ex,ev}$ at the evaporator exhaust becomes an input of the overall A/C loop model.

In Figures 3.65 and 3.66, the subscript *sf* stands for secondary fluid, typically air or glycol water.

Typical operating conditions of an A/C are illustrated through the example hereunder. The model is based on the flowchart of Figure 3.66.

Example 3.6 *Whole A/C Loop*

The following technical information regarding a vehicle air-conditioning system operating with R1234yf has been provided:

Compressor

Refrigerant supply pressure: 3 bar; refrigerant exhaust pressure: 21 bar; refrigerant exhaust temperature: 91°C; volumetric effectiveness: 60%; refrigerant mass flow rate: 190 kg h^{-1}; swept volume: 170 cm^3

Condenser

Air mass flow rate: 2350 kg h^{-1}; air supply temperature: 47°C; refrigerant exhaust subcooling: 9 K

Evaporator

Air mass flow rate: 550 kg h^{-1}; air supply temperature: 47°C; air supply relative humidity: 40%; air exhaust relative humidity: 90%; refrigerant exhaust superheat: 8 K

The following variables need to be computed:

- The compressor rotational speed [rpm]
- The cooling capacity (sensible and latent capacities) [kW]
- The evaporator air exhaust temperature [°C]
- The evaporator sensible heat ratio (SHR) [−]
- The condenser air exhaust temperature [°C]
- The coefficient of performance (COP) of the system [−]
- The conductance AU of the condenser [W K^{-1}]
- The isentropic effectiveness of the compressor [−]

Represent the refrigeration cycle in the pressure-specific enthalpy diagram of the refrigerant.

Analysis

The model of the A/C loop shown in Figure 3.66 is used here in a "parameter-identification" mode. This means that some outputs are imposed and some parameters are identified. If the model was run in the "simulation" mode, all parameters would be imposed, and all outputs would be computed.

In the model developed hereunder, pressure losses in the evaporator and condenser are neglected.

The air relative humidity at the evaporator exhaust is imposed, and the sensible heat ratio is computed. A more advanced model should be able to compute the humidity at the exhaust of the evaporator (and therefore the sensible and latent cooling capacities).

Solution

```
"1. INPUTS"
fluid$='R1234yf'
"1.1. Compressor"
"N_cp=30,1" "[Hz]" "Compressor rotational speed"

"1.2. Condenser"
T_a_su_cd=47 [C] "Air temperature at the condensor supply"
m_dot_a_cd=2350/3600 "[kg/s]" "Air mass flow rate through the
    condenser"
DELTAT_r_ex_cd=5 [K] "Refrigerant subcooling at the condenser
    exhaust"

"1.3. Evaporator"
T_a_su_ev=47 [C] "Air temperature at the evaporator supply"
RH_su_ev=0,4 [-] "Air relative humidity at the evaporator supply"
m_dot_a_ev=550/3600 "[kg/s]" "Air mass flow rate through the
    evaporator"
P_atm=1e5 [Pa] "Atmospheric pressure"
DELTAT_r_ex_ev=8 [K] "Refrigerant superheat at the evaporator
    exhaust"

"2. OUTPUTS"
"2.1. Compressor"
m_dot_r=190/3600 "[kg/s]" "Refrigerant mass flow rate" "imposed"
T_r_ex_cp=91 [C] "Refrigerant temperature at the compressor exhaust"
    "imposed"

"2.2. Condenser"
P_r_ex_cp=21e5 [Pa] "Refrigerant pressure at compressor exhaust"
    "imposed"

"2.3. Evaporator"
"SHR= 0,6691" "[-]" "Sensible heat ratio"
RH_ex_ev=0,9 [-] "Air relative humidity at the evaporator exhaust"
    "imposed"
"Q_dot_ev=3917" "[W]" "Cooling capacity"
P_r_su_cp=3e5 [Pa] "Refrigerant pressure at compressor supply"
    "imposed"

"2.4. Overall"
"COP=1,276"

"3. PARAMETERS"
"3.1. Compressor"
```

```
V_s_cp=170e-6 [m^3] "Compressor displacement"
epsilon_v_cp=0,6 [-] "Compressor volumetric effectiveness"
"epsilon_s_cp=0,6079" "[-]" "Compressor isentropic effectiveness"
   "identified"

"3.2. Condenser"
"AU_cd=379,3 [W/K]" "Overall conductance of the
   condenser" "identified"

"4. MODEL"
"4.1. Compressor »
"4.1.1. Mass flow rate displaced by the compressor"
m_dot_r=epsilon_v_cp*N_cp*V_s_cp/v_r_su_cp "=> N_cp"
v_r_su_cp=volume(fluid$;P=P_r_su_cp;T=T_r_su_cp) "Refrigerant
   specific volume at compressor supply"
T_ev=temperature(fluid$;P=P_r_su_cp;x=1) "Evaporating temperature
   defined at evaporator exhaust"
T_r_su_cp=T_ev+ DELTAT_r_ex_ev "Refrigerant temperature at
   compressor supply"

"4.1.2. Power consumpion"
W_dot_cp=m_dot_r*(h_r_ex_cp-h_r_su_cp) "Assuming the compressor to
   be adiabatic"
h_r_su_cp=enthalpy(fluid$;P=P_r_su_cp;T=T_r_su_cp) "Refrigerant
   specific enthalpy at compressor supply"

"4.1.3 Compressor discharge"
h_r_ex_cp=enthalpy(fluid$;P=P_r_ex_cp;T=T_r_ex_cp) "Refrigerant
   specific enthalpy at compressor exhaust"

"4.1.4 Isentropic effectiveness"
epsilon_s_cp=m_dot_r*(h_r_ex_cp_s-h_r_su_cp)/W_dot_cp
s_r_su_cp=entropy(fluid$;P=P_r_su_cp;T=T_r_su_cp) "Refrigerant
   specific entropy at compressor supply"
h_r_ex_cp_s=enthalpy(fluid$;P=P_r_ex_cp;s=s_r_su_cp) "Refrigerant
   specific enthalpy at compressor exhaust if isentropic
   compression"

"4.2. Condenser"
 "4.2.1. Energy balance on refrigerant side"
h_r_su_cd=h_r_ex_cp
h_r_ex_cd=enthalpy(fluid$;P=P_r_ex_cp;T=T_r_ex_cd) "Refrigerant
   specific enthalpy at condenser exhaust"
T_r_ex_cd=T_cd-DELTAT_r_ex_cd "Refrigerant temperature at condenser
   exhaust"
```

```
T_cd=temperature(fluid$;P=P_r_ex_cp;x=1) "Condensing temperature
   defined at condenser supply"
Q_dot_cd=m_dot_r*(h_r_su_cd-h_r_ex_cd) "Heat transfer rate in the
   condenser"

"4.2.2. Energy balance on air side"
Q_dot_cd=m_dot_a_cd*c_p_a*(T_a_ex_cd-T_a_su_cd)
c_p_a=cp(Air;T=T_a_su_cd)

"4.2.3. Performance"
 Q_dot_cd=epsilon_cd*m_dot_a_cd*c_p_a*(T_cd-T_a_su_cd)
epsilon_cd=1-exp(-AU_cd/(m_dot_a_cd*c_p_a)) "Condenser
   effectiveness"

 "4.3. Expansion Valve"
h_r_su_ev=h_r_ex_cd "Isenthalpic evolution"

"4.4. Evaporator"
"4.4.1. Energy balance on refrigerant side"
h_r_ex_ev=enthalpy(fluid$;P=P_r_su_cp;T=T_r_ex_ev)
T_r_ex_ev=T_ev+DELTAT_r_ex_ev "Refrigerant temperature at evaporator
   exhaust"
Q_dot_ev=m_dot_r*(h_r_ex_ev-h_r_su_ev) "Heat transfer rate in the
   evaporator"

"4.4.2. Energy balance on air-side"
Q_dot_ev=m_dot_a_ev*(h_a_su_ev-h_a_ex_ev)
h_a_su_ev=enthalpy(AirH2O;P=P_atm;T=T_a_su_ev;R=RH_su_ev) "Humid
   air enthalpy at evaporator supply"
h_a_ex_ev=enthalpy(AirH2O;P=P_atm;T=T_a_ex_ev;R=RH_ex_ev) "Humid air
   enthalpy at evaporator exhaust"

SHR=Q_dot_ev_sens/Q_dot_ev "Sensible Heat Ratio"
Q_dot_ev_sens=m_dot_a_ev*c_p_ha*(T_ha_su_ev-T_a_ex_ev) "Sensible
   cooling capacity"
c_p_ha=CP(AirH2O;P=P_atm;T=T_a_su_ev;R=RH_su_ev)
Q_dot_ev_lat=Q_dot_ev-Q_dot_ev_sens "Latent cooling capacity"

"4.5. Overall Performance"
COP=Q_dot_ev/W_dot_cp
residual=Q_dot_cd-Q_dot_ev-W_dot_cp "To check the overall model"
```

Results

The SHR value indicates that one-third of the cooling capacity is latent capacity.

The refrigeration cycle is represented in the pressure-specific enthalpy diagram Figure 3.67.

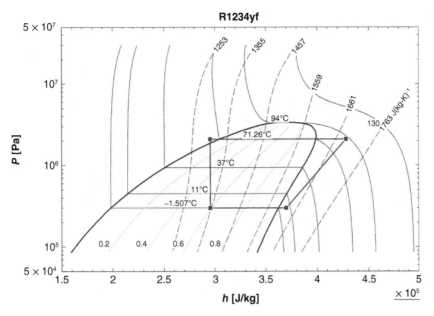

Figure 3.67 Representation of the refrigeration cycle in the P-h diagram.

```
AU_cd=379,3 [W/K]
COP=1,276 [-]
epsilon_s_cp=0,6079 [-]
N_cp=32,1 [Hz]
P_atm=100000 [Pa]
P_r_ex_cp=2,100E+06 [Pa]
P_r_su_cp=300000 [Pa]
Q_dot_cd=6989 [W]
Q_dot_ev=3917 [W]
Q_dot_ev_lat=1162 [W]
Q_dot_ev_sens=2756 [W]

RH_ex_ev=0,9 [-]
RH_su_ev=0,4 [-]
SHR=0,7035 [-]
T_a_ex_cd=57,64 [C]
T_a_ex_ev=29,94 [C]
T_cd=71,26 [C]
T_ev=-1,507 [C]
T_r_ex_cd=66,26 [C]
T_r_ex_cp=91 [C]
T_r_ex_ev=6,493 [C]
T_r_su_cp=6,493 [C]
W_dot_cp=3071 [W]
```

3.6 Production of Heating Capacity

Vehicles equipped with internal combustion engines typically use the heat released into the engine coolant loop for heating the cabin. As illustrated in Figure 3.29, the heater core, that is part of the HVAC unit, is connected to the engine coolant.

With the increase in powertrain efficiency, there is a deficit in heating capacity. Turbo-diesel direct injection engines show heating capacity largely reduced with respect to conventional diesel engines. This could lead to a higher difficulty in quickly achieving thermal comfort inside the cabin. Also, the engine warm-up time is increased. Additional heating systems can be used, such as PTC heaters or heating seats and a heating steering wheel. A deficit in heating capacity can also occur in very cold climates. A better match between the heating load and the heating capacity can also be obtained by decreasing the heating load. This can be achieved by a better cabin thermal insulation or by a higher recirculation ratio of stale air.

3.6.1 Heating with the Engine Coolant Loop

As shown in Figure 3.29, the heater core is integrated into the engine coolant loop in parallel with the radiator. This water-to-air heat exchanger is part of the HVAC unit (see Figure 3.19). It warms up the air before being supplied to the cabin. The maximal capacity of this heat exchanger is of the order of 5–9 kW. The heater core is supplied with water at a temperature close to 85°C when the engine has reached its steady-state temperature. The mechanisms used to control the heating capacity of the heater core have been described in Section 3.4.3.3.

Heater cores are typically tube-fin heat exchangers. Round tubes and flat tubes are used. Both mechanical and brazed assemblies are considered. An example of practical realization of tube-fin type is given in Figure 3.68. The design of this heat exchanger must ensure a good resistance to corrosion, vibration, temperature, and cycling pressure.

For full electric vehicles, the coolant reaches a maximal temperature much lower (of the order of 60°C). The smaller temperature difference with outdoor air does not allow for a large enough heating of the cabin.

Figure 3.68 Heater core. Source: Courtesy of Valeo.

3.6.2 PTC Heaters

For highly efficient engines or for electrified vehicles running in the electric mode, showing a deficit of heating capacity, a PTC heater, that is an electric resistor that heats by Joule effect, can be used

Figure 3.69   View of a PTC heater. Source: Courtesy of Valeo.

in series with the heater core on the air-side. Such a PTC heater is represented in Figure 3.69. The position of the PTC heaters in the HVAC unit is illustrated in Figure 3.29. This heating solution can be used as a supplementary heating system fed with low voltage for thermal engines, with a power of around 0.5 to 2 kW (among others for cold regions). It can also be used as the main heating system fed with high voltage in electrified vehicles, with a heating capacity of several kilowatts.

3.6.3   Heat Pump Systems

The heat pump cycle is similar to the vapor-compression air-conditioning cycle, but the useful effect is the heat rejected at the condenser. In electrified vehicles (full electric, plug-in hybrid, hybrid, etc.), vapor compression heat pumps can be used for heating the cabin during the electric mode when the vehicle does not benefit from the heat from the thermal engine. From the energy point of view, heat pumps show the advantage over PTC heaters to have a COP larger than 1. Improving the performance of the electric heating system in an electric vehicle increases the driving range of the latter. Heat pump systems will be described in more detail in Chapter 4.

3.7   Local Cooling and Heating Systems

As explained here above, comfortable cabin temperatures are typically achieved by pulsing air that has been thermally conditioned inside the HVAC unit, which is a *centralized* heating and air-conditioning system. During pull-down or warm-up, this acceptable level of temperature should be reached in a reasonable time.

By opposition to the centralized systems, spot (or *local,distributed*) cooling/heating aims at creating a micro-climate environment around the occupants. Rather than conditioning the entire cabin, the spot cooling/heating systems deliver a localized thermal comfort, by cooling/heating specific parts of the body, such as the face, the chest, the back, the hands, the neck, or the lap.

Localized comfort can be reached by convective, conductive, or radiative effects.

Conductive effect could be achieved by cooling or heating the seat and the steering wheel. On a seat, this solution is obtained by:

- supplying the component with treated air coming from the centralized heating and air conditioning system
- heating locally the seat with an electric layer,
- cooling/heating the seat by means of a reversible Peltier cooler/heat pump
- ventilating the seat with no thermal capacity generation.

Convective effect can be achieved by a set of nozzle generating air streams targeting different zones of the body under adequate temperatures and mass flow rates (Wang et al., 2014).

Radiative effect could be obtained by means of activated radiant panels, consisting of heated portions of the cabin wall. The temperature of the surface has to be limited to avoid any risk of burns for the occupants.

The major benefit of localized comfort is thermal energy saving: it allows for a reduction in the cabin cooling and heating loads. This reduction could be even more important if the spot cooling/heating system could detect the presence of occupants (Wang et al., 2014). Localized comfort also allows the slow pull-down and warm-up phases to be compensated, reducing largely the time to comfort (Ghosh et al., 2012). It also helps preventing any local discomfort by improving the thermal state of local body segments as well as the whole-body sensation.

3.7.1 Heated, Cooled, and Ventilated Seats

In driving conditions, with a conventional seat, the contact zones with the seat create a local thermal constraint related to the difficulty to heat the body zones in contact with the cold surfaces or to the difficulty to evacuate the sweat caused by the contact with the hot surfaces. These contact zones with the seat are felt colder or hotter than the rest of the body, which yields a local discomfort sensation that contributes to a more important global discomfort sensation.

As a solution, the seat can be equipped with a heating function and possibly a cooling function by conduction, allowing to heat or cool large skin surfaces, which could reduce the thermal constraint at the contact zones and reduce the global discomfort sensation.

These thermal functions can be achieved by equipping the seats in different ways:

- By a heating mat in the seat pan and in the backrest, with a temperature that can be regulated by means of a thermal switch,
- By a reversible Peltier system comprising thermoelectric Peltier cells integrated in the seat pan and in the backrest, by a ventilation system and a porous foam lining that allows to direct the air from the inner of the seat to the contact zones with the occupant.
- By connecting the seat to the air-conditioning module and by diffusing the treated air through the fabrics of the seats.

In the heating mode, the surface temperature is limited to roughly 40°C. The electric consumption of these intermittent localized heating/cooling functions is of the order of 50–100 W. They provide an appreciable gain in terms of comfort when entering the vehicle, especially when the latter is not preconditioned or for extreme outdoor temperatures. In addition, the cooling function of the seat allows for preventing sweating on contact zones with the seat pan and with the backrest.

3.7.1.1 Heated Seat with an Electric Mat

Heated seat with heated electric layers supplied with low voltage is an existing economic solution that brings a fast heating. The use of heated seats allows to reduce the blown air temperature. Such technical solution yields a decrease in the electrical consumption of the electric vehicle for heating.

For instance, Shonda (2013) showed that even with one electric seat consuming 40 W, a net decrease of 320 W could be achieved at −5°C regarding the consumption for cabin heating in a car originally heated with an electrical resistance. The use of thermal manikin allows measuring the equivalent temperature.

3.7.1.2 Seat with Peltier Cells

Seats equipped with a reversible heating/cooling system use Peltier cells (relying on thermoelectric effects). The system is composed of 2 Peltier cells (one for base and one for back), a fan, and 2 air heat exchangers on each Peltier cell (one to extract the heat on one side and the other one to reject the heat on the other side of the cell). Ducts introduced in the foam distribute the air. The system is supplied with low voltage electric source. The needed electric consumption does not exceed 100 W. The COP in the cooling mode is low, around 0.5, depending on the conditions. COP achieved with Peltier coolers are much lower than those achieved with vapor-compression systems. The thermoelectric solution is well adapted to supply cooling and heating in a confined system relatively far from a centralized system.

In hot conditions, the system generates a cooling capacity while heat is rejected in the cabin in a zone that has no influence on the comfort. Similarly, in cold condition, the system generates heating capacity, while cold is rejected in the cabin in a zone that has no influence on the comfort. The reversibility is simply obtained by inversion of the voltage. The objective allows to improve the reception comfort and the comfort during the driving in the winter or summer and to avoid the sweat.

Peltier effect is also used with low voltage in temperature control boxes to maintain cooled or heated liquid or food.

3.7.1.3 Ventilated Seat

A ventilated seat uses a fan that blows the air of the cabin in the seat (base and back) through holes drilled in the foam of the seat. The objective is to act against the sweat phenomenon and to give a cooling sensation on the body zone in contact with the seat.

3.7.2 Heated Steering Wheel

The use of a heated steering wheel can potentially reduce, in a cold environment, the local thermal discomfort at the hands level and improve the overall comfort. The heating of the contact zones is achieved by conduction. The steering wheel comprises a conductive wire integrated in the foam in the zones corresponding to hands positions. It is supplied with 12 V, and its electric consumption is of the order of 40 W.

The cooling of the steering wheel is more difficult because of its rotational displacement. The use of Peltier solution or of a heat pipe can be studied but are complex to integrate.

3.7.3 Electric Radiant Panels

In a cold environment, the body extremities are felt colder than the rest of the body, leading to a local discomfort that degrades the overall thermal comfort. Moreover, the thermal environment inside the cabin is characterized by a thermal heterogeneity due to the temperature of the side windows, which is an additional source of discomfort. The use of electric radiant panels for reducing the asymmetry of thermal fluxes and the thermal constraint at the level of the lower body extremities may improve the overall comfort. This innovative solution is quiet, because of the reduced air flow rates and can be personalized or decentralized to match the occupants' position in the perspective of future autonomous cars among others.

3.7.4 Head Cooling

In a perspective of optimization of thermal comfort in hot environment, the thermal state of the head can also be considered, given its importance in the judgments of perception of thermal comfort. This can be achieved by pulsing cold air in the vicinity of the head. The improvement of thermal comfort at this body segment should allow for decreasing the energy consumption of the convective system.

3.8 Thermal Energy Storage

Thermal energy storage can be used on all types of vehicles, among others for

- Accelerating the heating up of the thermal engine: the thermal storage is thus placed in the glycol water circuit.
- Accelerating the heating up or cooling down of the cabin.

Different technologies to store thermal energy are available. Major technologies are sensible and latent heat storages.

3.8.1 Sensible Thermal Energy Storage

The sensible thermal storage usually consists of a volume of glycol water. The heat storage is achieved by the direct heating of the glycol water that ensures both functions of storage and distribution of heat, which allows for a short response time. The internal configuration, including the inlet and outlet, must account for stratification inside the tank and for mixing phenomena with inlet fluxes that penalize the valorization of the stored heat.

The pre-conditioning of the whole heat transfer fluid contributes also to the function of thermal storage.

Example 3.7 *Sensible Heat Storage*
A sensible heat storage system consists of a 1-liter tank of glycol water (50% in mass of ethylene glycol) that is heated from 55°C to 95°C. During the heat discharge, glycol water feeds a heater core and comes back to the tank at 55°C. Compute the capacity of the heat storage in [Wh] as well as the energy density in [Wh L^{-1}] and in [Wh kg^{-1}]. If the heat transfer rate in the heater core is 350 W, compute the discharge time of the heat storage [min]. We assume that a plug flow is achieved, maintaining a perfect stratification in the tank. Also, we assume that there are no thermal inertia losses in the materials of the pipes and heat exchangers.

Analysis

The capacity of the storage can be determined based on volume, density, and specific heat of glycol water as well as temperature variation. The discharge time is determined by expressing the energy balance on the storage during discharge.

Solution

```
"1. INPUTS"
T_gw_i=95 [C] "initial temperature of glycol water (start of
   discharge)"
T_gw_f=45 [C] "final temperature of glycol water (end of discharge)"
```

```
Q_dot_hc=350 [W] "heat transfer rate in the heater core"

"2. OUTPUTS"
"Q_storage=?? [Wh]" "storage capacity"
"tau=?? [minutes]" "discharge time"

"3. PARAMETERS"
Vol_gw=0,001 [m^3] "volume of glycol water"
C_g=50 [%] "mass concentration of ethylene glycol in the aqueous
    solution"

 "4. MODEL"
"4.1. Energy capacity"
c_gw=cp(EG;T=(T_gw_i+T_gw_f)/2;C=C_g)
rho_gw=density(EG;T=(T_gw_i+T_gw_f)/2;C=C_g)
Q_storage=Vol_gw*rho_gw*c_gw*(T_gw_i-T_gw_f)/3600 "energy capacity"
m_storage=Vol_gw*rho_gw "mass of glycol water"
Q_storage_kg=Q_storage/m_storage "mass energy density"
Q_storage_L=Q_storage/Vol_gw/1000 "volumetric energy density"

"4.2. Discharge time"
tau=Q_storage*3600/Q_dot_hc/60
```

Results

```
c_gw=3544 [J/kg-K]
m_storage=1,034
Q_storage=50,87 [Wh]
Q_storage_kg=49,22 [Wh/kg]
Q_storage_L=50,87 [Wh/L]
rho_gw=1034 [m^3/kg]
tau=8,721 [min]
```

3.8.2 Latent Thermal Energy Storage

The latent energy storage uses the liquid–solid phase transition of an intermediate material to store or provide heat by fusion or solidification, respectively. The thermal energy exchange between the heat transfer fluid and the storage material requires a heat exchanger that increases the response time and introduces a temperature pinch point. However, this process allows for the thermal energy to be stored and recovered at an almost constant temperature, because of the constant temperature phase change of the material. The phase-change temperature of the phase change material must be selected according to the necessity of heating or cooling. A phase-change temperature of around 70°C is an example for heating the heat transfer fluid.

3.8.2.1 Phase Change Materials and Ice

Materials from the family of paraffins or fatty acids can be used for storing latent heat with phase change temperatures compatible with vehicle applications (roughly between 0°C and 100°C).

Materials should show a high energy density in [Wh kg^{-1}] or [Wh m^{-3}] (for comparison, energy density of ice is around 92.5 [Wh kg^{-1}]), a good thermal conductivity (not to limit the charging and discharging heat transfer rates), a good chemical stability (to prevent performance degradation with time), and industrial availability and allow for a low system cost.

For cold storage, water is an appropriate and efficient fluid because of its phase change temperature of 0°C at 1 atm. The storage is achieved under the form of ice in a small tank or in nodules. During regeneration, ice is built by a cooling production system, for instance, a vapor-compression system or a Peltier cooler. Implementing a cold storage is more complex than implementing a heat storage, since in the latter case regeneration by heat production is easier. A material from the family of paraffins (PCM) is also suitable, because of its phase change temperature of around 7°C.

3.8.2.2 Evaporator with Latent Thermal Energy Storage

Evaporators can be combined with phase change material (PCM) storage that can store cooling energy, under the form of latent heat of fusion and sensible heat. Dissociation in time of production and consumption of cooling energy allows for maintaining cabin comfort during idling or vehicle engine stop when a mechanical compressor is used. Better cabin comfort can thus be achieved during heavy traffic. For vehicles equipped with stop & start technology, the fraction of engine stop time increases yielding fuel consumption reduction. Indeed, with mechanical compressors, cooling production can only be achieved during operation of the engine.

Using evaporator with cooling energy storage allows for maintaining thermal comfort for around 50 seconds when the compressor is switched off, while this period is reduced to around 20 seconds without thermal energy storage.

3.8.3 Sorption Energy Storage

Sorption storage uses a reversible process where a gas (or a liquid) is fixed or released on the surface of a porous solid matrix to achieve sorption/desorption of the fluid. For instance, zeolite can be used as micro-porous solid (adsorbent bed) with water (adsorbate). The family of metal organic framework (MOF) made up of metallic ions connected by organic groups are relatively new solid materials, allowing for a potential significant improvement of performance in applications using adsorption. The adsorption reaction is exothermic. The regeneration (desorption) of the material is achieved by heating the material. This type of process avoids thermal losses from the storage over time and foreshadows, despite its complexity, volumetric energy densities larger than those obtained with sensible and latent energy storage. The working principle on an EV or PHEV consists of regenerating the thermal storage, for instance, during the charging period of the main electrical battery or possibly during the thermal mode with the heat delivered by the thermal engine.

3.8.4 Thermal Insulation

The thermal insulation of the thermal energy storage can be necessary to avoid losses over time, among others if the thermal energy is not quickly used. This could happen if the thermal storage is charged at vehicle standstill and even during driving with a discharge operation during the first minutes of the next driving or depending on the strategy of usage of the thermal storage. Insulation can be achieved by an external vacuum insulation or by insulating materials.

3.8.5 Energy Density

In overall, as it will be shown in Chapter 4, the thermal storage has a positive impact on the driving range in the electric mode. This range will be extended during short trips.

Table 3.6 Comparison of specific energy contents and energy densities of different storage solutions.

Thermal storage[a)]	Electrical storage			
	Glycol water with a $\Delta T = 50K$[b)]	Ice	PCM[c)]	Li-ion cells
Specific energy content [$Wh_{th}\ kg^{-1}$ or $Wh_e\ kg^{-1}$]	49	93	c. 56	c. 100
Energy density in [$Wh_{th}\ L^{-1}$ or $Wh_e\ L^{-1}$]	51	92	c. 42	>150
Associated components	Tank, insulation	Heat exchanger, insulation, tank[d)]	Heat exchanger, insulation, tank[d)]	Battery thermal management system, battery pack packaging

a) Specific energy contents and energy densities correspond to the fluid only.
b) refer to Example 3.7
c) considering a heat of fusion of 200 J g^{-1} and a density of 750 kg m^{-3}
d) depending on the type of container used for the storage material.

The major criteria allowing for the assessment of the potential of those solutions are:

– The energy density (or specific energy content) in Wh kg^{-1}
– The energy density in Wh L^{-1}
– The specific cost of energy storage in EUR Wh^{-1}

Typical specific energy content and energy densities of different storage solutions are given in Table 3.6. These densities do not account for the weight and volume of associated components (which are listed in the last line of the table).

References

Abraham, G.S., Ravikumar, A.S., and Shah, R.K. (2006). Design considerations for an integral-receiver dryer condenser. In: *SAE 2006 World Congress & Exhibition*. Paper 2006-01-0725.

ASHRAE (2002). Chapter 45: Refrigerant-control devices. In: *2002 ASHRAE Handbook—Refrigeration*. Atlanta: American Society of Heating, Refrigerating and Air-Conditioning Engineers, Inc.

ASHRAE (2003a). Chapter 33: Solar energy use. In: *2003 ASHRAE Handbook—HVAC Applications*. Atlanta: American Society of Heating, Refrigerating and Air-Conditioning Engineers, Inc.

ASHRAE (2003b). Chapter 53: Radiant heating and cooling. In: *2003 ASHRAE Handbook—HVAC Applications*. Atlanta: American Society of Heating, Refrigerating and Air-Conditioning Engineers, Inc.

ASHRAE (2005a). Chapter 8: Thermal comfort. In: *2005 ASHRAE Handbook—Fundamentals*. Atlanta: American Society of Heating, Refrigerating and Air-Conditioning Engineers, Inc.

ASHRAE (2005b). Chapter 30: Nonresidential cooling and heating load calculations. In: *2005 ASHRAE Handbook—Fundamentals*. Atlanta: American Society of Heating, Refrigerating and Air-Conditioning Engineers, Inc.

ASHRAE (2005c). Chapter 31: Fenestration. In: *2005 ASHRAE Handbook—Fundamentals*. Atlanta: American Society of Heating, Refrigerating and Air-Conditioning Engineers, Inc.

ASHRAE (2008). Chapter 37: Compressors. In: *2008 ASHRAE Handbook—HVAC Systems and Equipment*. Atlanta: American Society of Heating, Refrigerating and Air-Conditioning Engineers, Inc.

Aw, K.T. and Ooi, K.T. (2021). A review on sliding vane and rolling piston compressors. *Machines* 9 (6): 125. https://doi.org/10.3390/machines9060125.

Benouali, J., Clodic, D., and Malvicino, C. (2002). External and internal control compressors for mobile air-conditioning systems. International Compressor Engineering Conference. Paper 1614. http://docs.lib.purdue.edu/icec/1614.

Boewe, D., Bullard, C.W., Yin, J., and Hrnjak, P. (2001). Contribution of internal heat exchanger to transcritical R-744 cycle performance. *HVAC&R Research* 7: 155–168. https://doi.org/10.1080/10789669.2001.10391268.

Breque, F. (2017). Etude et amélioration d'une pompe à chaleur pour véhicule électrique en conditions de givrage. Energie électrique. Université Paris sciences et lettres. Français. ⟨NNT: 2017PSLEM083⟩. ⟨tel-03081279⟩.

Bullard, C.W., Yin, J.M., and Hrnjak, P.S. (2000). Transcritical CO2 mobile heat pump and A/C system. Experimental and model results. SAE Automotive Alternate Refrigerants Symposium, Scottsdale, Arizona, USA.

Choo, W.C. (2021). Development of multi-vane Revolving Vane (RV) compressor. PhD thesis. Nanyang Technological University. Singapore (NTU Singapore).

Cola, F., De Gennaro, M., Perocchio, D. et al. (2016). Integrated receivers with bottom subcooling for automotive air conditioning: detailed experimental study of their filling capacity. *International Journal of Refrigeration* 62: 72–84. ISSN 0140-7007.

Colinet, A. (1993). La climatisation automobile, Editions techniques pour l'automobile et l'industrie.

Cuevas, C., Winandy, E., and Lebrun, J. (2008). Testing and modelling of an automotive wobble plate compressor. *International Journal of Refrigeration* 31 (3): 423–431.

Daly, S. (2006). *Automotive Air Conditioning and Climate Control Systems*. Butterworth-Heinemann.

DuBois, D. and DuBois, E.F. (1916). A formula to estimate approximate surface area, if height and weight are known. *Archives of Internal Medicine* 17: 863–871.

Duffie, J.A. and Beckman, W.A. (1980). *Solar Engineering of Thermal Processes*. New York: Wiley.

Fanger, P.O. (1970). *Thermal Comfort Analysis and Applications in Environmental Engineering*. New York: McGraw-Hill.

Fanger, P.O. (1982). *Thermal Comfort*. Malabar, FL: Robert E. Krieger.

Fayazbakhsh, M. and Bahrami, M. (2013). Comprehensive modeling of vehicle air conditioning loads using heat balance method, SAE Technical Paper 2013-01-1507, doi:https://doi.org/10.4271/2013-01-1507.

Ghosh, D., Wang, M., Wolfe, E. et al. (2012). Energy efficient HVAC system with spot cooling in an automobile – Design and CFD analysis. *SAE International Journal of Passenger Vehicle Systems* 5 (2): doi: 10.4271/2012-01-0641.

Gillet, T. (2018). Étude expérimentale et modélisation numérique d'un système de climatisation multi-évaporateurs pour véhicule électrifié. Unpublished doctoral thesis. ULiège - Université de Liège.

Holmér, I. (2005). Evaluation of vehicle climate. In: *Environmental Ergonomics, The Ergonomocs of Human Comfort, Health and Performance in the Thermal Environment* (ed. Y. Tochihara and T. Ohnaka). Elsevier.

Huelle, Z.R. (1972). The MSS line – a new approach to hunting problem. *ASHRAE Journal* 43–46.

Ishii, N., Abe, Y., Taguchi, T., and Kitamura, T. (1990). Dynamic behavior of a variable displacement wobble plate compressor for automotive air conditioners. In: *Proceedings of the International Compressor Engineering Conference at Purdue*, 345–353.

ISO (2006). Ergonomics of the thermal environment – Evaluation of thermal environments in vehicles – Part 2: Determination of equivalent temperature. ISO 14505-2:2006.

ISO (2006). Ergonomics of the thermal environment – Evaluation of thermal environments in vehicles – Part 3: Evaluation of thermal comfort using human subjects. ISO 14505-3:2006.

ISO (2007). Ergonomics of the thermal environment – Evaluation of thermal environments in vehicles – Part 1: Principles and methods for assessment of thermal stress. ISO 14505-1:2007.

Klein, S.A. (2007). *TRNSYS 16 Program Manual*. Solar Energy Laboratory, University of Wisconsin: Madison, USA.

Knibbs, L.D., de Dear, R.J., and Atkinson, S.E. (2009). Field study of air change and flow rate in six automobiles. *Indoor Air* 19: 303–313.

Madsen, T.L., Olesen, B.W., and Christensen, N.L. (1984). Comparison between operative and equivalent temperature under typical indoor conditions. *ASHRAE Transactions* 90 (Part 1).

Marcos, D., Pino, F.J., Bordons, C., and Guerra, J.J. (2014). The development and validation of a thermal model for the cabin of a vehicle. *Applied Thermal Engineering* 66 (1–2): 646–656.

Moan, J. (2001). Visible light and UV radiation. In: *Radiation at Home* (ed. D. Brune). Outdoors and in the Workplace: Scandinavian Science Publisher.

Monforte, R. (2001). Identification of the numerical model for an automotive application thermostatic expansion valve. SAE Technical Paper 2001-01-1700. https://doi.org/10.4271/2001-01-1700.

Pomme, V. (1999). Improved automotive A/C systems using a new forced subcooling technique. SAE Technical Paper 1999-01-1192. https://doi.org/10.4271/1999-01-1192.

Pottker, G. and Hrnjak, P.S. (2012a). Effect of condenser subcooling of the performance of vapor compression systems: experimental and numerical investigation. International Refrigeration and Air Conditioning Conference. Paper 1328. http://docs.lib.purdue.edu/iracc/1328.

Pottker, G. and Hrnjak, P.S. (2012b). Designated vs non-designated areas for condenser subcooling. International Refrigeration and Air Conditioning Conference. Paper 1332. http://docs.lib.purdue.edu/iracc/1332

Rugh, J. and Farrington, R. (2008). Vehicle ancillary load reduction project close-our report. An overview of the task and a compilation of the research results. *Technical Report NREL/TP-540-42454*.

SAE J2234 (1993). Equivalent temperature. Surface Vehicle Information Report, sae.org, J2234.

Saint Gobain (2013). Thermal Comfort, Glazing that keeps the cool. Technical documentation.

Schnubel, M. (2016). *Today's Technicians™: Classroom Manual for Automotive Heating & Air Conditioning*, 6e. Boston, USA: Cengage Learning.

Seo, J. and Choi, Y. (2013). Estimation of the air quality of a vehicle interior: The effect of the ratio of fresh air to recirculated air from a heating, ventilation and air-conditioning system. *Proceedings of the Institution of Mechanical Engineers, Part D: Journal of Automobile Engineering* 227 (8): 1162–1172.

Shonda, O. (2013). Solutions for EVs: thermal management enabling a higher range. Paper presented at 3rd International Conference on Thermal Management for EV/HEV, 24–26 June 2013, Darmstadt, Germany.

Sullivan, R. and Selkowitz, S. (1990). Effects of glazing and ventilation options on automobile air conditioner size and performance. SAE Technical Paper Series, no. 900219. 15 pp. Society of Automotive Engineers: Warrendale, PA.

Torregrosa-Jaime, B., Bjurling, F., Corberán, J.M. et al. (2015). Transient thermal model of a vehicle's cabin validated under variable ambient conditions. *Applied Thermal Engineering* 75: 45–53.

Toyota Industries Corporation (2022. Continuous variable-displacement type compressors. https://www.toyota-industries.com/products/automobile/compressor/variable/index.html#internally (accessed 23 April 2022).

Wang, S.K. (2001). *Handbook of Air Conditioning and Refrigeration*. McGraw-Hill.

Wang, M., Wolfe, E., Ghosh, D., Bozeman, J. et al. (2014). Localized cooling for human comfort, *SAE International Journal of Passenger Cars: Mechanical Systems* 7(2): 2014, doi:https://doi.org/10.4271/2014-01-0686.

Yamanaka, Y., Matsuo, H., Tuzuki, K., Tsuboko, T. et al. (1997). Development of sub-cool system. SAE Technical Paper 970110. https://doi.org/10.4271/970110.

4

Thermal Energy Management in Hybrid and Electric Vehicles

4.1 Introduction

The needs in mobility of the population are evolving. Also, the automotive sector is largely changing in terms of uses, regulations, and technologies. The main issues for a clean transportation system are, among other things, the reduction of

- the emissions of the greenhouse gases, including CO_2,
- the pollutants in cities, with increasing stringent regulations,
- the dependency toward fossil fuels.

Regarding CO_2 emissions, major governments worldwide have established regulations to progressively decrease emissions of new passenger cars and light commercial vehicles. For instance, Figure 4.1 indicates the European fleet-wide average emission target for new passenger cars from 2000 to 2030. In 2019, the European average CO_2 emissions of newly registered cars were $122 \, g \, km^{-1}$ (ICCT, 2020). The average CO_2 emissions are progressively decreasing, but this decrease is currently slowed down for different reasons, including the larger fraction of SUV in the vehicles mix and the reduction in the sales of diesel vehicles.

The raise in awareness by our modern society has led to the seek for less polluting alternative energy sources and the emergence of a new generation of vehicles, including vehicles with electric powertrain, which is the topic of this chapter.

The growth of the electric vehicles (EVs) market seems inevitable in the current environmental context. The deployment of solutions that ease the electric mobility can be speeded up by investing in infrastructures and services (charging stations, vehicles maintenance, etc.), by defining local regulations and incentives (at city, regional, or country levels) that progressively restrict the use of vehicles with thermal engines (including diesel engines), and by a breakthrough societal demand.

Several technologies are part of the panel of solutions to be developed to go toward a "zero emission" mobility and to have zero emission vehicles (ZEVs):

- purely EVs with a battery that stores the whole embedded energy (battery electric vehicle [BEV]), which have already been mass-produced in markets for several years,
- purely EVs with a storage or production of hydrogen used to feed a fuel cell for electricity production (fuel cell electric vehicle [FCEV]),
- hybrid vehicles equipped with a mixed thermal and electric powertrain (hybrid electric vehicle [HEV]),
- HEVs equipped with a mixed thermal and electric powertrain and batteries that can be recharged (plug-in hybrid electric vehicle [PHEV]),
- smart and fast EV charging systems,

Thermal Energy Management in Vehicles, First Edition. Vincent Lemort, Gérard Olivier, and Georges de Pelsemaeker.
© 2023 John Wiley & Sons Ltd. Published 2023 by John Wiley & Sons Ltd.
Companion website: www.wiley.com/go/lemort/thermal

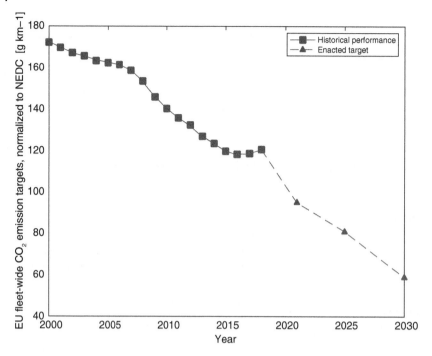

Figure 4.1 EU passenger car CO_2 emissions targets, normalized to NEDC. Source: Adapted from https://theicct.org/chart-library-passenger-vehicle-fuel-economy.

– contactless power transfer in charging systems,
– electricity production by means of renewable energies,
– other hybrid associations can emerge by combining electric, thermal, and fuel cell solutions, leading, for instance, to Range Extender solutions that increase the driving range of the vehicle (extended range electric vehicle [EREV]).

In parallel, a massive introduction of EV broadens the role of the vehicle itself. The latter is positioned at the center of an ecosystem that connects the automotive world, the information and communications technologies (ICT) sector, the energy industry, and the charging infrastructures. Actually, the embedded energy in a vehicle (for instance, 40 to 100 kWh) is a non-negligible source of energy with interesting perspectives at the individual, city, or territory scales. Moreover, the evolutions of the cost of lithium-ion batteries make their use in stationary applications more and more economically feasible.

The market introduction of EVs and their future evolution represent a triggering event on the automotive road map that forces to identify the major upstream consequences on the vehicle overall thermal energy management. These include the thermal control at mild or low temperature of new components such as the electric battery, the electric motor, the electronic components, or the battery charger. These components require specific needs in cooling or heating as a function of the temperature levels. These cooling and heating needs must allow for the optimization of the use of components and for the maximization of their lifespan. They must also account for the different phases of the use of the vehicle (driving, parking, and connection to the charging electricity grid), for the deficit in heat released by the electric powertrain and for the impact of the cabin thermal control in the electric mode on the driving range of the vehicle.

The heat rejection of the different components, mainly achieved in the front-end module of the vehicle, and their associated low levels of temperature require binding sizes of heat exchangers and a tradeoff with the design of the vehicle and its aerodynamic drag forces.

The air-conditioning system also evolves with the electrification of the compressor, which is controlled in speed, the reversibility of the refrigeration system to allow for the heat pump mode with an improved efficiency, the valorization of waste heat, and the use of the evaporation step to cool the battery or to store the coldness in phase change materials (PCM). The overall thermal management requires the imbrication and the control of the different fluid loops in a way to ensure the thermal control of the cabin and of the different components while limiting the energy consumption.

This chapter aims at describing the impact of vehicle electrification on the different thermal energy management systems. The first section of the chapter recalls the classification of electrified vehicles. The second section addresses the impact of electrification on the cabin climate control. The next two sections cover aspects specific to HEVs and EVs: the battery thermal management and the e-motor and power electronics cooling. Finally, the thermal management of the HEVs or EVs as a whole is discussed.

4.2 Classification of Electric and Hybrid Electric Vehicles

The two main families of electrified motorizations are the full EVs and the hybrid vehicles. The former ones are working only in the electric driving mode, while the latter ones combine a heat engine and an electric motor allowing for both thermal and electric driving modes.

The main options for the architecture of the vehicle platform are as follows: the Conversion Platform (an ICE platform is adapted to the BEV with a minimum of modifications), the Modular Platform (the platform is adapted to receive an ICE vehicle, a HEV or BEV), and the BEV Platform (new platform developed specifically for a BEV).

4.2.1 Electric Vehicles

In EVs, the distinction can be made between BEVs and FCEVs.

4.2.1.1 Battery Electric Vehicles

A BEV is a vehicle using a battery to store the electricity and whose propulsion is fully electric. In such vehicles, wheels are driven by an electric motor. More than one electric motor could possibly be used. This solution allows to benefit from the assets of an electric motor, including the good efficiencies, the absence of pollutant emission from the vehicle, and a noiseless operation.

The evolution from an ICE vehicle to an EV involves three major changes. First, in terms of energy, the consumption metrics is no more the liter of fuel, but the kWh of electricity. Second, with regard to charging location and duration, the options range from a charge of a few minutes in a charging station to a charge during part of the night at home for instance. Third, in terms of energy utilities, from oil industry to electricity industry, which is even today less familiar with mobility.

Electric motors are smaller, lighter, and more efficient, but batteries show an energy density smaller than that of fuel tanks (40 L of gasoline correspond to 380 kWh, and the tank filling rate is roughly $50\,L\,min^{-1}$).

The energy flow diagram of the powertrain of a battery electric vehicle is represented in Figure 4.2. Electricity is stored in a battery pack. The battery can be externally charged from the

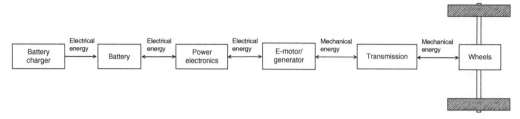

Figure 4.2 Energy flow representation of the powertrain of a battery electric vehicle.

electricity grid. The battery supplies the electric motor with electrical energy. Power electronics components, inserted between the battery and the electric motor, convert the electricity from one form to another. These components are typically DC–DC converters to modify the voltage of direct current, inverters to convert direct current to alternating current, and rectifiers to convert alternating current into direct current. The electric motor converts the electrical energy into mechanical energy, which is finally supplied to the wheels through a transmission system.

During regenerative braking, kinetic energy is recovered, and energy flows in the opposite direction. Mechanical energy is converted into electricity by means of the electrical motor working as a generator. The power electronics has to ensure that electrical energy is in a suitable form to feed the battery.

The success of electric vehicles will largely depend on their driving range (limiting the problem of "range anxiety") and on the charging time, which are today the major constraints. In 2017, the battery capacity typically ranged from 15 to 40 kWh, with a driving range up to 400 km, and the electric motor power from 50 to 100 kW. Announced driving ranges correspond to the early stages of the life of the battery. The latter one degrades over the life cycle due to aging that is function of the temperatures at standstill and during operation. In the future, the battery capacity is expected to progressively grow to increase the range of the vehicle. The battery pack can be externally charged by plugging into the electricity grid. Inconveniences include the charging time and the lack of access to charging stations over long journeys and full depletion of battery. Fast charging with higher electric power transfer is currently developed to reduce the time needed to recover the full capacity or an intermediate acceptable capacity of the battery. Moreover, induction technology should allow, in a more or less short term, to recharge the batteries while parking or driving on dedicated portions of roads (the latter option being more ambitious).

4.2.1.2 Integration of EVs in Electricity Grids

Taking benefit of the maturity of ICT, smart charging infrastructures will have to be developed. Deployment of EVs is carried out in parallel with the development of smart grids. The "smart-grid" is an electricity grid that promotes the transfers of information to control in real time the electricity fluxes and hence better manage the grid. Smart grids allow for smart charging, for interactions between the vehicle and the electricity grid ("vehicle-to-grid"), and for projects of second life batteries.

The "smart-charging" modulates the charging of EVs as a function of the needs of the user and of the available electricity on the grid, in such a way as to maximize the use of renewable energy and benefit from the lowest electricity rates (during off-peak hours to participate to the electrical grid management). The user indicates the requested state of charge and the time of use of the vehicle.

The charging can be bidirectional between the vehicle and a stationary charging point. The onboard battery can be used to store electrical energy produced for instance by the conversion of a renewable energy source by photovoltaic panels or wind turbines. This stored energy can be re-injected in the electric grid, such as an electricity micro-production device, during evening peak demand hours. Hence, EVs can act as temporary storage units. The EV would therefore contribute decarbonizing the electricity by, inter alia, limiting the peak consumption. The CO_2 balance could become negative.

The EV battery, at the end of its vehicle lifetime, still shows a substantial storage capacity and could be reused in a second life in less stringent environments than automotive.

This second life of batteries before being recycled when they do not meet the very demanding needs of an automotive use (driving range, acceleration rate) is proposed to transform them into energy storage units and feed:

– Green electricity distribution grids,
– Office buildings, industrial sites, and residential buildings (single and multi-family buildings),
– Charging stations for EVs limiting the cost of connection to the grid,
– Back-up solutions in case of grid failure by providing an energy reserve.

In households equipped with PV panels, the solar electricity produced during daytime or the electricity at low cost provided by the grid during off-peak hours can be stored in these recycled batteries and sent back during peak hours by means of smart electricity meters. They allow the fraction of produced electricity that can be self-consumed to be increased rather than being sent back to the grid. This stored electricity can also be used to charge EVs. Their cost is decreased by sharing the latter with the stationary application and by the optimized life cycle with an additional use in the order of 5 to 10 years after the use in EVs that is in the order of 8 to 10 years.

4.2.1.3 Fuel Cell Electric Vehicles

Another category of EVs is the FCEV. This type of EV is equipped with a self-generation of electricity by means of a fuel cell fed by hydrogen stored onboard the vehicle. In such a propulsion system, electricity and water are simultaneously produced by the fuel cell from hydrogen and oxygen of the air and directed toward an electric motor to propel the vehicle. This technical solution allows for overcoming the bulkiness and mass of the battery of a BEV. Its charging time is similar to that of a vehicle with conventional fuel, provided that a large enough network of hydrogen supply exists.

Several uses of the fuel cell are possible in the transportation sector. The full power application consists of powering the vehicle with a large-power fuel cell, a large quantity of on-board hydrogen, and a small-capacity battery used during transient phases (acceleration and battery charging). The auxiliary power unit (APU) application, or the range extender system, works as an auxiliary charger onboard a BEV. The fuel cell power and the onboard quantity of hydrogen are lower than those in the previous application. The drive is ensured by the battery of a large capacity. This approach allows for a reduction of the costs of the fuel cell system due to its smaller size and for a limitation of the constraints on the hydrogen pressure storage, thermal systems, and bulkiness. An intermediate application between the full power and the APU range extender, associating hydrogen and the battery, can offer functionalities similar to those of a PHEV. Finally, an alternative to hydrogen storage consists of equipping the vehicle with a reformer that produces hydrogen onboard the vehicle from a conventional fuel. This solution allows for overcoming the limitation of the infrastructures and of the investments, even though the emitted CO_2 cannot be captured and stored.

Theoretically, FCEVs and BEVs reject neither pollutants nor greenhouse gases from the tank-to-wheels. Also, their drive is soundless and ensures a driving pleasure. However, the well-to-wheels CO_2 emissions mainly depend on the production mode of hydrogen as well as the modes of transportation, storage, and distribution of hydrogen to the refueling stations.

4.2.1.3.1 *Hydrogen Properties and Production*

Hydrogen is very abundant on Earth, but almost always combined with other atoms such as carbon and oxygen (hydrocarbons, water), meaning that it has to be produced. Hydrogen has a simple atomic structure with a nucleus composed of one proton and one orbiting electron. It is very light with a molar mass of 2.016 g mol^{-1} and with a density of 0.08235 kg m^{-3} at $25\,°$C and $101\,325$ Pa. At standard atmospheric pressure ($101\,325$ Pa), the melting point of hydrogen is $-259.14\,°$C and its boiling point is $-252.87\,°$C. Hydrogen shows a very large energy density (120 MJ kg^{-1}, i. e. $33, 33$ kWh kg^{-1}), which is approximately three times that of hydrocarbons. In gaseous phase, hydrogen is flammable with concentrations in air larger or equal to 4%. Hydrogen is odorless, invisible, environmentally neutral, nontoxic, and noncorrosive, but it is a highly reducing agent. It is highly diffusible and mixes easily with air. A low viscosity and its small size ease the leaks. Combining hydrogen with oxygen O_2 produces water.

Hydrogen production and fuel cells have been known from one century. Hydrogen can be produced from methane (natural gas) ($CH_4 + H_2O \rightarrow 3H_2 + CO$ [$700 - 1100\,°$C]) or by reforming liquid hydrocarbons. It can also be produced from the gasification of coal that produces a mixture of H_2, CO, and CO_2. Hydrogen can also be produced without emitting CO_2 by the electrolysis of water from a renewable source of electricity (wind, hydraulic, or solar energy), which transforms water and electricity into hydrogen H_2 and oxygen O_2. Hydrogen and electricity show a good match because of the reversibility of the fuel cell/electrolysis concepts.

In the energy transition, hydrogen, as a green energy vector produced by the electrolysis of water ($2\,H_2O \rightarrow 2\,H_2 + O_2$) from a renewable source of electricity, can be seen as a potential alternative to fuel oil. This is due to its high energy density, its reaction that only produces water, and its promising capacity to convert and store intermittent renewable electricity in excess ("power-to-gas" as a mean to contribute to the management of electricity grids) and to be transported in distribution networks.

However, the technology is still complex and costly. Moreover, many technological limitations need to be overcome, including limitation associated with the infrastructure and the cost of production, transportation, and storage. Moreover, this technology must operate with outdoor temperatures of 40–50°C in hot climate and $- 10$ to -20°C in cold climate, while addressing the risk of freezing of the produced water. For large quantities, hydrogen can be transported in the liquid phase at cryogenic temperatures with semi-trailers. For very large quantities, a transportation network can be used. One of the specific issues of the hydrogen filling stations is to ensure a fast refueling of the vehicle tank. The required compression of hydrogen to reach 700 bar to feed an intermediate buffer tank and the vehicle tank is quasi adiabatic and heats the gas. As a consequence, either a hydrogen source at a higher pressure (>800 bar) is used so that there is no need of cooling or hydrogen is cooled down before being filled in the tank. Also filling liquid hydrogen at very low temperatures requires caution and a quick and sophisticated connection. These elements, in addition to social acceptance and security, constitute the main issues of the design engineers.

4.2.1.3.2 *Hydrogen Storage*

The main technologies of storage of hydrogen in vehicles are based on physical storage solutions (gas compression, liquefaction, and physisorption on porous materials) or chemical storage solutions (adsorption in chemical hydrides).

While hydrogen has a mass energy density three times larger than that of hydrocarbons ($120\,MJ\,kg^{-1}$ versus $45\,MJ\,kg^{-1}$ for gasoline), hydrogen is a very light component, with a density equal to 0.069 53 that of air at 25 °C and 101 325 Pa. Hence, its volume energy density is relatively low. This leads to the need to store gaseous hydrogen at very high pressure (350 to 700 bar). The density of gaseous hydrogen at 700 bar and 300 K is only $0.039\,kg\,L^{-1}$, while it is only $0.072\,kg\,L^{-1}$ for liquid hydrogen at 10 bar and 20 K. Hence, assuming a need of hydrogen of 6 kg to ensure a driving range of roughly 500 km for a FCEV, the volume of hydrogen itself would be around 83–154 L. To this volume, must be added the volume of the envelope, the valves, and other accessories. The weight of the tank could also become a drawback, because it must sustain the very large pressures of the compressed gas or, for a liquid fuel, it must ensure a highly efficient thermal insulation to limit the vaporization due to the heating and a smart management of hydrogen leaks due to boiling. The risk factors associated with the high pressure and the explosivity of hydrogen must also be taken into account. The costs associated with those technologies represent a critical issue and must be reduced while maintaining the security level. Most of the studies are oriented toward a gaseous storage of 700 bar or possibly 350 bar in the case of light tanks (around 40 kg) in carbon fiber-reinforced composite materials. Moreover, storage under high-pressure compressed gas or under liquid has a large energy cost. As illustrated in Example 4.1, the compression of hydrogen from 1 to 700 bar represents around 10% of its high heating value.

Example 4.1 *Hydrogen Compression*
Hydrogen is compressed in a 2-stage compressor from 1 bar and 25°C to 700 bar. Both compression stages are assumed adiabatic and reversible. At the outlet of the first compression stage, hydrogen is cooled down in an intercooler down to 25°C. Compute the ratio of the compression work to the heating value of hydrogen.

Solution

```
"1. INPUTS"
fluid$='hydrogen'
LHV=119,96e6 [J/kg] "Low heating value of hydrogen"
HHV=141,80e6 [J/kg] "High heating value of hydrogen"
P_1=1e5 [Pa] "Pressure at inlet of first compression stage"
T_1=25 [°C] "Temperature at inlet of first compression stage"
P_4=700e5 [Pa] "Pressure at outlet of second compression stage"
"2. OUTPUTS"
"w_rel"
"3. PARAMETERS"
"both compression stages are assumed isentropic"
"4. MODEL"
"4.1 First stage compression (1=>2)"
P_2=sqrt(P_1*P_4)
h_1=enthalpy(fluid$;P=P_1;T=T_1)
s_1=entropy(fluid$;P=P_1;T=T_1)
h_2=enthalpy(fluid$;P=P_2;s=s_1)
T_2=temperature(fluid$;P=P_2;s=s_1)
"4.2 Intercooling (2=>3)"
T_3=T_1
```

```
P_3=P_2
"4.3 Second stage compression (3=>4)"
h_3=enthalpy(fluid$;P=P_3;T=T_3)
s_3=entropy(fluid$;P=P_3;T=T_3)
h_4=enthalpy(fluid$;P=P_4;s=s_3)
"Total compression specific work"
w=h_2-h_1+h_4-h_3 "Specific compression work"
w_rel=w/HHV "Relative specific compression work"
```

Results

```
P_2=2,646E+06 [Pa]
w=1,388E+07 [J/kg]
w_rel=0,09787 [-]
```

The energy consumption for liquefying hydrogen down to 20 K represents roughly 30% of its heating value.

Chemical storage solutions with a reformer onboard to produce hydrogen have also been tested. Methanol CH_3OH is interesting due to its intrinsic mass fraction of hydrogen of 12.5% and due to the fact that it is liquid at room pressure and temperature. But this transformation path is debatable because of the toxicity of methanol. Research is currently conducted on storage processes less constraining that liquefaction or compression, such as low-temperature physisorption on different forms of nano-structured materials, including organometallic materials called metal–organic frameworks (MOF) and chemisorption in metallic hydrides or in complex hydrides offering large theoretical possibilities of hydrogen storage.

4.2.1.3.3 Fuel Cell Operation A fuel cell is an electrochemical generator, which is an apparatus that produces electricity from a chemical reduction–oxidation reaction that converts the chemical energy of the component into electricity. The fuel cell theoretically operates provided that it is fed in fuel (hydrogen) and oxidizing agent (oxygen). This reaction occurs inside an elementary cell or membrane electrode assembly (MEA), which is made up of two porous electrodes separated by an electrolyte. The first electrode, the anode, is negatively charged and is fed by hydrogen. The second electrode, the cathode, is positively charged and is fed by oxygen.

Different types of fuel cell exist and can be distinguished by the nature of the electrolyte. The two major types of fuel cells are described hereunder: the proton exchange membrane fuel cell (PEMFC) and the solid oxide fuel cell (SOFC).

In a PEMFC, at the anode, a platinum catalyst is used to dissociate hydrogen into protons (positively charged hydrogen ions) and electrons. The protons H^+ move toward the cathode through the electrolyte. The electrons are transferred from the anode to the cathode through an external circuit, which generates electricity. At the cathode, a catalyst allows for the re-combination of protons, electrons, and oxygen to form water. Globally, the fuel cell consumes hydrogen and oxygen to produce electricity, water, and heat. The electrolytic membrane material must show interesting properties of proton conductivity, resistivity against the flow of current and robustness. The active layer comprises catalysts based on noble metals from platinum group. Its optimization consists of reducing the quantity of noble metals and in the increase of the active surface by reducing the diameter of the nanoparticles in a way to reduce the cost and the risk of limited available resources and corrosion. Low-temperature PEMFC require a minimum temperature of 70–80 °C to operate, while this temperature is 160°–180 °C for high-temperature PEMFC (Wang et al., 2014).

The elementary cells are assembled in a stack as a function of the power that is needed. Water electrolysis work in the opposite way. An electric current passes through the water, the molecule of which is split into hydrogen and oxygen.

PEMFC show several assets:

– A good efficiency of electricity production: an efficiency of 33% based on natural gas HHV has been measured by Davila et al. (2022) on a commercialized stationary fuel cell,
– The relatively constant efficiency of a PEMFC coupled to an electric motor of a vehicle during intermittent urban use,
– The absence of production of polluting gases (nitrogen oxides, ozone, etc.) and particles, independently from the pollution resulting from the production of hydrogen.

However, assessing the global efficiency in a well-to-wheels approach, considering the use of renewable energies, the efficiency for a BEV from the electricity production to the use onboard the car is of the order of 70–80%. By contrast, for a FCEV, with the production of hydrogen from electricity in an electrolyzer and with the production of electricity onboard the vehicle with the fuel cell, the global efficiency is of the order of 25–30%.

The other type of fuel cell under industrial development and commercialization is the SOFC, which works with a solid ceramic electrolyte conductor of negatively charged ions O_2^-. In a SOFC, at the cathode, oxygen from air is split into oxygen atoms. The latter ones take two electrons to become oxygen ions. These ions migrate through the electrolyte to reach the anode. At the anode, hydrogen is split into protons and electrons. The electrolyte being an electric current insulator, electrons pass through an external circuit. Protons and oxygen ions react to form water.

The SOFC has to work at a high temperature (600–1000°C) to ensure a high enough conductivity of the electrolyte to the oxygen ions. At such temperatures, the SOFC allows for internal reformation of natural gas. This fuel cell technology offers high efficiencies. An efficiency of 60% based on natural gas LHV has been measured on-site for residential applications (Paulus and Lemort 2022). However, due to the high operating temperature, it is difficult to implement inside a vehicle with short-duration and intermittent trips with frequent start-up phases. Indeed, such profiles of use of the vehicle are not compatible with the time needed to start-up the SOFC. As a consequence, it is currently mainly used in stationary and steady-state applications of cogeneration of electricity and heat.

4.2.1.3.4 *Fuel Cell Thermal Management*

The fuel cell system comprises the stack, and its surroundings are composed of components necessary for its operation (tanks, compressor, heat exchangers, inverters, etc.) and include air supply, water production, electricity production, and cooling systems. The air supplied to the fuel cell is cooled down. The water produced by the reaction is rejected outside the vehicle. Electricity produced by the fuel cell is either directly supplied to the motor or stored into batteries after conversion.

The cooling system is of paramount importance. It must ensure the thermal management of the system at temperatures lower than those of a thermal engine, which can make more difficult the evacuation of heat in hot climates. The first prototypes and pre-development programs of hydrogen vehicles have only appeared recently. Also, stationary fuel cell electricity generators have been arriving in the market for the last few years. It is however important to ensure that all phases of their life cycle (manufacturing, operation, and recycling) as well as their cost, security and lifespan are well mastered. Major technological, scientific, and societal issues must also be overcome with a good control of the car accident risks and of the infrastructure. The major challenges of the system are the large reduction of the costs (MEA, tank, compressor to increase the FC stack power density, thermal management system); the thermal management and the integration of the heat exchanger and fan components, which are critical on a full power system, as defined previously, since most of the heat released by the fuel cell is injected in the coolant loop at low temperature; and the cleanness of the heat exchangers for the non-contamination of water and gases, which is stringent; the filtration of the air at the supply of the cathode that impacts the lifespan of the fuel cell; the manufacturing process and the mass of the tank; the geometrical integration of the system and its reservoir inside the vehicle without any trade-off with the interior room; collision scenarios; the management of the air infiltrations in the fuel cell air supply and the impact of the aerodynamics and the external design, the compactness, the efficiencies of the components and their improvements.

Dedicated research and development programs must be carried out to prepare the future generations of technologies in terms of production, hydrogen storage, and development of new non–fluorine platinum-free materials for the electrodes of the fuel cell.

4.2.2 Hybrid Electric Vehicles

A HEV, or simply « hybrid vehicle », is a vehicle that can be propelled by fossil fuel and electricity source. As a function of the definition of hybridization, both energy sources can be used simultaneously or alone to propel the vehicle. Hybrid EVs cope with the drawbacks of ICE vehicles and electric vehicles. In terms of energy efficiency, HEV show the advantages of optimization of the ICE operation and performance and recovery of kinetic energy during braking (regenerative braking). There are numerous configurations of hybrid powertrain systems. Technical and scientific literature typically classifies the different hybrid systems according to the degree of hybridization and the hybrid powertrain architecture.

4.2.2.1 Classification According to the Degree of Hybridization

Hybridization, or the electrification of ICE, has become essential in the use of engines and vehicles to reduce CO_2 emissions. The range of electrification of powertrain is very large and is among other things function of the battery voltage and capacity. It mainly encompasses the principles of:

- *Stop & start*, which allows for the shutdown of the thermal engine during the standstill phases of the vehicle (traffic jam, traffic light, etc.) before its restart.
- *Regenerative braking*, which allows producing electricity during deceleration or engine braking phases.
- *Power assist*, which uses the electrical energy from the battery to supplement or boost the thermal motorization.
- *Stop and start sailing*, allowing to shut down the ICE at relevant times and to reduce the time of use of the ICE driving mode.
- *EV driving mode* (or ZEV mode) with an electric propulsion of the vehicle limited to the capacity of the battery.

Hybrid vehicles can be classified as follows:

Micro-hybrid vehicles show the lowest level of hybridization. It integrates the Stop & Start function. For sedan vehicles, the electric motor provides a power of around 2.5 kW to start the ICE (Chan 2007). The battery capacity is typically around 0.1–0.2 kWh. Regenerative braking can possibly be implemented in micro-hybrid.

Mild-hybrid vehicles are equipped with a more powerful electric motor (around 10–20 kW for sedan vehicles (Chan 2007)). Mild-hybrid vehicles allow for Stop & Start operation and boost (power assist system) during acceleration and regenerative braking. Battery capacity of mild-hybrid vehicles typically ranges from 0.8 to 1.5 kWh.

Full-hybrid vehicles show the functions of the mild-hybrid vehicles as well as the EV driving mode and have electric motors with power larger than 20 kW and typically around 50 kW for sedan vehicles (Chan 2007). Such vehicles can be partially driven by the electric engine. The battery capacity typically ranges from 1.5 to 3 kWh.

In *PHEV*, electrification is heavier. The battery pack (the capacity of which is around 5–15 kWh) is externally rechargeable by plug-in it to the stationary electricity grid. This charging mechanism of the battery allows for getting a larger driving range, in electric mode (ZEV), for instance 20–60 additional kilometers depending on the size of the battery. It also yields CO_2 emission reductions, but with a large extra cost. When the electric battery is empty, the working mode is similar to a classical hybrid vehicle. Electric motor shows a power ranging from 20 to 100 kW. The vehicle can be propelled solely by the electric engine. The interest for a PHEV lies in its capacity to be regularly plugged for charging the battery. In the opposite case, the PHEV is like a hybrid vehicle, and its additional weight can yield fuel overconsumption.

The range thus extends from small mild hybrids in 12 or 48 V (which allow for relatively small fuel saving of the order of 5%) to full hybrids (which can slightly exceed 20% of gain with respect to ICE engines in an optimal case). The reduction of the weight and the drag coefficient SCx is still an important means to decrease the CO_2 emissions. For instance, a full-hybrid SUV can have a similar impact in terms of consumption to that of an ICE sedan vehicle, because of the increase in the weight and SCx of a SUV and of the weight of the hybrid technology.

Beyond the Full Hybrid, the PHEV has involved almost all car manufacturers to achieve large CO_2 savings, but required an important investment on the technology, which was difficult to implement. The last stage is now the full electrification of the vehicle without ICE. Intermediate solutions could be considered, including for instance a PHEV at a reduced cost or Range Extenders.

In order to meet the future CAFE regulations, largely electrified vehicles, such as PHEV and BEV, appear to be increasing necessary in the mix of vehicles of the manufacturers.

ICEs without electrification are expected to disappear in regions such as Europe where regulation is stringent.

In a hybrid motorization, the controller selects between e-driving or ICE driving. The energy management is complex and integrates many parameters such as the request by the driver at the accelerator pedal, the state of charge of the battery, and the level of use of accessories. This energy management must also ensure several fundamental principles:

1. Optimize the function of electric regenerative braking to recover as much electric energy as possible while ensuring the braking function.
2. Stop the ICE when possible, by replacing it by the e-drive mode, during for instance driving conditions at low load and low speed where the ICE shows lower efficiency (because of the relatively high share of friction losses that do not depend on the load). For a similar work, the ICE is thus used during a shorter time duration, but at higher load in order to limit constant frictional losses.
3. Use the ICE on better efficiency operating points in boost mode or with a power production larger that the driver's demand. The surplus of energy is thus converted into electricity and added to battery charge to initiate electric modes and ICE stops.

The number of operating hours of the ICE engine is thus two to three times lower than with a conventional ICE, but re-starts of the ICE are roughly three times more frequent and the average load of the operating ICE roughly two times larger. Because of this lower cumulated time of use of the engine, the climatization compressor and the glycol water circulating pump are electrically driven. Moreover, the lubrication systems must account for the more frequent engine restarts.

4.2.2.2 Powertrain Architectures

The distinction is usually done between series hybrid, parallel hybrid, and series–parallel hybrid and complex hybrid configurations.

In *series hybrid* configuration, the wheels are only driven by an electric motor. As shown in Figure 4.3, the ICE drives a generator that produces electricity. The latter can either charge the batteries or supply the electric motor. A series hybrid vehicle has less mechanical components and can be seen as an EV with an ICE to extend the driving range (Chan 2007).

The different operating modes of the series hybrid powertrain are (Chan 2007):

– Propulsion by the electric motor fed by the battery pack
– Propulsion by the electric motor fed by the ICE/generator
– Propulsion by the electric motor fed by both the battery pack and the ICE/generator
– ICE used to both propel the vehicle and charge the battery ("power split mode")
– Static charging (when the vehicle is parked)
– Regenerative braking

In *parallel hybrid* configuration, the wheels can be driven by either the ICE, or the electric motor or both. As shown in Figure 4.4, a mechanical transmission system ensures the mechanical coupling between both engines and the wheels. The parallel hybrid vehicle can be seen as an ICE vehicle assisted by an electric engine to decrease fuel consumption and emissions.

The different operating modes of the parallel hybrid powertrain are (Chan 2007):

– Propulsion by the ICE alone
– Propulsion by the electric motor alone
– Propulsion by both the ICE and the electric motor
– ICE used to both propel the vehicle and charge the battery ("power split mode")
– Static charging
– Regenerative braking

The series–parallel configuration combines the features of the two former configurations and is shown in Figure 4.5.

A fourth typical configuration is the complex hybrid one but is not presented here.

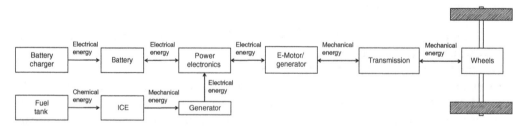

Figure 4.3 Energy flow representation of the powertrain of a series hybrid configuration.

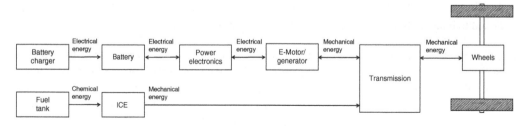

Figure 4.4 Energy flow representation of the powertrain of a parallel hybrid configuration.

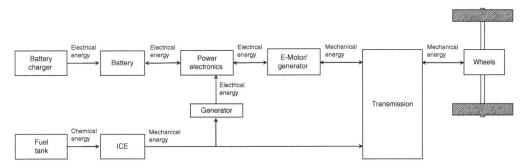

Figure 4.5 Energy flow representation of the powertrain of a series–parallel hybrid configuration.

Hybrid powertrain architectures have an impact on the increase in the volume and on the packaging. One of the main difficulties of installation concerns the zone of the battery and of the fuel tank, for which numerous architectures are proposed. A trade-off must often be found as it is the case for PHEV. This trade-off affects the driving range in relation with the battery capacity and with the volume of the tank, the volume of the trunk, and the comfort at the rear seats.

4.3 Cabin Thermal Control in HEVS and EVs

The needs for the thermal comfort in transient regime and in steady-state regime and the management of mist and frost on glazing leads to thermal loads profiles. As shown in Chapter 3, these profiles indicate that:

- The levels of heat transfer rate required for cabin heating are (according to the geographical zone) larger than those required for cooling,
- The levels of heat transfer rate during transient periods leading to the steady state regime (to keep the comfort) are larger than those during the steady-state regime by 50% to 100%,
- The air recycling rate has a large impact on, among other things, the heating demand in steady-state regime. For instance, the heating load can be reduced by 50% when implementing a full recirculation rather than a full air renewal,
- The dehumidification function that relies on simultaneously air cooling and heating leads to additional important loads,
- The energy balance depends on the climatic conditions of the concerned region, which determine the rate of use of heating and cooling functions.

In a vehicle equipped only with a thermal engine, the thermal losses of the engine can cover the cabin heating load. The compressor of the air-conditioning system and the pump of the coolant circuit are driven mechanically by the engine through the serpentine accessory belt drive. The upcoming electrification of vehicles forces technologies to evolve and new technologies to emerge.

4.3.1 Technical Challenges Associated with Cabin Thermal Control in Electrified Vehicles

4.3.1.1 Vehicles with Stop & Start Functions

The Stop & Start function allows for the consumption of micro-hybrid vehicles to be decreased by switching off the thermal engine when the vehicle is at standstill. The air-conditioning compressor and the coolant circuit pump, driven by the thermal engine, are thus also stopped, yielding a decrease of the means to achieve or maintain the thermal comfort. Only the thermal capacity of the glycol water stored in the heater core can be used for heating. The available low voltage (14 V for instance) does not allow for the use of an electric air-conditioning compressor and limits electric heating to a power of roughly 2 kW (larger powers would yield too high current intensity, cable section and Joule losses).

For the heating needs, an electric pump of small electric capacity can be used to ensure the circulation of hot glycol water through the heater core. Another solution consists of using a thermal storage to compensate the lack of heating production and increase the period of time when the thermal engine is switched off without affecting the thermal comfort objective. A solution uses a sensible storage tank filled with hot glycol water during the driving phase when temperature is high enough. This reservoir shows an efficient insulation layer to use it also straight after engine start up and hence speed up the temperature rise of the engine or cabin. This solution is investigated in Chapter 3. To increase the heat storage capacity (in [Wh L^{-1}]), latent energy storage solutions using PCM or sorption energy storage are currently investigated.

For the cooling needs, the evaporator can be equipped with a volume filled of PCM, generally of paraffin type, that stores cooling energy during the use of the air-conditioning system. This PCM is usually introduced in plates and exchanges heat directly with refrigerant and air. The quantity of PCM and their thermal capacity determine the duration of thermal conditions maintenance following a stop of the air-conditioning system as a function of ambient conditions (temperature, humidity, and radiation). The thermal engine is usually started again when the air temperature upwards the evaporator has reached up a temperature of around 12–15°C, in such a way to avoid the evaporation of condensates accumulated on the surface of the evaporator and associated odors released inside the cabin.

4.3.1.2 Vehicles with Regenerative Braking

The regenerative braking function allows for the battery to be charged during deceleration or braking phases. The recovered capacity is, however, not high enough to use an electric compressor during all phases of use. A hybrid compressor composed of a part driven mechanically by the thermal engine and a part driven electrically by an electric motor has been developed. It allows for a sporadic use of electricity, but this solution has not largely come to market.

4.3.1.3 Vehicles with Electric Driving Mode

HEVs and PHEVs (equipped with a thermal engine) and BEVs and FCEVs (full electric) have a high voltage and have an electric driving mode. Auxiliaries are not driven by a belt, and the air-conditioning compressor is generally an electric scroll machine (see Chapter 3). Rotating compressors can possibly be used.

4.3.1.3.1 Reduction of Driving Range
The analysis of the use profile of full electric vehicles or electrified vehicles working in electric mode show a large occurrence of short-distance trips and of short-duration trips characterized by transient regime associated with vehicle start. This leads to a larger relative weight of these transient conditions. This weight is even more important if the vehicle is "soaked" after outdoor parking with cold or hot conditions rather than indoor parking in a climatized place. Steady-state conditions can possibly not be met at the end of the trip.

Moreover, in the electrical mode, the large efficiency of the powertrain (roughly 95% for the electric motor and the power electronics), its cooling achieved at a low temperature (lower than 50°C), and the limited thermal losses (a few hundreds of W in urban and peri-urban driving) make the waste heat valorization more difficult to meet the cabin heating load (a few kW with a pulsed air temperature of 50–55°C). As a consequence, an additional solution for heating is necessary. In general, this solution yields a non-negligible electric consumption.

Another consequence of the large powertrain efficiency, combined to the capacity of energy recovery during driving, is its low average consumed electrical power during urban/peri-urban driving. On an urban driving cycle, it does not exceed 3.5 kW. As a consequence, the consumption associated with thermal comfort quickly reaches the same order of magnitude that the consumption associated with propulsion in the electrical mode. Hence, the driving range is highly affected by the thermal system, and the penalty depends more on the duration of use than of the traveled distance.

On trips characterized with larger powers of the powertrain, the discharge period of the battery is shorter, and the relative penalty of the thermal system is weaker.

Hence, with traditional techniques, in the electric mode, the electric consumption associated with the thermal comfort leads to a large decrease of the driving range. This decrease is a function of the climatic conditions, of the duration of the use of the thermal systems over the trip, and of the power delivered by the powertrain. During urban driving, this decrease of the driving range may reach 50% with electric heating when outdoor temperature is 0°C and 30% with electric air-conditioning when outdoor temperature is 35°C. This decrease is close to 25% as a yearly average over the European climate according to typical use, i.e. roughly 300 km of driving range instead of 400 km achieved with an electric vehicle when auxiliaries are not used. The driving range of EV is a major factor affecting their success. Hence, it appears necessary to minimize this impact on the driving range over a range of temperatures as large as possible. This could be achieved by developing solutions for the thermal management of the cabin that largely reduce the electric consumption

- Of heating, targeting mainly the range [−10°; 5°C],
- Of cooling, targeting mainly the range [25°, 35°C],
- Of dehumidification, which requires a combined production of cooling and heating for dehumidifying the air and post-heating it. This mainly occurs in the range [5°, 20°C],

The objective in summer conditions is easier to achieve than that in winter conditions, especially in regions where the yearly average temperature is lower than 15°C. This is mainly due to the high efficiency of the air-conditioning system because of the use of vapor-compression systems and to the larger potential of air renewal at high temperatures.

4.3.1.3.2 Deficit of Thermal Losses for Cabin Heating   In the electric driving mode, the cabin heating function can only valorize low thermal losses, which leads to the need of an electric heating. The latter one can deliver heat directly to the cabin or indirectly through a glycol water loop equipped with an electric pump. Electric heating solutions are facing significant changes with the evolution of electrification of vehicles and available electric voltage whose trend is to increase to enable fast charging of the battery. Available voltage reaches values of 400–800 V and yields a rate of generated heat ranging from 5 to 10 kW to meet the heating needs at very low outdoor temperature. Electric heating by PTC (electric resistances) converts almost entirely the electric energy in heat. Thermodynamic heating by means of a heat pump, reversing the air-conditioning loop, is an interesting alternative since its coefficient of performance (COP) is higher than one, which allows for a signification decrease of its impact on the driving range of the vehicle.

4.3.1.3.3 Need for a Global Thermal Management Approach   As explained previously, the functions of heating, air-conditioning and dehumidification of the cabin and thermal control of the powertrain and of the battery have a significant impact on the driving range of the vehicle in electric mode. This impact can be reduced by considering a global analysis of the needs and available means and a system-level approach for the design of the thermal management where synergies between different thermal functions (thermal comfort, thermal control of powertrain, and battery) are

sought. The objective aims at identifying and demonstrating the feasibility of solutions associated together in such a way to largely decrease the electric consumption of the whole vehicle thermal management.

The analysis allows for the identification of the following needs:

- Ensure the thermal, aeraulic and acoustic comforts as well as the visibility and air quality in the cabin.
- Increase the availability and the lifespan of the battery pack and maintain the temperature of the powertrain in the appropriate range.

The first step consists of identifying the means allowing for a *decrease of the cabin thermal load*. This could be achieved by pre-conditioning the cabin and optimizing on the one hand the insulation of opaque walls and glazing and on the other hand the management of air inside the cabin. The objective is to limit:

- the temperature gap (with comfort temperature) when the vehicle is started by handling the cabin air during the vehicle parking phase by means of the thermal management system,
- the solar gains,
- the radiative fluxes inside the cabin from the surfaces (roof lining, dashboard, and glazing) impacted by the outdoor environment (solar radiation and temperature),
- the conductive heat transfers with components (seat, steering wheel, etc.) impacted by the outdoor environment (solar radiation and temperature),
- the fresh air flow rate by optimizing the air recycling rate inside the cabin,
- the use of heating and cooling mechanisms consuming energy by storing thermal energy during parking phases without consuming electricity from the battery.

The second step consists of identifying the means of *local thermal production* that allow for a reduction of the global consumption associated with thermal comfort. As seen previously, the thermal comfort sensation results from a combination of convective, radiative, and conductive heat exchanges. It is thus possible to act on the comfort by using other levers than the air management. For instance, the following mechanisms can be used to treat locally the thermal comfort and ensure positive stimuli:

- the contact temperatures (seat and steering wheel),
- the radiation from the walls,
- the air flows handled locally around the passengers.

The third step consists of identifying the means allowing for the *improvement of the efficiency of the thermal production systems*, by among other things:

- optimizing the air-conditioning loop by the electrification of the compressor, the choice of the refrigerant, the energy recovery during the refrigerant expansion, the heat transfer between the high and low pressures (through the internal heat exchanger), the technology of throttling device, the management of the outdoor flow, and the control of the system,
- optimizing the heating system by a thermodynamic solution,
- valorizing the residual thermal losses from the powertrain,
- optimizing the production using a thermal storage system,
- optimizing the production using a mutualized heat transfer loop that allows for energy transfers and recovery of losses.

This analysis must also take into account the good repartition of air inside the cabin by wise positioning of air inlets and outlets, the purification of air by means of filtration systems, and the good visibility through the glazing.

Finally, the full system used to achieve thermal comfort must be *judiciously integrated inside the vehicle global thermal management system*, which takes into account the thermal control needs of the different components of the vehicle (battery, powertrain, etc.). The glycol water loop can then get more complex by mutualizing the thermal transfers of the different components.

This latter step aims at achieving the following objectives:

– the optimization of charging/discharging cycles of the battery,
– the maintenance of the battery temperature in its operating range with cooling/heating solutions through a fluid loop (air, liquid, oil, or refrigerant) or electric resistances,
– the maintenance of the powertrain temperature in its operating range with cooling solutions through a fluid loop (air or liquid).

4.3.1.3.4 Synthesis of Relevant Thermal Management Technical Developments Among the technological solutions that address this analysis, one can mention:

– A better air management through an optimized partial air recycling allowing to increase the rate of cabin air recycling. As shown in Chapter 3, a minimum outdoor air flow rate depending on the number of passengers in the vehicle cabin is necessary in order to prevent, especially at low outdoor air temperature, any risk of mist formation in the cabin. This constraint is the main factor of limitation of air recirculation, which would be really useful at low outdoor temperature in electrified vehicles and mainly in BEVs that do not have any thermal powertrain that could deliver during some driving modes waste heat at a sufficient temperature to contribute to thermal comfort. An important gain on the driving range is then possible when the optimal recirculation rate is large especially when one single passenger is in the cabin. The increase of the number of passengers yields a decrease of this gain. The recirculation rate must hence be well managed in such a way to prevent the use of the air/conditioning loop to dry the air, which would lead to an additional electric consumption. However, in heating mode, even if the optimal recirculation rate is lower than in the case of high outdoor temperatures, in the electric mode, it could provide a significant gain in terms of consumption and driving range, because the usual recirculation rate is low or even null due to risk of mist formation.
– Heat pump systems with an efficiency much larger than that of an electric resistance.
– Seat heating, which brings a solution for local thermal comfort.
– Innovative glazing with high thermal performance, such as the heated windshield with a surface treatment with reflective and electrical conductivity properties. Such properties reduce the solar gain in summer and improve the availability and security of the vehicle and the electric consumption during demisting and defrosting phases of the windshield. Another type of innovative glazing shows an insulating treatment at low emissivity that reduces the solar gains in summer and traps the heat inside the vehicle in winter (and reduces the sensation of cold wall), reducing the thermal demand at iso comfort.
– The thermal pre-conditioning of the cabin achieved when the vehicle is at a standstill in the parking mode, especially if it is connected to the stationary electricity grid. The heating or cooling solutions are activated as a function of the needs in such a way to reach or to get close to the thermal comfort and to reach a good visibility through the glazing before occupants use the vehicle. Pre-conditioning can be activated through a mobile phone, a schedule, or when getting close to the vehicle.
– The thermal energy storage systems that can be charged by means of electricity from the stationary grid during parking phases, which allows not to consume electricity from the battery during

driving phase. This thermal energy can be otherwise produced and stored when conditions are favorable to obtain a good efficiency and released during less favorable phases to improve the operating point and the efficiency of the system.

- The air stratification with two independent air circuits (fresh dry air pulsed in the upper zone of the cabin in contact with the glazing, recirculated more humid air in the lower zone), allowing, for the same humidity ratio in the vicinity of the glazing, to minimize the air renewal and hence to increase the recycling rate without any risk of mist formation despite a larger average humidity ratio. This solution shows the advantage of decoupling the management of the temperature, of the humidity and of the air renewal, but also of decoupling the management in flow rate and temperature of the air flows to the heads and to the feet.
- The containment of the thermal comfort to the zones occupied by a passenger, for instance the sole driver zone if the driver is the only occupant of the cabin. The comfort localized to the driver zone can be achieved with largely reducing the flow rates to the other passenger seats. Among the risks incurred, one can mention the movements inside the cabin of air masses at temperatures far from the thermal comfort, for instance during vehicle turns.
- The recovery and the valorization of thermal losses of the powertrain and of the battery to heat the cabin, either directly or by using a heat pump to upgrade the temperature level.
- The use of CO_2 refrigerant in an indirect water/water compact loop, which eases the assembly and maintenance of the loop and which operates at a very low temperature with an improved efficiency.

The refrigeration circuit can therefore be complexified by the new function of heat pump and by the cooling of the battery by means of a specific evaporator on air or on glycol water circuit (chiller). In the front-end module, the condenser that was in direct contact with outdoor air can evolve toward an indirect condensation with a coolant circuit that can be shared with other thermal systems of the vehicle. Also, the original condenser can evolve toward a reversible heat exchanger (condenser in the A/C mode and evaporator in the heat pump mode). The evaporator section of the A/C circuit can also evolve to ensure the cooling of the battery or the valorization of the heat losses (in the heat pump mode) or integrate phase change materials for cold storage useful during periods when the A/C compressor is switched off. These evolutions allow for the use of the A/C loop in the global thermal management with the objective to reduce the energy consumption while ensuring the thermal comfort and the thermal control of the components.

Additional gains could be achieved by minimizing the objectives in terms of thermal comfort and by modifying the habits of occupants.

The function of thermal comfort will have to adapt to the coming evolution toward concepts of electric or highly electrified, autonomous, and connected vehicles with passengers whose characteristics are identified and showing different locations and positions in the cabin. The thermal comfort will have to be personalized, more efficient and adaptive all along the journey.

The concepts of air distribution and heating/cooling by convection, conduction, and radiation are also subject to change.

Presence sensors (infrared, contact sensor integrated in the seat, etc.), CO_2 and humidity sensors, solar radiation sensors, heart rate sensors to assess the stress or risks, alertness sensors (camera) will be possibly introduced in the cabin to optimize the thermal comfort in the different zones of the cabin as a function of the presence, the position and the activity of passengers.

With those evolutions, the air-conditioning system will be increasingly integrated in the global thermal management of the new powertrain and mobility technologies in such a way to limit the energy consumption and optimize the reliability of components.

4.3.2 Heat Pump Systems

In comparison to conventional vehicles with internal combustion engines, EVs suffer in the heating mode and in the dehumidification mode from a lack of waste heat at an appropriate temperature to heat the cabin. Electric resistance heaters (PTC) can be used (to produce directly hot air or hot water), but such a heating mode could contribute to drive range anxiety or to increase the size of the battery pack. According to Feng and Hrjnak (2016), using electric resistance heaters can decrease the driving range up to 60% depending on the outdoor air temperature and the usage conditions. This is due to the low performance of electric resistance heaters, which exhibit maximal coefficient of performance of 1. A better alternative from the energy point of view is to use a heat pump. In cold weather conditions, the driving range could be increased by roughly 20–30% when using a heat pump instead of electric resistances (Feng and Hrjnak, 2016), which also decreases the variability of the driving range linked to the cabin heating system.

Also, the dehumidification function achieved with the air-conditioning loop by condensing the water and hence drying the air, followed by an electric resistance heating to compensate for the cooling and meet the comfort, consumes a lot of energy. A reversible air-conditioning/heat pump loop, which is wisely designed, allows for the simultaneous use in the Air Conditioning Module of the evaporator to cool the air and of the condenser to heat the air with a minimized electric consumption.

With a heat pump system and an optimized management of the cabin thermal needs, the yearly average electric consumption, for the European climate, of the thermal functions at isothermal comfort can be reduced by approximately 50% depending on usage conditions. This gain, which largely varies with the climatic conditions, allows for a reduced variability of the driving range. For instance, for a passenger car:

- At −18°C, the gain is larger than 3 kW, which is roughly 50% of additional driving range by combining a controlled air renewal rate, a heat pump, heated seat, and low-emissivity glazing.
- Between −5 and + 15°C, a largely occurring temperature range, this gain reaches 1.4 kW, which represents 30% of additional driving range.
- At +35°C, gains are close to 400 W, with a better control of air renewal, new glazing, cooling seats, the improvement of the A/C loop (with a better control of expansion and heat exchanges), leading a gain in driving range of 5 to 10%.
- Between −20 °C and + 5 °C, an additional gain in the driving range of around 25% can be reached by the pre-conditioning of the cabin and by a heat storage (approximately 3 L) charged during standstill, in direct relation with a use of the vehicle on short trips, for instance lower than 20 minutes.

In a heat pump mode, the compressor must accept pressure ratios up to approximately 12 to explore a large range of operating conditions. The energy efficiency optimum is obtained by combining the heat pump with an electric resistance heating system, which allows for the pressure ratio to only rarely exceed 8. Also, the minimum suction pressure is limited to 1.1 bar to ensure tightness and prevent any risk of non-condensable gases and humidity infiltration due to sub-atmospheric conditions. The compressor swept volume is defined to be adapted to the targeted range of heating power, while limiting large rotational speeds, and associated noise, at large heating power. A thorough mechanical validation for suction temperatures and pressures particularly low (−20°C and 1 bar) must be conducted to check, among other things, the appropriate lubrication while meeting constraints on refrigerant/oil miscibility and good oil return to the compressor.

More and more vehicles are equipped with heat pumps. There is a large panel of configurations. Similar to that done in building applications, heat pumps have been classified here according to the nature of the heat source and heat sink. The distinction can be done among air-to-air heat pumps, air-to-water heat pumps and water-to-water heat pumps. The system can thus deliver heat directly into the air through an air-cooled condenser located inside the HVAC unit or indirectly through a glycol-water-cooled condenser connected to the heater core by means of the glycol water loop. In most designs, the heat source of the heat pump is the outdoor air. Alternatively exhaust air from the cabin or heat generated by the power electronics and electric motor can also be used. These alternative heat sources present the advantage of being at higher temperatures, yielding to better heat pump performance. However, their heating capacity is limited.

In automotive applications, the same machine combines the functions of cooling (A/C) and heating (HP). Hence, heat pumps are said to be "reversible." However, the reversibility of the machine does not always rely on the use of a four-way valve as in stationary applications. As shown hereunder, switching from cooling to heating production can be achieved by using three-way or two-way valves, the cost of which is lower than that of four-way valves. The heat pump has also to be able to dehumidify the air entering into the cabin for windshield demisting. It will be shown in the next sections how the design of the heat pump can be adapted to fulfill such a function. Finally, heat pump systems have to be integrated into larger thermal management systems ensuring the cabin climate control but also the thermal conditioning of the battery, power electronics, and electric motor.

4.3.2.1 Air-to-Air Heat Pumps

Air-to-air heat pumps, or "direct heat pumps," use air as both the heat source and the heat sink of the heat pump. A configuration using three heat exchangers currently installed in several EVs is depicted in Figure 4.6. In this configuration, the outer heat exchanger (OHEX) is located in the front-end module. In subcritical cycles, it works as a condenser in the cooling mode and as an evaporator in the heating mode. The HVAC unit houses both the inner condenser (ICD) (or inner gas cooler) and the inner evaporator (IEV). The inner condenser replaces the heater core of a traditional heating system. PTC heaters are used in series with the ICD to increase the heating capacity. When using R-744 as a refrigerant, the outer condenser and the inner condenser become gas coolers.

The circuit can also ensure the battery cooling when it is necessary.

According to the selected operating mode, the expansion valves can be either used or deactivated by means of 2-way or 3-way valves.

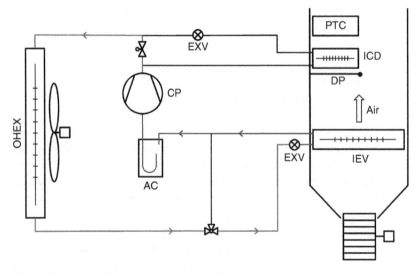

Figure 4.6 Air-to-air heat pump in the cooling mode.

The accumulator upwards the compressor allows for variations of refrigerant charge to be absorbed in the different operating modes and allows for the compressor to be protected against liquid slugging.

4.3.2.1.1 Use of the Heat Pump in the Cooling Mode

Figure 4.6 illustrates the cooling mode. Cold air is produced in the inner evaporator. The position of the damper (DP) prevents air from flowing through the inner condenser. On the refrigerant side, a three-way valve allows refrigerant to flow successively through an electronic expansion valve (EXV) and through the IEV. At the outlet of the compressor, a two-way valve prevents refrigerant from flowing through the ICD.

As illustrated in Figure 4.7, some designs do not prevent refrigerant flow through the ICD in cooling mode. This yields to a lower A/C system performance, because of air leakage through the ICD, resulting in undesired heating-up (which must be compensated by additional cooling capacity).

Figure 4.7 also indicates how the coupling with the battery cooling could be achieved. The two-way valve 2-WV3 opens to feed the battery cooling evaporator (producing cold air or chilled water).

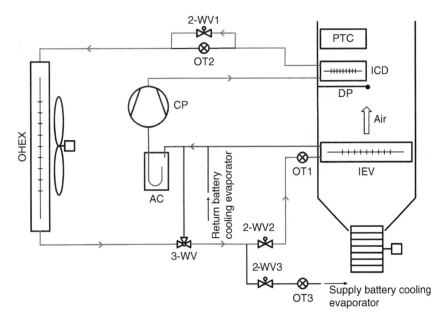

Figure 4.7 Air-to-air heat pump in the cooling mode.

4.3.2.1.2 Use of the Heat Pump in Heating Mode

Figure 4.8 illustrates the operation in the heating mode. When switching from the cooling (Figure 4.7) to the heating mode, the two-way valve 2-WV1 closes allowing the fluid to flow through the orifice tube (OT2). Also, the damper is actuated to drive the air through the ICD.

In the configuration presented in Figure 4.7 and Figure 4.8, the refrigerant always flows in the same direction through the OHEX regardless of the operating mode. The OHEX must be designed accordingly. Inversion of the direction of flow could be made possible by using additional valves. This would offer a better match between the heat exchanger geometry and the fluid density. If flow is reversed in the OHEX, using an accumulator at the compressor supply rather than a receiver associated to the OHEX could ease the operation of the OHEX in both modes.

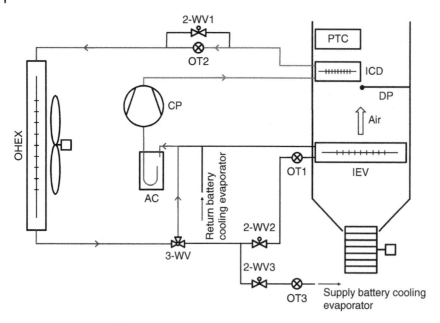

Figure 4.8 Air-to-air heat pump in the heating mode.

Figure 4.9 Inner condenser: extruded tubes and fins (Courtesy of Valeo).

The use of additional valves (2-way, 3-way, and expansion valves) contribute to increasing the cost of a heat pump in comparison to a traditional A/C system.

Figure 4.9 shows an example of the inner condenser. This heat exchanger is a fin-tube one with a U-flow (2 passes) configuration on the refrigerant side. The top-down pass is on the rear side of the heat exchanger in the direction of the air flow, such as it is in contact with hotter air.

4.3.2.1.3 *Use of Heat Pump in the Dehumidification Mode* This configuration with three heat exchangers allows for the dehumidification mode. In such an operating mode, depicted in Figure 4.10, the inner evaporator and the inner condenser (or inner gas cooler) are simultaneously used. Air entering the ACM is first cooled down in the IEV underneath its temperature dew point to allow for water condensation. Dehumidified air is then post-heated in the ICD before being pulsed into the cabin. Figure 4.10 shows a configuration with electronic valves (EXV) for the purpose of illustrating the large range of technical solutions.

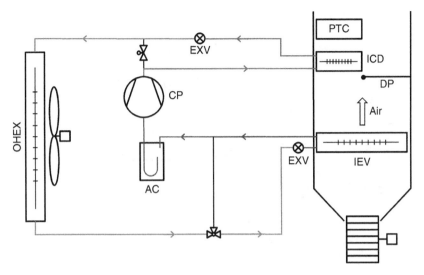

Figure 4.10 Air-to-air heat pump in the dehumidification mode.

4.3.2.1.4 Features of the Outdoor Heat Exchanger The OHEX, or "evaporator-condenser," is a key component of the heat pump. Its design must ensure a compromise to satisfy the two operating modes. Its heat transfer capacity must be maximized and the temperature pinch point (difference between the air and refrigerant temperatures) minimized in such a way to optimize the heat pump performance and the management of frost formation by delaying it and making easier its evacuation by means of a higher surface temperature.

A "down-flow" circuit where tubes are put vertically limits the risk of frost formation, while easing the drainage of condensates along the tubes, but penalizes the packaging with a larger number of tubes and longer fluid boxes. A "cross-flow" circuit where tubes are set horizontally offers a better packaging but is less adapted to the management of successive frosting-defrosting cycles. The behavior of the heat exchanger is sensitive to external conditions of integration (air side) and to the internal conditions (refrigerant pressure, upstream condenser sub-cooling in heat pump mode) that modify the homogeneity of the distribution of the air and of the refrigerant and the evolution of the quality inside the heat exchanger, and consequently the repartition of the exchanges and the internal pressure losses. In the heating mode, the heat exchanger is located on a low-pressure branch, leading to high flow speeds for a given mass flow rate and to a pressure drop that must be minimized in such a way to limit the pressure ratio and the power consumed by the compressor. This pressure drop also yields a drop of the evaporating temperature between the inlet and outlet of the heat exchanger, and thus a variation of the temperature difference between the refrigerant and the air, which limits the heat transfer capacity.

In the heat pump mode, the frost starts to form on the outdoor heat exchanger in contact with outdoor air in the vehicle front-end module when the heat exchanger surface temperature is lower than 0°C. Frost tends to block the air passage and consequently decrease the air flow rate and the performance. As a consequence, the heating system must be sized to have enough capacity during this phase. Frost formation can be exacerbated by the snow, the hygrometry and the speed of the vehicle. Occurrences of frost formation are however limited, and the impact of frost can be controllable if it can be detected.

The management of the risk of frost formation or the achievement of a quick and efficient defrost process consists of maintaining a high enough surface temperature or in melting the frost

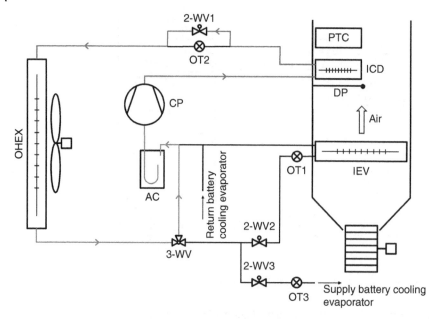

Figure 4.11 Air-to-air heat pump in the OHEX defrosting mode.

accumulated on the surface. Defrosting can be achieved by reversing the cycle and operating the heat pump as an A/C unit, with some caution not to cool the cabin. Alternatively, an example of modified ("degraded") cycle is proposed in Figure 4.11. The refrigerant remains in the vapor phase. The compressor works between a medium and a high pressure. Through the outdoor heat exchanger, the vapor is cooled down from high temperature to a medium temperature.

Another technique consists of using grille shutters to limit the air flow rate. Also, when the vehicle is at a standstill, depending on the type of the heat pump system, heat can be provided by another heat source than from the cabin, as for instance, a glycol water loop connected to the battery or to a heat storage system.

To delay frost formation, a hydrophilic coating of the surface of the outdoor heat exchanger or of the evaporator inside the HVAC unit can ease the evacuation of water. Hydrophobic coatings can also delay the frost formation. These treatments must account for the exposure to dust and more constraining splashing than on the HVAC unit evaporator. The fin spacing is also a factor that impacts the drainage of condensates and the management of frost formation.

It seems more favorable to place the outdoor heat exchanger behind the low temperature radiator that, compared to the high temperature radiator of a thermal vehicle, shows a low thermal load leading to an acceptable air heating for the performance in *air-conditioning* mode. *In heating* mode, it thus recovers part of the heat evacuated in the air by the radiator, which improves the performance and delays the risk of frost formation. The heat exchanger is also protected from splashing water, snow and dust, which reduces risks of frost formation for low ambient temperature and high moisture content.

4.3.2.2 Air-to-Water Heat Pumps

Air-to-water heat pumps, also called "indirect heat pumps," use a water-cooled condenser (or gas cooler), instead of an air-cooled condenser in heating mode (Figure 4.12), that heats a secondary water circuit. Glycol water is used instead of pure water because of its anti-freezing properties. This architecture concerns not only pure EVs but also hybrid vehicles having a full electric mode.

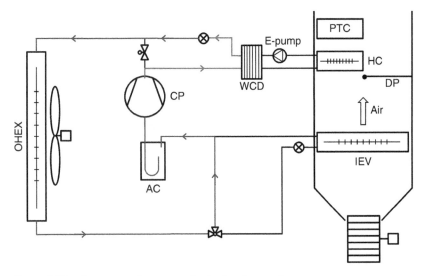

Figure 4.12 Air-to-water heat pump in the heating mode.

In hybrid vehicles, the glycol water circuit is shared by the heat pump and by the engine cooling and is connected to a heater core inside the HVAC unit. The same coolant loop can be used to cool down both the engine and the condenser. Produced hot glycol water is then driven to the heater core by a pump. In hybrid vehicles, it is necessary during thermal engine mode to heat the cabin with the engine coolant without using the heat pump.

However, this system shows a larger thermal inertia due to the secondary fluid loop, which slows down the warm-up process after vehicle start. Also, the energy efficiency is lower because of this thermal inertia, of the additional temperature pinch point due to the use of an additional heat exchanger that increases the condensing temperature and of the thermal losses of the additional water circuit under the hood. This yields a degradation of the COP, with respect to the COP of a direct air-to-air heat pump, depending on the usage and the sizing of the secondary loop. The glycol water flow rate feeding the condenser and the heater core is limited in such a way to obtain a water temperature at the condenser outlet that is high enough and close to the refrigerant temperature. This water temperature is much lower than the temperature obtained with a thermal engine.

Air-to-water heat pumps are more compact units than air-to-air heat pumps, limiting the charge of refrigerant and the risk of leaks. The condenser is a compact heat exchanger, such as a brazed plate heat exchanger. This architecture also allows keeping the heater core and preserving the traditional design of the HVAC unit, leading to a cost reduction because of a joint manufacturing with thermal vehicles from a same platform showing larger volumes.

The indirect configuration shows the advantage of a better control of the stability, among other things during frosting phases of the outdoor front heat exchanger without inducing a significant decrease in performance.

4.3.2.3 Water-to-Air Heat Pumps

In a water-to-air heat pump, the evaporator is heated by a glycol-water coolant. This coolant is itself heated by air in a heat exchanger integrated in the front-end module. Adding an outdoor secondary fluid loop in the front-end module largely degrades the energy performance of the heat pump. At low temperature, with a glycol water circuit at temperatures lower than 0°C, one can observe a large increase of the glycol water viscosity and of the pumping power. This yields

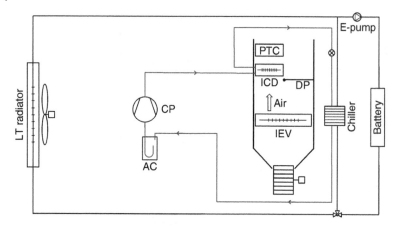

Figure 4.13 Water-to-air heat pump for battery waste heat recovery.

a drop in the flow rate and in the heat transfer capacity. Moreover, the heat pump evaporating temperature is 5 to 15 K lower and the evaporating pressure 0.5 to 1 bar lower than the values obtained with a heat pump directly connected to outdoor air. As a consequence, the COP is largely degraded due to the increase of the compression ratio but also to the increased pressure drop in the low-pressure line. Moreover, the objective of not achieving sub-atmospheric conditions in the refrigerant loop (the evaporating pressure must be larger than 1.1 bar) to ensure the tightness reduces the heat pump operating range at very low temperature. Finally, the defrosting of a front-end heat exchanger fed by a secondary fluid loop is more difficult to achieve, because the heating capacity and dynamics are limited when the sensible heat of the water circuit is used.

One possible configuration of a water-to-air heat pump is illustrated in Figure 4.13. It is part of a more complex thermal management system. The heat source here is the heat generated by the battery pack. The evaporator, named "chiller," is heated by the battery pack cooling loop. In the absence of heat demand onboard, the battery pack cooling loop can also be cooled by a low temperature radiator in the front-end module. Heat not only from other electrotechnical components (electric motor and power electronics) but also from a thermal storage or from cabin exhaust air can be valorized.

4.3.2.4 Water-to-Water Heat Pumps

This configuration provides the opportunity to consider a compact and possibly standard (nonreversible) refrigerant loop, the reversibility being ensured by the glycol water circuits.

A simple configuration of water-to-water heat pump, inspired by Leighton (2015), is represented in Figure 4.14 and Figure 4.15. The evaporator and condenser are heated and cooled, respectively, by glycol water loops. In the cooling mode (Figure 4.14), the evaporator is connected to the cooler core (CC) and the condenser to the OHEX. When switching to the heat pump mode (Figure 4.15), by using valves (not represented in those Figures), the evaporator is connected to the outdoor heat exchanger and the condenser to the heater core (HC). Other heat sources and heat sinks can be connected to the low-temperature and high-temperature glycol water loops such as batteries, electric motor, and power electronics.

This configuration of heat pump shows some advantages in comparison to configuration previously introduced and also shows the drawbacks of water-to-air heat pumps. The absence of cycle reversal prevents from any problem associated with refrigerant and oil migration (Leighton 2015).

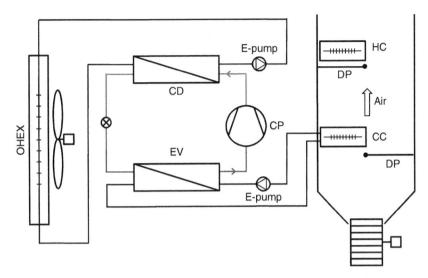

Figure 4.14 Water-to-water heat pump in the cooling mode.

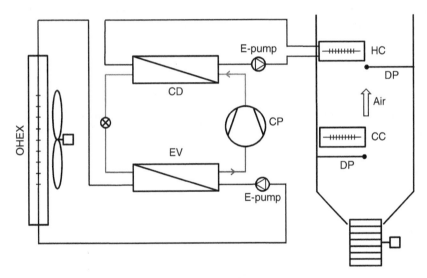

Figure 4.15 Water-to-water heat pump in the heating mode.

Also, the evaporator and condenser always work in the same mode, which simplifies their design (Leighton 2015). The refrigerant loop is much more compact, limiting the risk of leaks and the charge of refrigerant.

4.3.2.5 Back-Up Electric Resistance Heating System

Nevertheless, the heat pump systems described here above must be completed by an electric resistance heating system that works simultaneously with the heat pump system, in such a way to limit the heat pump thermal capacity and limit the compressor pressure ratio to maximize its efficiency and optimize the overall system COP. This situation occurs when the heat load is important, for instance, during cold start phases or when the outdoor temperature is very low, resulting in system efficiency decrease.

Using a back-up electric resistance heating system also allows the frost formation to be slowed down by increasing the evaporating temperature in the outdoor front heat exchanger. Electric resistance heating can be achieved:

- On the air in the HVAC unit with a PTC resistance, which allows a maximal efficiency, a short response time, and a de-coupling between the thermodynamic heating system and the electric resistance heating system to optimize the global efficiency of the system. However, having a large power under a high voltage imposes integration and security constraints. Also, PTC cannot be used to heat the water circuit, since it is on the air side.
- On the water, limiting the impact on a standard HVAC unit and allowing to deliver a high power. However, this solution suffers from the thermal losses and inertia of the water circuit during transients.

Example 4.2   *Air-to-Air Heat Pump with Back-up Electric Resistance*

An electric passenger car is heated with an air-to-air heat pump equipped with a back-up electric resistance of 1800 W (12 V PTC). At −5°C outdoor temperature, during warm-up phase, the heat demand of the cabin is 4 kW. The maximum rotational speed of the heat pump compressor is 8000 rpm. Compute the optimal power of the electric resistance and the overall COP of the heating system. At −5°C outdoor temperature, the heat pump capacity and COP are given as a function of the compressor rotational speed by polynomials given in the "3. Parameters" section.

Analysis

The global COP, accounting for both the compressor and electric resistance consumptions, is first expressed. The MIN-MAX function of EES is then used to maximize this COP as a function of the compressor rotational speed.

Solution

```
"1. INPUTS"
Q_dot_cabin=4000 [W] "Cabin heat demand"
{rpm_cp=2000 [rpm] "Compressor rotational speed"}
"2. OUTPUTS"
"W_dot_el_PTC [W]" "Power of the electric resistance"
"3. PARAMETERS"
Q_dot_cd=329,798077 + 0,368139607*rpm_cp + 0,00000971632187*rpm_cp^2
   "Heat pump capacity"
COP=2,48202787 - 0,000173483763*rpm_cp + 4,27326745E-09*rpm_cp^2
   "Heat pump COP"
"4. MODEL"
"4.1. Power of the electric resistance"
W_dot_el_PTC=Q_dot_cabin-Q_dot_cd
"4.2. Power of the electric compressor"
COP=Q_dot_cd/W_dot_cp
"4.3. Overalll COP"
Q_dot_heat=Q_dot_cd+W_dot_el_PTC
COP_global=Q_dot_heat/(W_dot_cp+W_dot_el_PTC)
```

Results:

```
COP=1,503 [-]
COP_global=1,376
N_cp=6773 [rpm]
Q_dot_cabin=4000 [W]
Q_dot_cd=3269 [W]
Q_dot_heat=4000 [W]
W_dot_cp=2175 [W]
W_dot_el_PTC=731 [W]
```

Results are represented graphically in Figure 4.16. It can be observed that the heat pump capacity (heat transfer rate delivered at the condenser) increases with the rotational speed, while the COP decreases. For a compressor speed below approximately 4500 rpm, the heating system is not able to cover the cabin heating demand, because the back-up resistance saturates at 1800 W. The rotational speed maximizing the overall COP is 6773 rpm. At that speed, the electric resistance provides 731 W and the heat pump 3269 W.

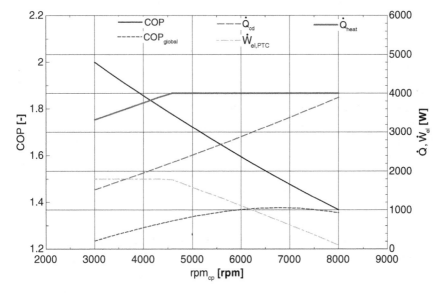

Figure 4.16 Evolution of the powers and COP with the compressor rotational speed.

4.3.3 Local Heating Systems

As described in Chapter 3, the heating and cooling systems traditionally used in automotive industry are based on heat transfers by convection by adjusting the mass flow rate and temperature of air pulsed inside the cabin, used as heat transfer fluid. The interest of a localized solution is to improve the thermal state of local body segments as well as the whole-body sensation and to reduce the use of the centralized convection-based system. The solutions are, among other things, efficient during transient ("convergence") phases when the local thermal constraint is strong by providing a thermal sensation quicker than the sensation provided by a centralized convection-based system. Actually, in the latter case, the thermal sensation depends on the thermal inertia of the heating production system itself and of the cabin air.

The local heating solutions bring a complement to the centralized heating function to satisfy the thermal comfort of passengers by combining global and local comfort. It also modifies the thermo-physiological behavior of the human being by stimuli through contacts as well as through the field of speeds and temperatures of air inside the cabin. For a vehicle in the electric mode that does not beneficiate from enough waste heat from the thermal engine, the local heating solutions (especially heated seats) in addition to PTC electric resistances can allow for reducing the air heating load and the overall energy consumption for a given comfort as well as the pulsed air temperature.

Among the different local heating systems described in Chapter 3, the following ones are relevant for extending the EV driving range: heated seats, heated steering wheel, and electric radiant panels.

As described in Chapter 3, some technologies of glazing aim at decreasing the energy consumption by improving the thermal comfort in the cabin and by countering mist formation. The electric reflective thin layer inserted in the windshield offers a gain against solar radiation constraints in a hot environment and in regards of convective needs during defrosting/demisting phases of the glazing in cold and humid weather (due to Joule effect), also allowing for consolidating the strategies of partial air recycling while limiting mist formation. Also, glazing with low emissivity of the inner surface shows benefits in both winter and summer as explained in previous sections.

4.3.4 Thermal Energy Storage

As shown in Chapter 3, the thermal storage can be used to participate to cabin temperature control. This can provide means of improving the energy performance of cabin climate control in electric vehicles. However, thermal storage solutions must be compared to electric batteries in terms of volume, mass, cost and capability to be used as a function of weather conditions. Both storage solutions can possibly be coupled.

The advantages of electric batteries are the possibility to store energy that can be used for a large range of usages, including non-thermal usages and the negligible losses over time. Hence, in a hot environment, the hot thermal storage will not be used while the electric battery can be used for the powertrain or to supply the air-conditioning system. Consequently, it does not allow for reducing the sizing of the powertrain electric battery, which must ensure the autonomy in any circumstance.

The charging of a thermal storage can be achieved during parking phases, simultaneously with the charging of an electric battery. The potential strategy of a heat storage on a trip in the electric mode would be:

- Charging of the thermal storage at a high storage temperature.
- Use of the thermal storage in direct heating mode on the glycol water loop of the heater core inside the HVAC unit. The storage can allow for a quick warm-up of the cabin in the first minutes of use of the vehicle and cover the peak of heating demand without using electric energy.
- Use in combination with another thermodynamic or resistive heating mechanism, if the heat transfer rate delivered by the storage is too limited.
- Use in indirect heating mode, to be valorized by means of a heat pump when the temperature of the storage is not high enough for a direct use, while being higher than ambient temperature.

The thermal storage can therefore be valorized on an extended range of temperature of the fluid, theoretically down to the ambient temperature, which allows for limiting the thermal losses at low temperature and for exploiting the major part of the stored energy. Similarly, the quickest use of the thermal storage allows for limiting thermal losses to the outdoor and for exploiting at most the stored energy.

In such a way to optimize the global thermal system, the thermal storage can also be used during the driving cycle for producing more heat from heating device and storing this surplus of thermal energy when conditions are favorable with a high efficiency of the whole energy chain of heating from electrical battery to heat production and discharge this stored energy when conditions are unfavorable.

4.4 Battery Thermal Management (BTM)

An electric battery or accumulator is an electrochemical energy storage device that allows for the reversible conversion of chemical energy into electrical energy. Because of its capacity to instantaneously deliver large powers, it is used in numerous applications including:

- Energy storage systems and industrial applications and, among others, in the management of the use of intermittent renewable energy sources (wind, photovoltaic, etc.) in such a way to account for production variability constraints.
- The development of electrical transportation (EV, PHEV, HEV, etc.) to ensure a better environment.

4.4.1 Description of a Battery

4.4.1.1 Battery Pack, Modules and Cells

Electric vehicles use packs of rechargeable batteries. A battery pack (illustrated in Figure 4.17) is made up of several modules connected in parallel or in series. A module is composed itself of several cells connected in parallel or in series. The cell is the smallest form of battery and is characterized by a voltage of a few volts (around 3.6 V for a lithium-ion cell). Electrochemical reactions occur in cells and produce electric current during battery discharging or store energy during battery charging or during vehicle braking (regenerative braking).

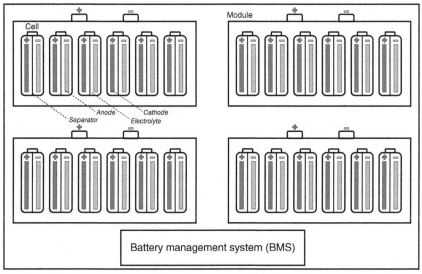

Figure 4.17 Cells, modules, and battery pack.

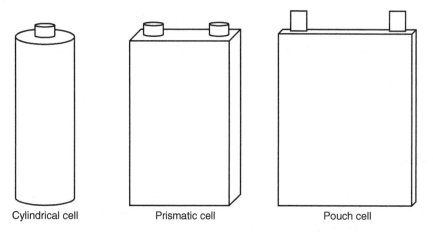

Cylindrical cell   Prismatic cell   Pouch cell

Figure 4.18 Different geometries of cells.

There exist different shapes of integration of cells (Figure 4.18):

- Pouch cell (thin soft casing)
- Prismatic cell (hard casing)
- Cylindrical cell (hard casing)

Because of their high surface area to volume, pouch cells seem to be the most adapted for effective thermal management.

The electrical traction of a vehicle requires a high voltage obtained by associating in series a large number of battery cells, whose individual voltage is low. The battery pack is therefore made up of a number of modules containing a small number of cells arranged in a parallel or series electric configuration. The parallel configuration increases the battery pack capacity (in Ah) and the series configuration increases the overall voltage. The geometry of the battery pack must also meet the architecture constraints.

4.4.1.2 Operating Principle of Lithium-Ion Battery Cells

Electrified vehicles make use of an electric energy storage device that must combine a good autonomy and a low cost with an acceptable compactness. The chemical family of lithium-ion batteries constitutes the technology currently preferred in the domain of electric powertrain, due to its high energy and power densities and its lifespan. Another technology of rechargeable battery is the nickel-metal hydride (NiMH).

However, its cost, which could reach 30% of the cost of the vehicle in the period 2010–2020, and its total capacity, which reflects the driving range of the vehicle, are two major weaknesses that demand future developments. Moreover, the lifespan of the battery pack does not reach that of a traditional vehicle, which is 10 to 15 years. The understanding and management of aging are necessary. The drawbacks are the risks in the case of leakage, overheating, and development of a crystalline structure of lithium between electrodes. The battery management system (BMS), described hereunder, is a monitoring system that allows for preventing such risks.

A lithium-ion battery cell is made up of porous electrodes: a negatively charged one (anode) able to give electrons and a positively charged one (cathode) able to catch electrons. Both electrodes are connected and immerged in an electrolytic solution that conducts ions. Lithium ions Li+ can be exchanged between the anode and the cathode by traveling through the electrolyte by diffusion. A separator ensures an electric insulation and prevents any contact between electrodes that would yield short-circuit. During operation, ions are inserted into and extracted from electrodes

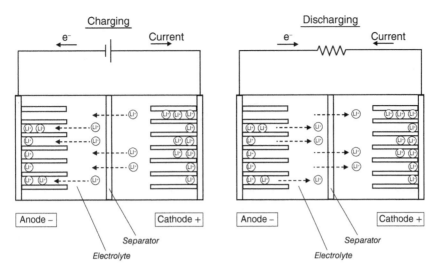

Figure 4.19 Representation of a Lithium-Ion battery in discharging and charging modes. Source: Adapted from Voelker (2014).

("intercalation" and "de-intercalation" processes). Electrons travel in an external circuit, in the opposite direction of current. During discharge, the electrical voltage created by the electromotive force generates a current of discharge between the two electrodes. The material of the negative electrode (reducing agent) is oxidized and transfers its electrons to the positive electrode (oxidizing agent) which undergoes a reduction reaction. Movement of lithium ions and electrons are depicted in Figure 4.19 for discharging and charging modes.

4.4.1.3   Battery Technical Characteristics
A battery is generally characterized by:

- its capacity [Ah] (also called "coulometric capacity")
- its energy content [Wh] (also called "energy capacity")
- its specific energy [Wh kg^{-1}]
- its energy density [Wh L^{-1}]
- its specific power [W kg^{-1}]
- its power density [W L^{-1}]

The capacity and energy capacity represent the total Ah and Wh when the battery is fully discharged from a fully charged state. Figure 4.20 shows the evolution of the cell Open Circuit Voltage as a function of the discharged ampere hours. When the cell voltage reached the cut-off voltage, the cell is considered empty, which allows defining the cell capacity. The area underneath the curve represents the energy capacity [$V \cdot Ah = Wh$] of the cell.

4.4.1.4   State of Charge (SOC)
The state of charge (SOC) is the ratio of the remaining battery capacity to the maximum capacity. It can be seen as fuel gage in % (Figure 4.21). The SOC is estimated by the BMS and is used on the display device of the dashboard and is also useful for different internal calculations (maximal powers, driving range, etc.). The depth of discharge (DOD) is the fraction of the maximum capacity that has been discharged.

However, both capacity and energy capacity are defined for a specified C-rate. The C-rate [h^{-1}] is used to indicate the current or power at which the battery is charged or discharged safely. It is

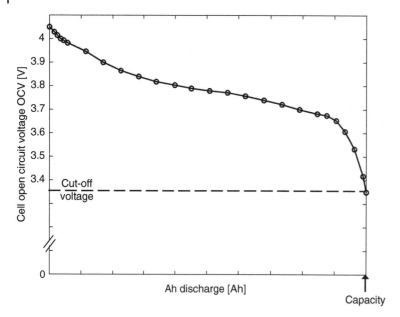

Figure 4.20 Open Circuit Voltage as a function of the ampere-hours discharged. Source: Data from Buller et al. (2005).

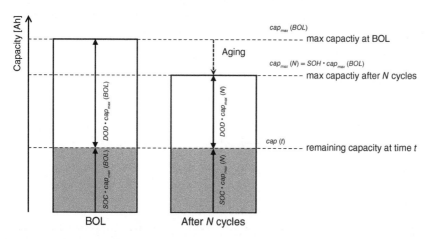

Figure 4.21 Illustration of the state of charge (SOC), depth of discharge (DOD) and state of health (SOH).

defined as the ratio of the (dis)charge current [A] to the battery capacity [Ah]. Multiplying both numerator and denominator by the battery voltage allows defining the C-rate as the ratio of the (dis)charge power [W] to the battery energy capacity [Wh]. Hence, the C-rate quantifies the rate at which the battery is charged or discharged, independently of the capacity of the battery. A C-rate of 1C indicates that the current or power allows for the battery to be (dis)charged in one hour. For instance, if a 50 kWh battery is discharged at a power of 100 kW, the C-rate is 2C.

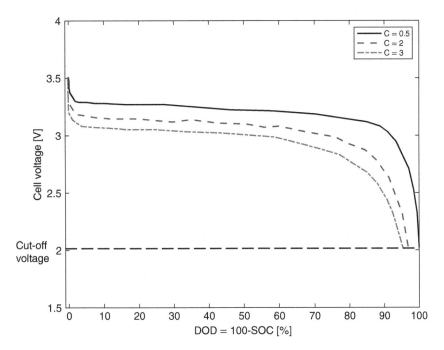

Figure 4.22 Evolution of cell voltage as a function of DOD for different C-rates.

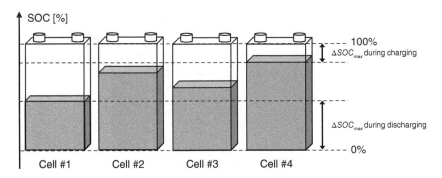

Figure 4.23 Schematic illustrating capacity loss during charging and discharging due to cells unbalance.

Figure 4.22 illustrates the evolution of the cell voltage with the DOC for different C-rates. Increasing the C-rate increases the discharge current. This yields a larger voltage drop through the internal resistance of the cell.

The cells imbalance causes, among other things, a difference of SOC between the most charged cell (cell #4 in Figure 4.23) and the least charged cell (cell #1 in Figure 4.23), which corresponds to a capacity loss during charging (as soon as the most charged cell has reached the maximal voltage, the charging is stopped) and to a loss during discharge (as soon as the least charged cell reaches the minimal voltage, discharging is stopped). A system of cell balancing is necessary for balancing cell voltages.

Batteries used in full hybrid, plug-in hybrid, and EVs have voltage larger than 200 V, power larger than 20 kW and energy content tending to 100 kWh.

- Mild hybrid electric vehicle (M-HEV): <1 kWh
- Full hybrid (F-HEV) and PHEV electric vehicle: 1 to 10 kWh
- EREV: 5 to 20 kWh
- BEV: 15 to around 100 kWh and more in the future

As one example, the evolution of battery pack energy density of the Renault Zoe electric vehicle is given hereafter:

- 2011: 23.3 kWh, 280 to 320 Wh L^{-1}
- 2016: 41 kWh, 450 to 480 Wh L^{-1}, autonomy close to 400 km on NEDC cycle (not including thermal comfort), improvements on cell chemistry (lithium-manganese oxide [LMO]-based to pure lithium-nickel-manganese-cobalt [Li-NMC]) and on cell engineering (better space utilization)
- 2019: 52 kWh, autonomy of 395 km on WLTP cycle (not including thermal comfort)

In the future, the evolution will be quick, and EVs will propose up to 800 km autonomy with consumptions on WLTC cycle of around 12 kWh (100 km^{-1}) and a possible charging power of around 250–350 kW and then reach even larger charging powers for a quick charge (up to 80% SOC in 15 minutes).

The major issues for next-generation batteries are the increase of the energy density and specific energy (which impacts the driving range), the increase of the lifespan, the decrease of the cost in such a way to make the EV affordable, the security, the impact of charging/discharging cycles, fast charging, storage conditions and temperature, the CO_2 footprint, and the increase in the power (high power e-powertrain).

One of the best ways to increase energy density and specific energy in batteries is to reduce the overall mass and volume of the thermal management components of the battery pack.

4.4.2 Battery Charging

Battery charging is a major issue associated with the development of BEVs. The stationary plug-in charging of the battery is possible in both slow charging and fast charging modes. The slow charging is achieved by using a low charging power of the order of 2–7 kW (8–32 A in single-phase power supply) over a rather long duration, for instance from four to eight hours, generally during nighttime with a standard domestic electric plug.

The fast charging is achieved by using a high charging power of the order of 20–50 kW (32–73 A in three-phase supply) available on specific charging stations. The battery can, for instance, be charged within less than one hour. An ultra-fast charging with an even larger power (50–350 kW and more [direct current]) could for instance ensure a partial charge in five minutes and a full charge in 15 minutes. It is therefore necessary to develop a charger of 200 kW to theoretically charge in 15 minutes a battery with a capacity of 50 kWh or 360 kW to charge in 15 minutes a battery with a capacity of 90 kWh capable of providing a driving range of the order of 800 km without accounting for energy use associated with thermal comfort.

The main difficulty associated with ultra-fast charging of a battery pack is the inability to remove heat as fast as it is generated. During ultra-fast charging, the cell performance and reliability are mainly determined by the cell temperature and the cell temperature gradient.

On a PHEV vehicle, the battery can be charged by the electrical grid, by regenerative braking or by the thermal engine. It is discharged by the electric motor.

The major issues associated with charging are:

– The development of faster and faster charging of batteries, while the latter show larger and larger capacities to increase their driving range.
– The expected line at the charging station due to the charging time.
– The development of residential energy supply to EV batteries.

Several alternative solutions to plug-in static charging could be considered:

– The replacement (quick drop) of the discharged battery by a new preloaded battery. This solution imposes to have a higher number of batteries (in vehicles plus charging stations) than vehicles on the road, limits the packaging possibilities and imposes standardization. This solution can allow for time saving by replacing the discharged battery by a charged battery by means of a robotic system (estimated time of the order of five minutes), rather than waiting for the charging of the battery that is in the vehicle. The discharged battery is then recharged for future users. The battery could be rented with an access to batteries exchange stations.
– The contactless charging at vehicle standstill (static mode),
– The contactless charging with driving vehicle (dynamic mode).

The wireless dynamic charging uses the induction technology. Inductive coils buried in the ground are fed by alternating current and emit an electromagnetic field. The latter is caught by the nearby moving vehicle and converted back into electricity by an onboard system. For instance, in urban zones, an electric vehicle driving 25% of its time on a road equipped with such a technology may not need to be charged at a charging point. The coupling with solar collectors integrated or not inside the road is also considered as a promising future solution. The issue is to more massively deploy the EV with an increased autonomy:

– Transferring the specific problem of driving range on the infrastructure rather than increasing the vehicle onboard energy would allow for decreasing the volume and weight of the battery.
– Addressing the limited number of classical charging points as well as the required time for a full charging would allow for a simplified charging process, more accessible and with a less impacting duration. Indeed, with such a dynamic charging, fast charging becomes less or not necessary.

4.4.3 Battery Aging

4.4.3.1 Calendar and Cycling Aging

As the years pass by and with thousands of kilometers traveled, the autonomy and power of the powertrain battery decrease and the charging time increases. Battery aging is linked to electro-chemical reactions and physical degradations of cells.

One can distinguish two modes of aging of Li-ion batteries, which can lead to a battery replacement, characterized by a progressive decrease of the battery capacity and an increase of its internal resistance (see Figure 4.24), that is, a decrease of the available energy reserve and of the maximal power at which it can be delivered:

– Calendar aging occurs when the battery is at rest in the "parking" mode. This mode of aging depends on the storage temperature (a high temperature accelerates the parasitic chemical reactions and the expansion of the solid electrolyte interface [SEI]) and on the SOC of the battery (a high SOC yields premature degradations of the cells).
– Cycling aging occurs when the battery is used with a current that flows through the cell in either charging or discharging. This mode of aging is more severe since it integrates other parameters of the degradation, such as the current regime, the quantity of exchanged electricity, the state of charge window (ΔSOC) and the regime of utilization (cycling).

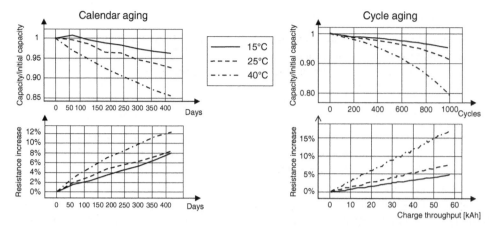

Figure 4.24 Thermal behavior of lithium-ion battery: calendar and cycle aging. Source: Adapted from Kolp et al. (2016).

Both modes of aging are accelerated with the increase in temperature, since it impacts the growth of SEI layer (and therefore its resistance), which builds at the contact between electrodes (mainly the cathode) and electrolyte.

It is therefore important to develop strategies allowing for predicting the behavior and the aging of the battery, for extending its lifespan as a function of the vehicle usage and of the charging modes, for providing an estimation of the autonomy and for controlling the charging. The most classical technique for characterizing the aging consists of tracking the decrease of the capacity during the life duration of the battery. This capacity is measured by a charging followed by a full discharging of the battery.

The tracking of the increase of the battery internal resistance R_{in} also gives an image of its aging. The resistance can be measured by electrochemical impedance spectroscopy (EIS) or using current pulses.

4.4.3.2   State of Health (SOH)

The state of health (SOH) represents the state of degradation of the battery, due to aging, in comparison with its original state at beginning of life (BOL). It is therefore the ratio of the actual maximal capacity of the battery to the original maximal capacity ("rated capacity"). The SOH can be seen as the size of tank. It reflects the decrease in the available energy and the decrease in power during charging/discharging due to, among others, the increase of the internal resistance of the cell. An illustrative comparison between SOC and SOH is proposed in Figure 4.21.

The SOH at time t can also be defined as function of the internal resistance R_{in}. Actually, with aging (and also degradation), the internal resistance increases and therefore the voltage drop across the cell increases for a given current I. Considering the resistance $R_{in}(t)$ at time t, the resistance $R_{in}(t = 0)$ at BOL and the resistance $R_{in}(t = t_{EOL})$ at end of life (EOL), Eq. (4.1) can be established. Noura et al. (2020) suggest to consider $R_{in}(t = t_{EOL}) = 1.6 \times R_{in}(t = 0)$.

$$SOH(t) = \frac{R_{in}(t = t_{EOL}) - R_{in}(t)}{R_{in}(t = t_{EOL}) - R_{in}(t = 0)} \times 100\% \tag{4.1}$$

4.4.4 Battery Management System (BMS)

The battery management system (BMS) is an electronic system that aims at managing the high voltage battery in an electric or hybrid vehicle. It is placed within the battery pack. It ensures performance and security of the battery as well as security of passengers. It manages a system where all physical variables to monitor/compute/control are interacting. It ensures:

– the management of the energy quantity (cell balancing in such a way to equalize the voltage across each cell),
– the management of the maximal values of powers during the discharging and charging (power of discharging and power of charging during driving and during standstill),
– the thermal management from the temperature of cells (for performance, durability, and security reasons),
– the diagnoses,
– the battery protection,
– the assessment of several state indicators (SOC, SOH, available energy, and temperature of the battery).

All these actions are ensured using measurements (time [calendar], current, voltage, temperature, isolation of high voltage circuit, assessment of SOC, assessment of maximal power, etc.).

The battery management system, among other functions, calculates the discharging rate, the charging rate during driving and the charging rate at a standstill as a function of the temperature for limiting the degradation or ensuring the security of the battery. It can impose a temporary derating of the powers for cells temperatures of the order of 50–60°C with the current technologies. Also, it can directly drive actuators (cooling or heating) to bring back the battery temperature inside an ideal zone.

Battery thermal management system is installed in mild-hybrids, full-hybrids, plug-in hybrids, BEVs, and FCEVs.

4.4.5 Energy Balance Across a Battery Cell

The energy balance across a battery cell can be expressed as follows:

$$m_{cell}\, c_{cell} \frac{dT_{cell}}{dt} = \dot{Q}_{gen} - \dot{Q}_{amb} - \dot{Q}_{BTM} \qquad (4.2)$$

where

m_{cell} is the mass of the cell, [kg]
c_{cell} is the specific heat of the cell, [J kg^{-1}K^{-1}]
T_{cell} is the temperature of the cell (assumed uniform), [K]
$\dot{Q}_{gen}$ is the heat rate generated inside the cell, [W]
$\dot{Q}_{amb}$ is the heat exchange rate with the ambient, [W]
$\dot{Q}_{BTM}$ is the heat transfer rate related to the battery thermal management, [W]

4.4.5.1 Heat Generation inside the Cell

Batteries generate heat during charging and discharging. Heat generation inside the cell is due to both the Joule effect (related to the internal resistance) and to the entropy change resulting from the electrochemical reactions (Karimi and Li, 2012). In a simplified approach, the rate of heat generated inside the cell $\dot{Q}_{gen}$ is the sum of the Joule effect heat rate $\dot{Q}_{Joule}$ (irreversible process) and the reversible entropic heat generation rate (or reaction heat rate) $\dot{Q}_{rev}$ (Liu et al., 2014). It can be written as

$$\dot{Q}_{gen} = \dot{Q}_{Joule} + \dot{Q}_{rev} \tag{4.3}$$

The Joule effect heat rate can be expressed as (Liu et al., 2014)

$$\dot{Q}_{Joule} = I_{cell}(U_{OCV} - U) = I_{cell} R^2_{in,cell} \tag{4.4}$$

where

I_{cell} is the current through the cell, [A]
U_{OCV} is the open circuit voltage, [V]
U is the cell voltage, $[V]$
$R_{in,\,cell}$ is the internal resistance of the cell, $[\Omega]$

The reversible entropic heat generation rate is given by

$$\dot{Q}_{rev} = -I_{cell}\, T_{cell} \frac{\partial U_{OCV}}{\partial T_{cell}} \tag{4.5}$$

This term corresponds to the rate of heat generation or consumption due to reversible entropy change associated with cell electrochemical reactions (Pesaran et al., 1997). The partial derivative of the open circuit voltage with the temperature is the temperature coefficient (Pesaran et al., 1997). When multiplied by the absolute temperature $T_{cell}[K]$, the term is the effective entropic potential that varies with the SOC and battery chemistry (Liu et al., 2014). The general and complete development of the energy balance for a battery and the justification of simplifications can be found in Bernardi et al. (1985).

As explained previously, the internal resistance $R_{in,\,cell}(SOC, SOH, T_{cell})$ of the cell is function of the state of charge, aging, and temperature. This resistance increases as the age of cell increases and the temperature decreases.

The rate of heat generated is relatively small (typically lower than 1 kW for a compact passenger car) during driving, but it increases significantly with the need for fast charging. Including the rate of heat generated by the electric motor and power electronics, the total rate of heat generated by an electric vehicle is still lower than 10 kW, while that of a thermal engine vehicle can exceed 100 kW. The implementation of a cooling system can be more an issue because of the low level of coolant temperature, because of a temperature difference with hot ambience that could be as low as 15–20 K (what represents roughly 1/3 of the temperature difference obtained with a thermal engine) and, for the battery, because of the necessity of using the air-conditioning system to prevent derating.

4.4.5.2 Heat Exchange with the Ambient and with the Heat Transfer Fluid of the BTMS

The heat transfer rate to the ambient $\dot{Q}_{amb}$ comprises both convective and radiative contributions

$$\dot{Q}_{amb} = A_{cell}\, h_c\, (T_{cell} - T_a) + A_{cell}\, h_r\, (T_{cell} - T_w) \tag{4.6}$$

where

A_{cell} is the heat transfer area of the cell, $[m^2]$
h_c is cell convective heat transfer coefficient with the ambient, [W m^2 K^{-1}]
h_r is cell radiative heat transfer coefficient with the ambient, [W m^2 K^{-1}]
T_{cell} is cell surface temperature (assumed equal to the cell temperature), [°C]
T_a is the air temperature around the cell, [°C]
T_w is the mean radiant temperature around the cell, [°C]

In the case of a PHEV or HEV, following the location of integration of the battery, besides the Joule effect due to its operation, the battery can be subject to other solicitations that condition its sizing:

– If the battery is located under the car body, there is convection and radiation from the exhaust gas line. When the engine is in operation, the mass flow rate of gas and its temperature can be high.
– If the battery is located in the rear trunk, there is the solar radiation during the day that generates a temperature in the trunk of the order of 65°C when the outdoor temperature and the solar radiation are high.

4.4.6 Undesired Effects of Battery Operating Temperature

During charging and discharging cycles, heat is generated within the battery. This phenomenon leads to an increase in the temperature and potentially to an acceleration of the aging process. The cell temperature is a function of the ambient temperature, the cooling mechanism, and the vehicle operation.

Different undesirable effects occur when the cell temperature is beyond acceptable levels and when the temperature is not uniform within the cell. These phenomena lead to a decrease in performance (potentially irreversible) and a decrease in lifespan and could yield safety issues.

4.4.6.1 Cell Temperature Level

Undesirable phenomena can be classified according to the level of temperature of the cell. The optimal operating temperature is around 25°C. Given this reference, low temperature, high temperature and very high temperature operation can be distinguished.

During its life, the cell, because of ambient conditions and undergone electric loads, is subject to variations of temperature. The ideal operating temperature lies around 15–35°C (Pesaran et al., 2013), temperature which is achieved with the thermal control.

4.4.6.1.1 Low Temperature At low temperatures, significantly below −10°C (Bandhauer et al., 2011), the energy that can be extracted from the battery is reduced. Actually, the cell internal impedance increases since electrolyte conductivity decreases (Ma et al., 2018). This causes the ion mobility in the electrolyte to be reduced. Moreover, the reactions of intercalation and de-intercalation at the electrodes are slower. These phenomena lead to a decrease in energy capacity and power.

Lithium plating can also decrease in an irreversible way the performance of the cell. Lithium plating is the formation of metallic lithium on the graphite anode. This phenomenon mainly appears during charging at high current, low temperature, and high SOC (Petzl et al., 2015). Therefore, at low temperatures of the battery and high SOC, the charging power and the SOC at the end of charging must be limited. Moreover, dendritically formation of metallic Ion can lead to short circuits when penetrating the separators (Ma et al., 2018).

It is thus necessary to heat the battery during extreme cold conditions so that it recovers its capacity and its electric power, and it increases its availability.

4.4.6.1.2 High Temperature At high temperature, cell degradation leads to battery energy (capacity) and power fade, as well as lifespan decrease. This phenomenon appears both when the battery is stored or cycled at high temperature. It is related to the increase in the SEI layer thickness and internal resistance of cells. For lithium-ion batteries, the upper threshold is approximately 50°C at the outer of the cell. The electrolyte of a Li-ion battery is currently limited to 60°C. For higher temperatures, parasitic reactions occur, and the electrolyte is degraded. Irreversible chemical phenomena also occur leading to a loss of available Li ions. In such a way to improve the lifespan and avoid limiting the power (derating), the cell temperature during charging and discharging must be controlled at the least energy cost.

Battery capacity is significantly decreased when it is stored for a long period at high temperature. Given the small value of the self-discharging rate, 0.44% per day at 60°C (Bandhauer et al., 2011), self-discharging is not an issue for HEVs and EVs frequently driven.

4.4.6.1.3 *Very High Temperature*

Above a critical temperature that depends on the cell chemistry, SOC and abuse event (Bandhauer et al., 2011), exothermic reactions can occur within the cell. For lithium-ion batteries, such reactions are triggered over 100°C. Released heat yields further increase in temperature and additional reactions. Overheating is then propagated in an uncontrolled way to the other surrounding cells. Hence, thermal runaway can occur even if only a portion of the battery reaches the critical temperature.

The influence of temperature on battery behavior is summarized in Figure 4.25, showing qualitatively the evolution of the power with the temperature during charging and discharging operations. High temperature operation occurs during boost function in hybrids, high outdoor temperature, fast charging in EV, and regenerative braking.

4.4.6.2 Battery Temperature Gradient

Besides maintaining the battery temperature within an appropriate range, it is important to limit the temperature gradient across the cells.

Cells are typically assembled in series in order to increase the useful voltage. There is a dispersion of the characteristics of the cells, which is related to their manufacturing tolerance, to their temperature (as explained hereunder) and to their aging. If one of the cells is characterized by a lower

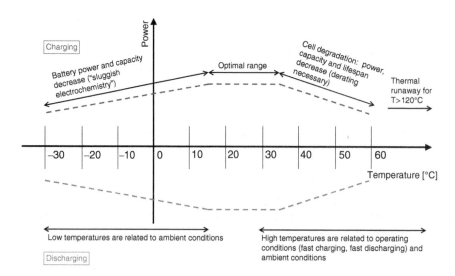

Figure 4.25 Influence of temperature on battery behavior. Source: Adapted from an original drawing of Smith et al. (2008) and Pesaran et al. (2013).

capacity, the charging and discharging would have to be stopped before the battery has reached a SOC of 100 and 0% respectively. Hence, the battery would store or retrieve less energy than the available energy content.

There is a risk that the cells with the smallest capacity are overcharged or over-discharged, while the cells with the largest capacities are only partially cycled. Overcharged of cells could yield thermal runaway. Battery electrical balancing allows the cells with the largest capacity to be fully charged/discharged and weakest cells to be protected.

Capacity is a function of cell temperature. Hence, temperature gradients across the battery would cause some cells to be overcharged. Hence, for safety issues, it is important to maintain a uniform temperature within the battery. There are more related safety risks with lithium-ion batteries than with nickel-metal hydride batteries (Bandhauer et al., 2011). Also, maintaining a uniform temperature between the cells could simplify the electrical balancing schemes (Bandhauer et al., 2011). The homogeneity of the cells inside the pack must be inferior to 3–5 K.

4.4.6.3 Battery Thermal Inertia

The battery shows a very large thermal inertia, while the heat density generated during driving is comparatively low. The advantage is that its temperature slowly increases (10–30 K during a full discharging cycle depending on the C-rate). However, it requires a large energy to be heated: for instance, around 10 kW during 12 minutes for heating it from −20 to 20 °C.

4.4.7 Battery Thermal Management Systems (BTMS)

The thermal management of powertrain battery thermal management (BTM) largely participates in maintaining its performance and lifespan and in preventing security risks.

Practically, the role of the battery thermal management system (BTMS) is to maintain the battery temperature within given limits and to ensure a uniform battery temperature. The optimal battery temperature range is 15–35°C. A heat transfer fluid is used to either heat or cool the battery pack.

The BTM maximizes the available capacity during charging/discharging, limits internal losses, maximizes the power, ensures the availability in a cold environment, and prevents derating in a hot environment (which consists in limiting the powertrain power and the charging power in such a way to master the thermal aspects and protect the chemistry of the battery). As a consequence, the battery thermal management participates in maintaining the efficiency, thereby decreasing the consumption and increasing the driving range of the vehicle and also contributes to decrease the depreciation cost of the battery strongly related to its average operating temperature, to the maximal temperature peaks during its life duration and to its reliability. An abnormal rise in battery average temperature of 10 K (at rest and during operation) can lead to a division by 2 of its lifespan.

It is also recommended to keep a maximum temperature difference of 3–5 K between the coldest and hottest cells.

The solutions for thermally controlling a battery in such a way to maintain it in optimal thermal conditions must consume as less as possible energy and impact as less as possible the architecture of the thermal management system.

Many configurations of BTMS have been proposed. They first differ by the heat transfer fluid (air, glycol-water, a dielectric fluid, or a refrigerant) that is used to heat or cool the battery. As shown latter, a refrigeration circuit can be used to cool down the heat transfer fluid. However, a battery cooling system that only uses the refrigeration circuit would lead to a use of the compressor in climatic conditions where a direct cooling by an outdoor radiator would be sufficient (outdoor temperature <20°C).

The main objectives of the thermal control of the battery are:

– Cool down in a homogeneous way the battery pack and slow down the heating process or reduce the temperature in such a way
  o to ensure a good level of performance and of autonomy,
  o to preserve the capacity of the battery in time and its durability
  o to optimize the charging time
  o to prevent phases of derating.
  During the charging phase at a standstill, an appropriate solution would consist in sufficiently cooling the battery to compensate both the heat released during charging and the heat that will be later released during next driving. This will allow for the optimization of the driving range by limiting the use of the cooling system during driving.
– At a standstill, when the battery has a low temperature, heat up quickly the latter in such a way to ensure the next driving or to allow a fast charging, downgraded but acceptable.

The thermal system can be used during the driving phase, the charging phase (slow or fast), and the pre-conditioning phase. It must be efficient to reject heat from the battery core to the interfaces of the battery and to optimize the cooling/heating system of the vehicle (outside the battery).

To cool down or heat the battery, a heat transfer fluid (air, glycol-water, a refrigerant, or a dielectric fluid) is used. This fluid flows either around the components of the battery or inside plates to which the battery components are attached. As illustrated in Figure 4.26, the thermal problem can then be split into a thermal problem of conduction in a heterogeneous medium inside the cells (or modules) (thermal zone 1), of heat transfer from the external surface of the cells (or modules) to the heat transfer fluid (thermal zone 2) and of heat transfer by the heat transfer fluid from the battery pack to the vehicle outdoor (defined as thermal zone 3).

The factors that impact the choice of the thermal solution are the battery itself (technology, capacity, internal architecture, thermal resistance, location in the vehicle), the powertrain

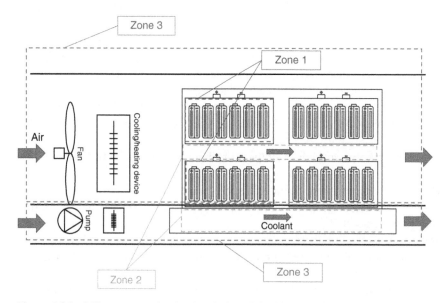

Figure 4.26 Different zones for the description of the heat transfer problem of battery thermal management (top: air-based cooling system; bottom: coolant-based cooling system).

(installed power), the strategies of charging/discharging (power, charging duration), the conditions in terms of ambient temperature for the country of commercialization, the use pattern (daily driving distance, number of charging cycles per day, solicitation in terms of power). Also, reducing the overall mass and volume of the thermal management system is a way to increase energy density/specific energy of the battery. The thermal needs can reach several hundreds of W to a few kW during driving. With the increase in the storage capacity (in such a way to propose larger driving ranges) and with the shorter fast charging times (to get close to the usages obtained with conventional fuels), the charging powers in parking conditions will become huge in the future and could reach several hundreds of kW, which will dissipate through the battery inertia and cooling systems several tens of kW of heat.

When there is no cooling capacity, or if the latter is not sufficient, the maximal allowable current can be limited ("derating") so that the battery does not reach a critical temperature. Without any thermal management solution, temperature differences appear in the pack and require balancing the cells current. In such a way to maintain the vehicle driving range, cooling solutions that consume as little as possible energy must be used. The use of the air-conditioning system to cool the heat transfer fluids (air or glycol water) to reach a low enough temperature to ensure cooling of the battery, which is necessary in a hot environment and especially when combined with a fast charging, should be limited. It allows to maximize the vehicle driving range.

4.4.7.1 Air-Based Systems

4.4.7.1.1 Configurations of Air-Based BTMS When cooling the battery pack with air, the latter is circulated around the modules to evacuate heat generated inside the cells. The air can circulate in a closed loop through the battery pack, as illustrated in Figure 4.27 (with 100% of recirculated air) and be cooled down by a cooling system (vapor-compression air-conditioning, Peltier cooler, etc.).

Air can also circulate through an open loop. Either *outdoor* air or *cabin* air can be used for thermally conditioning the battery pack. Also, the distinction can be done among *passive* and *active* systems.

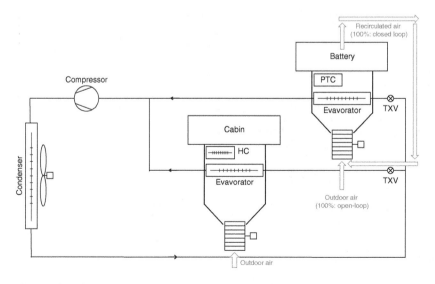

Figure 4.27 Active outdoor air-based system for battery thermal management.

In a *passive outdoor air* configuration, the outdoor air is neither heated nor cooled down before being routed to the battery. In an *active outdoor air* system, a heating system and a cooling system (the electric energy consumption of which limits its use during driving) are used to control accurately the air temperature before reaching the battery pack. Figure 4.27 shows such a configuration using a vapor-compression refrigeration unit for cooling production. The unit comprises two evaporators refrigerant/air in parallel: one for cabin air-conditioning and one on a secondary A/C loop for preparing cold air for the battery thermal management. Both evaporators are fed with outdoor air, or with a mix of outdoor air and recirculated air. In this specific example, electric resistance heaters (PTC) are used for heating the air, for instance, for countries with extremely cold weather. Other technologies can be used for heating and cooling, such as thermo-electrical systems.

Another example shown in Figure 4.28 illustrates an active outdoor air-based battery thermal management system associated to a heat pump system. The role of the different valves has been explained previously when describing Figure 4.7 and Figure 4.8.

In a cabin air system (passive or active), a dedicated blower can also route the air through the battery pack from car cabin.

A *passive cabin air* system, represented in Figure 4.29, uses air from the cabin without any additional heating or cooling to condition the battery.

In an *active cabin air* system, depicted in Figure 4.30, the cabin air is further cooled down or heated up before being routed to the battery.

A passive air system, with outdoor air or cabin air, shows some advantages (weight, security, low cost) and drawbacks (bad control of cooling power, limited range of cooling power from 0 to roughly 600 W, as a function of outdoor temperature, volume, and noise).

An active air system, with recirculated air or not, also shows some advantages (cooling performance and large range extending from roughly 0.5–2 kW) and drawbacks (limited cooling capacity for ultra-fast charging, volume of the BTM system, noise). In the air recirculation mode, the air

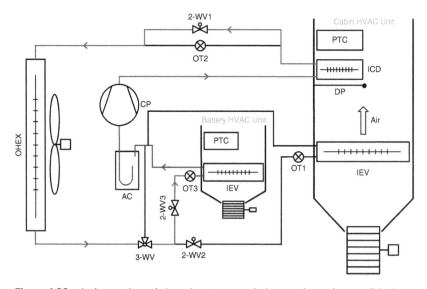

Figure 4.28 Active outdoor air-based system coupled to an air-to-air reversible heat pump.

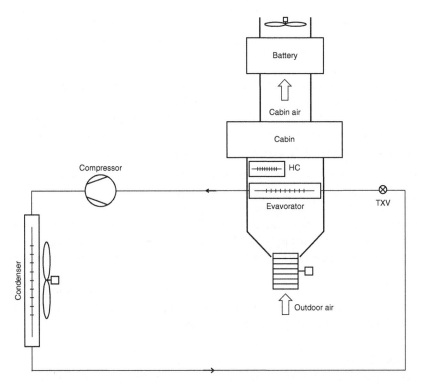

Figure 4.29 Passive cabin air-based system for battery thermal management.

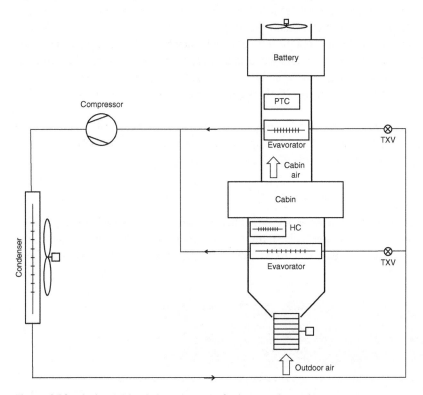

Figure 4.30 Active cabin air-based system for battery thermal management.

cooling system is a closed loop that offers better performance. The battery evaporator is mainly used during battery charging. In the driving mode, the use of the battery evaporator impacts the driving range and the cabin air-conditioning mode.

4.4.7.1.2 Advantages and Drawbacks of Air-Based BTMS The air-based cooling system is a simple solution. The major advantages of using air as the heat transfer fluid are the absence of a liquid fluid inside the battery pack (Girmscheid 2014), the non-flammability and non-electrical conductivity of the air, and the low cost of this cooling solution. However, air-based systems show some drawbacks with respect to liquid-based systems (Bandhaur et al., 2011; Girmscheid, 2014):

- They require an advanced and bulky system (larger ducts than with liquid hydraulic system in such a way that all cells are cooled down in a homogeneous way).
- More energy is needed by the fan to drive the air than by the pump to circulate the liquid.
- Lower convective heat transfer coefficients are achieved with air than with liquids, leading to a larger temperature difference between the heat transfer fluid and the battery surface (Bandhauer et al., 2011). Their performance in cooling and heating mode is therefore limited in comparison with liquid-based solutions.
- The lower capacity flow rate (in $[W\ K^{-1}]$) achieved with air yields larger temperature gradients inside the battery pack. Even with an optimized aeraulic design, it is difficult to ensure a homogenization in temperature under large heat transfer rates.

Example 4.3   *Air-Based Battery Cooling*

It is proposed to simulate the cooling by air of a battery described by Pesaran et al. (1997). The battery pack is made of 30 modules arranged in 3 columns of 10 modules (Figure 4.31). Each module contains 6 cells. Cooling air flows in the x direction. The air mass flow rate through the battery pack is $0.08\ kg\ s^{-1}$ and it enters at 25°C (ambient temperature).

Each module has a width of 11 cm, a length of 16 cm, and a height of 15 cm. The plastic casing has a thickness of 2 mm and a thermal conductivity of $0.25\ W\ (m\text{-}K)^{-1}$. The convective heat transfer coefficient between the module casing and the air is assumed to be $35\ W\ (m^2\text{-}K)^{-1}$. The mass of each module is 8 kg and its overall heat capacity is $900\ J\ (kg\text{-}K)^{-1}$. The rate of heat generation in a cell is 5.83 W.

If all the modules are initially at ambient temperature, compute the evolution of the module temperature for five hours. We assume that the temperature is uniform within a module. Compute also the evolution with time of the heat transfer rate on the air side.

Analysis

It is assumed that in each row, the 3 modules show identical temperatures. The problem is therefore discretized in the X direction only. This is also a transient problem. For each row of modules, the transient energy balance is expressed and solved by the "integral" function of EES as in previous examples.

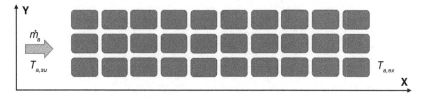

Figure 4.31 Configuration of the 30 modules of the battery pack.

Solution

```
"1. INPUTS"
m_dot_a=0,08 [kg/s] "mass flow rate of air through the battery pack
    in the x direction"
Q_dot_gen_cell=5,83 [W] "heat generation rate per cell"
T_a_su=25 [C] "air temperature at the supply of battery pack"
T_amb=25 [C] "ambient temperature"
"2. OUTPUTS"
"T_mod=?? [C]" "temperature of the modules with time"
"Q_dot_air_=?? [W]" "heat transfer rate on the air side with time"
"3. PARAMETERS"
m_mod=8 [kg] "mass of one module"
N_x=10 [-] "number of modules in x direction"
N_y=3 [-] "number of modules in y direction"
N_cells_mod=6 [-] "number of cells per module"
c_mod=900 [J/kg-K] "module heat capacity"
W_mod=0,11 [m] "module width"
L_mod=0,16 [m] "module length"
h_mod=0,15 [m] "module height"
e_plas=0,002 [m] "plastic thickness"
k_plas=0,25 [W/m-K] "plastic thermal conductivity"
h_out=35 [W/m^2-K] "convective heat transfer coefficient at the
    module outer surface"
"4. MODEL"
"4.1. Geometric variables"
N_mod=N_x*N_y "number of modules in the battery pack"
A_mod=2*(W_mod+L_mod)*h_mod+W_mod*L_mod
"4.2. Computational parameters"
tau_1=0 [s] "lower limit of time integration"
tau_2=5*3600 [s] "upper limit of time integration"
"4.3. Energy balances at module level"
c_p_a=CP(air;T=T_a_su)
T_a[1]=T_a_su
T_a[N_x+1]=T_a_ex
1/AU_mod=1/(A_mod*h_out)+e_plas/(A_mod*k_plas) "overall heat
    transfer resistance between the module content (assumed
    homogeneous) and the cooling air [K/W]"
NTU_cool=AU_mod/(m_dot_a_mod*c_p_a)
epsilon_cool=1-exp(-NTU_cool) "semi-isothermal heat exchanger"
m_dot_a_mod=m_dot_a/N_y "air mass flow rate around one module
    [kg/s]"
Duplicate j=1;N_x
"4.3.1. Initialization"
    T_mod_i[j]=T_amb
"4.3.2. Energy balance on module"
    Q_dot_gen_mod[j]=N_cells_mod*Q_dot_gen_cell "heat generation rate
    per module [W]"
```

```
    Q_dot_cool[j]=epsilon_cool*m_dot_a_mod*c_p_a*(T_mod[j]-T_a[j])
    "rate of cooling of each module [W]"
    m_mod*c_mod*dT_moddtau[j]=Q_dot_gen_mod[j]-Q_dot_cool[j]
    "Transient energy balance on each module"
    deltaT_mod[j]=integral(dT_moddtau[j];tau;tau_1;tau_2)
    "numerical solving of differential equation"
    deltaT_mod[j]=T_mod[j]-T_mod_i[j] "calculation of T_mod[j] at
    time Tau"
"4.3.3. Energy balance on air side"
    Q_dot_cool[j]=m_dot_a_mod*c_p_a*(T_a[j+1]-T_a[j]) "rate of
    heating of the air [W]"
End
"4.4. Global energy balance"
Q_dot_air=m_dot_a*c_p_a*(T_a_ex-T_a_su) "calculation of the
    temperature of air leaving the battery"
Q_dot_gen=Q_dot_gen_cell*N_cells_mod*N_x*N_y "heat generation rate
    of the battery pack [W]"
$integraltable tau:60 T_mod[1]; T_mod[2]; T_mod[3]; T_mod[4];
    T_mod[5]; T_mod[6]; T_mod[7]; T_mod[8]; T_mod[9]; T_mod[10];
    Q_dot_air
```

Results are presented in Figure 4.32. When reaching steady state, the temperature gradient between the 3 modules of the first row and the 3 modules of the last row is around 12 K. It can be observed that the heat transfer rate on the air side converges, in steady state, toward the heat generation rate of the full battery pack ($\dot{Q}_{gen} = 1049\ W$).

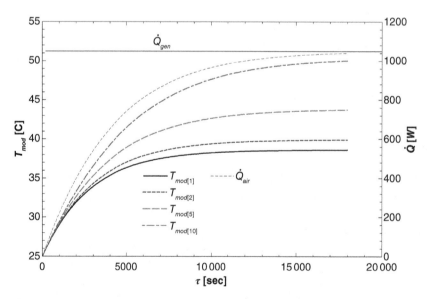

Figure 4.32 Time evolution of temperature of modules in rows 1, 2, 5, and 10 and of the heat transfer rate on the air side.

4.4.7.2 Liquid-Based Systems

4.4.7.2.1 Configurations of Liquid-Based BTMS In liquid-based systems, a liquid coolant loop is used to cool or heat plates ("liquid plates") to which the modules are fixed (Figure 4.33 a,b). Another

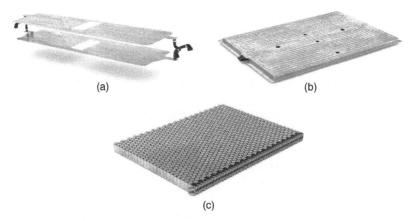

(a) (b)

(c)

Figure 4.33 Different configurations of liquid-based battery cooling systems. a: Modules are sandwiched between two liquid cooling plates. b: Modules are placed on a liquid cooling plate. c: Extruded U-tubes are nestled between rows of cells. Source: Courtesy of Valeo.

solution consists in cooling the cells through a direct contact with tubes through which liquid coolant flows (Figure 4.33 c). The coolant is usually an aqueous solution of glycol. The coolant loop temperature is typically comprised between 20°C and 40°C (Pischinger et al., 2014), but it can reach 60°C. A direct cooling mechanism using plates or tubes inserted between cells is much more efficient than an air-based system, but this solution shows a risk in terms of security related to the direct contact between the water jacket and the cells or the modules. One intermediate solution consists in cooling by means of conduction plates fixed at their extremities to liquid plates located on the side of the cells.

In the solution of passive cooling of the coolant loop, water heated by the battery must flow in an additional low temperature circuit through an air/water low temperature (LT) radiator ("battery cooler") located in the vehicle front-end module in such a way to be cooled down by outdoor air (Figure 4.34). Passive cooling can be achieved if the outdoor air temperature is low enough.

In the case of hot outdoor climate, the temperature difference between outdoor air and water is smaller and limits the thermal exchange in the radiator and hence the cooling of the battery with a risk of derating.

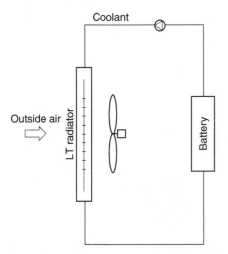

Figure 4.34 Passive cooling of the coolant loop.

Active cooling of coolant loop is achieved by a water-heated evaporator ("water chiller") of a refrigeration machine. This evaporator exchanges heat with a coolant loop that thermally manages battery cells (for example, through battery liquid plates) and allows in that case of achieving the targeted cooling of the water and of the battery. As illustrated in Figure 4.35, the BTMS and the air-conditioning system can share the same refrigeration machine, the latter having several evaporators in parallel.

Chillers are usually plate heat exchangers showing counter-flow configurations and assembled with their expansion valve. Such a plate heat exchanger chiller is represented in Figure 4.36. It can be observed that the thermostatic expansion valve of the chiller is equipped with a solenoid valve to fully close the thermostatic valve. On the left of the figure, a liquid cooling plate with modules is represented. Figure 4.37 shows an example of integration of the battery pack in the vehicle.

If the vehicle outdoor temperature is low enough, passive cooling of the coolant loop can be achieved through the low-temperature radiator ("battery cooler") in the front-end module or any air-to-coolant heat exchanger (Figure 4.34). In that situation, the air-conditioning loop is

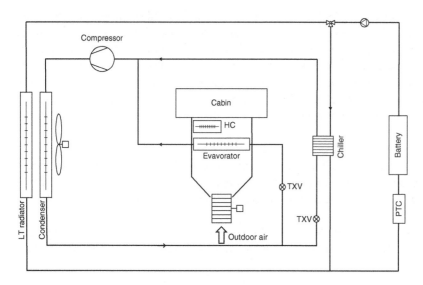

Figure 4.35 Active and passive battery cooling with the coolant loop.

Figure 4.36 View of a chiller (right) and its associated water-cooling plate (left). Source: Courtesy of Valeo.

Figure 4.37 Illustration of the integration of a liquid-based battery cooling system inside a vehicle. Source: Courtesy of Valeo.

switched-off. Switching from active to passive cooling can be done by means of a three-way valve (depicted in Figure 4.35), or simply by a two-way valve on the air-to-coolant line.

Power electronics components, such as DC–DC converters, can also be included in the same coolant loop.

The limited operating ranges of the battery can require a heating system in order to ensure the availability of the battery during low-temperature ambient conditions, for instance in very cold climates. Heating of the liquid coolant is achieved either with electric heaters (PTCs) or by means of a heater core.

4.4.7.2.2 Advantages and Limitations of Coolant-Based BTMS Liquid coolers cope with some of the drawbacks of air-based systems listed hereunder, because they are better at conducting heat away from the battery. Especially, they can ensure good cell temperature homogeneity within the battery (the gradient is lower than 4 K) and are very effective at reducing peak temperatures. For passenger cars, heating and cooling capacities of up to around 4 kW can be achieved. Such large cooling capacity is more suitable for fast charging.

Moreover, the good thermal performance of a coolant-based cooling allows for only using the active mode with the A/C loop for a higher outdoor temperature. Hence, the passive mode is more frequently used, among others during driving.

However, coolant-based BTMS requires higher safety effort because of the use of a conductive fluid in high electrical voltage environment. The limited contact surface with cells also limits the efficiency.

4.4.7.3 Refrigerant-Based Systems

In refrigerant-based systems, the batteries are directly fixed to an evaporator (refrigerant plate) in contact with the cells, which is illustrated in Figure 4.38. Similar to liquid-based systems, the battery evaporator is installed in parallel with the air-heated evaporator of the air-conditioning system on a secondary loop of the refrigerant circuit (Figure 4.39). The absence of a secondary coolant loop

Figure 4.38 Refrigerant plate system allowing for the cooling of 6 battery modules. Source: Courtesy of Valeo.

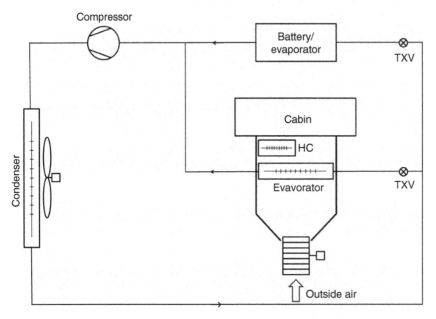

Figure 4.39 Refrigerant-based system for battery cooling.

makes the system simpler, more compact, lighter and quiet. Low battery temperature gradient and larger cooling capacities (1–4 kW) can be achieved with lower thermal inertia. These heat transfer rates are compatible with fast charging. However, this configuration requires safety precautions. Also, for limited cooling needs, this solution does not allow for using a passive solution that uses the outdoor air as the heat sink in order to limit the energy consumption. Finally, this solution allows for cooling, but not for heating (except if a reversible heat pump loop with an evaporator-condenser is considered).

4.4.7.4 Dielectric Fluid-Based System

In both the coolant-based system and the refrigerant-based system, there is a physical barrier between the heat transfer fluid and the cell. This barrier introduces a thermal resistance. It could be removed provided that the fluid does not conduct electricity not to interact with the electric components inside the battery (cells, electronics, contactors). By removing this barrier, the fluid can access a larger heat transfer surface closer to the hot spots (Solai et al., 2022). Such fluids, called dielectric fluids, could be for instance oils or HFO fluids. The battery cells can be submerged in the fluid (immersion cooling), but spray (or "jet") cooling can also be considered. The dielectric fluid can stay liquid or undergo a liquid–vapor phase change, which allows for higher cooling heat fluxes (van Gils et al., 2014). The hot liquid or the mixture of liquid and vapor or the pure vapor leaving the battery pack is then routed to a dedicated cooling heat exchanger.

Contact cooling with dielectric fluids offers different advantages as explained by Daccord and Wang (2021). It can ensure a better homogeneity of temperature within the cells and across modules. It also limits the propagation of cell thermal runaway and the risk of battery fire. Finally, such a solution does not require a coolant or refrigerant plate, which decreases the weight of the BTMS.

The high heat transfer fluxes and the good temperature homogeneity make this technique promising for high C-rates and ultra-fast charging.

4.4.7.5 Mutual Impact of Cabin Climate Control and BTM

The use of an additional evaporator on air or on the glycol water circuit, in parallel with the evaporator of the HVAC unit allows for a maximal cooling capacity delivered by the air-conditioning system during fast charging phases. However, the energy consumption associated with the use of the air-conditioning system during this phase impacts the charging power of the battery. Also, if the battery requires to be cooled down during very hot ambient conditions during the discharge phase of the battery in driving conditions to prevent, among others, derating, then the additional evaporator can be used, but this impacts the driving range and the cabin cooling operation. As a consequence, the sizing of the air-conditioning system has to account for the simultaneous cooling needs of the cabin in such a way to ensure the thermal comfort and the cooling of the battery, which compete during simultaneous cooling.

4.4.7.6 Coupling of Battery Modules on Coolant/Refrigerant Plates

When the battery module/cell is fixed on the plate, a thermal interface material (TIM), such as a thermal grease, is inserted between the battery and the plate (Figure 4.40). The use of the TIM aims at filling any air gap between the battery surface and the coolant channel wall that results from surfaces irregularities.

The rate of heat $\dot{Q}_{bat}$ transferred from the battery to the coolant can be expressed, in steady-state conditions, by:

$$\dot{Q}_{bat} = \frac{T_{bat,surf} - T_{cool}}{R_{c,bat} + \frac{e_{TIM}}{k_{TIM} \cdot A_{TIM}} + R_{c,cool} + \frac{1}{h_{cool} \cdot A_{cool}}} \tag{4.7}$$

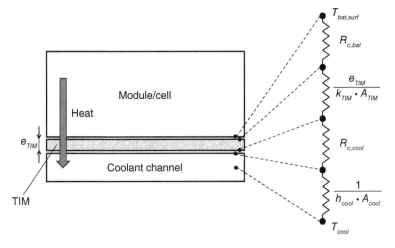

Figure 4.40 Connection of the battery to the plate.

where

$T_{bat,surf}$ is the batter lower surface temperature, [°C]

T_{cool} is the coolant temperature, [°C]

$R_{c,bat}$ is the thermal contact resistance between the battery surface and the TIM, [K W⁻¹]

$R_{c,cool}$ is the thermal contact resistance between the TIM and the coolant channel wall, [K W⁻¹]

e_{TIM} is the thickness of the TIM, [m]

k_{TIM} is the thermal conductivity of the TIM, [W m⁻¹ K⁻¹]

A_{TIM} is the area of the surface between the TIM and the battery, [m²]

A_{cool} is the heat transfer area between the coolant and the channel wall, [m²]

h_{cool} is the convective heat transfer coefficient between the coolant and the channel wall, [W m² K⁻¹]

The energy balance on the coolant side can be expressed as:

$$\dot{Q}_{bat} = \dot{m}_{cool}(h_{cool,ex} - h_{cool,su}) \tag{4.8}$$

If there is no phase change on the coolant side, the previous equation becomes:

$$\dot{Q}_{bat} = \dot{m}_{cool}\, c_{p,cool}(T_{cool,ex} - T_{cool,su}) \tag{4.9}$$

The contact between the battery wall, the TIM and the coolant channel wall is illustrated in Figure 4.41. The contact resistances $R_{c,bat}$ and $R_{c,cool}$ depend on the surface roughness and flatness, the contact pressure, the TIM elasticity and the "wetting" of the TIM and surfaces. The contact pressure also impacts the TIM thickness and can be increased to reduce the TIM thermal resistance. Deformation of the TIM can also overcome the thermal expansion and contraction of the battery cells during operation.

The TIM should present, among others, the following properties: appropriate operating temperature range, high thermal conductivity, minimum thickness, low electrical conductivity, deformability at low contact pressure, chemical compatibility with materials in contact, and minimum cost.

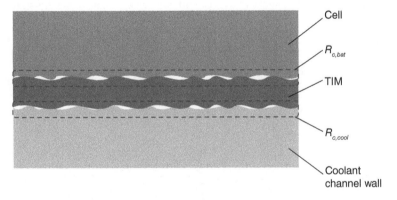

Cell

$R_{c,bat}$

TIM

$R_{c,cool}$

Coolant channel wall

Figure 4.41 Visualization of the contact between the battery, the coolant channel wall, and the TIM.

4.4.7.7 Comparison between Air Cooling and Glycol-Water Cooling Solutions

The comparison between physical properties of air and glycol water indicates that water is most advantageous, but is complex to be used if good performance, reliability, and security have to be ensured. Actually, the thermal conduction from the water plate to the cells is the bottleneck, since it can substantially affect the performance of the system and it depends on the shape and dimensions of the pack.

Table 4.1 Comparison of thermal properties of glycol water and air.

Thermal properties	Glycol water (50% in mass of EG at 85 °C)	Air (1 atm, 20 °C and RH = 50%)	Glycol water/air
Specific heat [J (kg-K)$^{-1}$]	3599	1018	3.536
Mass flow rate [kg h^{-1}]	800	360	2.22
Thermal conductivity [W (m-K)$^{-1}$]	0.4286	0.02517	17.03
Convective heat transfer coefficient [W (m^2-K)$^{-1}$]	1000–10000	10–100	100
Density [kg m^{-3}]	1023	1.199	853

For a same mass flow rate, a fan has an envelope volume 10 times larger than a pump with a head pressure 10 times smaller. Pipes cross-sectional areas are 30 times more important to limit the pressure losses in comparison to those associated with the component to cool down. Also, the electric consumption of the fan is not comparable with that of a pump. These constraints thus impose that the mass flow rate of air is smaller than that of a water-based system. When considering values of the water flow rate of 800 kg h^{-1} and of air of 360 kg h^{-1}, the main differences of thermal properties between an air-based cooling and a water-based cooling are summarized in Table 4.1.

Water has a capacity 100 times more important to extract heat to cool a hot plate (heat transfer coefficient).

Water has a capacity 8 times more important to transport heat (mass flow rate multiplied by specific heat).

However, the water-based cooling system requires a radiator for rejecting the heat to the outdoor air. But, it does not always need to activate the front-end module fan during driving, even at low vehicle speed where the rate of heat generated by the battery is low. During a charging cycle in static conditions, the system activates the existing front-end module fan supplied by the stationary grid (through the vehicle battery). The electric pump is therefore the only energy consumer of the system during driving conditions.

4.4.7.8 PCM and Other Technologies

A thermal storage made up of PCM can be integrated in the battery. The PCM surrounds the battery cell.

This PCM has the capacity to modify its physical state in a narrow fusion temperature range $T_{f,1} - T_{f,2}$. It allows for storing heat at an almost constant temperature, due to its high enthalpy variation in this fusion temperature range. If the temperature of the PCM is varied around this fusion temperature range, it presents a larger energy density than a sensible storage (for instance liquid water) operating with the same temperature range.

PCMs allow, by working as a damper, to shave the peaks of temperature, to homogenize and increase the average temperature in the battery, and to let the system working more often in passive mode.

When the PCM is heated:

- If $T_{PCM} < T_{f,1}$: the PCM in solid phase is heated up, because of its sensible heat
- If $T_{PCM} \in [T_{f,1}; T_{f,2}]$: the PCM stores heat through a phase change while its temperature is increased
- If $T_{PCM} > T_{f,2}$: the PCM in the liquid phase is heated up, because of its sensible heat

However, this solution requires to evacuate the energy stored in the PCM. This can be achieved by natural air convection around the PCM. However, with fast-charging and discharging, a larger heat generation rate from the battery has to be dissipated. Hybridization of PCM cooling with active air cooling and glycol-water cooling (see Luo et al. (2022)) can be a promising solution.

Because of its low thermal conductivity, it is useless to have a thick PCM for the needs in terms of charging and discharging of the battery. Indeed, with a thick PCM, during discharging, a fraction of the solid PCM will not be melted. Most of ongoing research aim at increasing the conductivity of the PCM by adding thermally conductive components inside the PCM.

To determine the amount of thermal energy $Q[J]$ that can be stored in the PCM, it is necessary to know its thermodynamic properties. For a PCM, it can be shown that $du \approx dh$ because of the limited variation of the specific volume on the considered temperature ranges (Dechesne et al., 2014). Therefore, if T_i and T_f are the initial and final temperatures of the PCM, respectively, the heat Q stored in the PCM (and associated heat transfer rate $\dot{Q}\,[W]$) can be determined by

$$Q = \int \dot{Q}dt = \Delta U_{PCM} = m_{PCM} \cdot \Delta u_{PCM} \approx m_{PCM} \cdot \Delta h_{PCM} = m_{PCM} \cdot \int_{T_i}^{T_f} dh_{PCM}(T) \qquad (4.10)$$

Solving the previous equation requires to know the enthalpy–temperature curve of the PCM. This curve can be measured by calorimetric methods such as DSC or T-history methods. An example of curve is proposed in Figure 4.42. It should be mentioned that this curve does not indicate the hysteresis effect: actually, the melting curve and freezing curve do not superpose.

Example 4.4 *Use of PCM for Battery Thermal Management*

It is proposed to cover one module of Example 4.3 with 5 mm of PCM, the enthalpy–temperature curve of which is represented in Figure 4.42. Initially, the air flow rate through the battery pack is reduced by a factor 10, and the convective heat transfer coefficient is assumed to be $10\ W\ m^{-2}K^{-1}$. After 2 hours, the air flow rate is increased back to its nominal value, and the convective heat transfer coefficient is $35\ W\ m^{-2}K^{-1}$. Follow the time evolution of the temperatures of the module, the module surface, the PCM and the air at the outlet of the module.

Analysis

In this problem, two thermal capacities are considered: that of the module and that of the PCM (Figure 4.43). Therefore, two transient energy balances must be written and numerically solved.

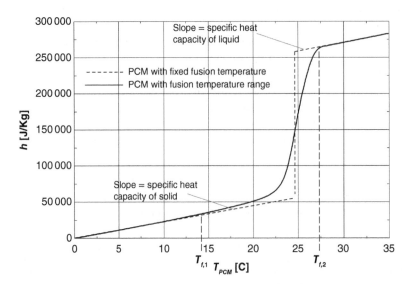

Figure 4.42 Example of enthalpy-temperature curve. Source: Adapted from Dechesne et al. (2014).

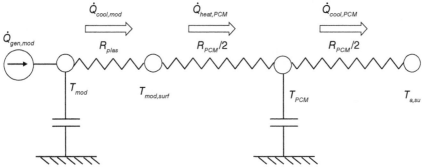

Figure 4.43 Heat transfer resistances and thermal capacities to describe the thermal behavior of one module covered by PCM.

Solution

```
"1. INPUTS"
"1.1. Battery Module"
T_mod_i=20 [C] "initial temperature of the module"
T_a_su=20 [C] "ambient air temperature"
m_dot_a_mod_min=0,02667/10 [kg/s] "minimum air flow rate around
    1 module"
m_dot_a_mod_max=0,02667 [kg/s] "maximum air flow rate around
    1 module"
m_dot_a_mod=if(tau;7200;m_dot_a_mod_min;m_dot_a_mod_min;
  m_dot_a_mod_max) "after 7200 sec, the air flow rate switches from
    the minimal to the maximum value"
"1.2. PCM"
T_PCM_i=20 [C] "initial temperature of the PCM"
"2. OUTPUTS"
"Time evolutions of T_mod; T_mod_surf; T_PCM; T_a_ex"
"3. PARAMETERS"
"3.1. Battery Module"
m_mod=8 [kg] "mass of one module"
c_mod=900 [J/kg-K] "module heat capacity"
W_mod=0,11 [m] "module width"
L_mod=0,16 [m] "module length"
h_mod=0,15 [m] "module height"
Q_dot_gen_cell=5,83 [W] "heat generation rate per cell"
N_cells_mod=6 [-] "number of cells per module"
"3.2. Plastic casing"
e_plas=0,002 [m] "plastic thickness"
k_plas=0,25 [W/m-K] "plastic thermal conductivity"
"3.3. PCM"
e_PCM=0,005""1e-12 [m]" "PCM thickness" "the value of 1e-12
    corresponds to no PCM"
rho_PCM=810 [kg/s] "PCM density"
h_out=if(tau;7200;10;10;35) "[W/m^2-K]" "convective heat transfer
    coefficient at the module outer surface"
"4. MODEL"
"4.1. Computational parameters"
```

```
tau_i=0 [s] "lower limit of time integration"
tau_f=3*3600 [s] "upper limit of time integration"
"4.2. Energy balance on module"
A_mod=2*(W_mod+L_mod)*h_mod+W_mod*L_mod "External surface are a of
    module covered by PCM, [m^2]"
R_plas=e_plas/(A_mod*k_plas)
Q_dot_gen_mod=N_cells_mod*Q_dot_gen_cell "Heat generation rate in
    one module, [W]"
m_mod*c_mod*dT_moddtau=Q_dot_gen_mod-Q_dot_cool_mod "Transient
    energy balance on each module"
Q_dot_cool_mod=(T_mod-T_mod_surf)/R_plas
deltaT_mod=integral(dT_moddtau;tau;tau_i;tau_f) "numerical solving
    of differential equation"
deltaT_mod=T_mod-T_mod_i "calculation of T_mod at time Tau"
"T_mod_surf=40" "guess before introducing the following sections"
"4.3. Energy balance on PCM"
"4.3.1. PCM heating power"
R_PCM=e_PCM/(2*k_PCM*A_PCM) "PCM heat transfer resistance"
k_PCM=interpolate3("'PCMcurve"';'k';'temp';temp=T_pcm)*c_natural
c_natural=2 "to account for the effect of natural convection"
Q_dot_heat_PCM=(T_mod_surf-T_PCM)/R_PCM
"T_PCM=20" "guess before introducing the following sections"
Q_dot_heat_PCM=Q_dot_cool_mod "remove guess on T_mod_surf"
"4.3.2. PCM cooling power"
rho_a=density(Air_ha;T=T_a_su;P=1e5)
c_p_a=cp(Air_ha;T=T_a_su;P=1e5)
A_PCM=2*(W_mod+2*e_PCM+L_mod+2*e_PCM)*(h_mod+e_pcm)+(W_mod+2*e_PCM)*
    (L_mod+2*e_PCM) "External surface are of PCM, [m^2]"
R_a=1/(h_out*A_PCM)
C_dot_a=m_dot_a_mod*c_p_a
Q_dot_cool_PCM=epsilon*C_dot_a*(T_PCM-T_a_su) "to determine the heat
    transfer rate"
epsilon=1-exp(-NTU)
NTU=AU/C_dot_a
AU^(-1)=R_a+R_PCM "total resistance between air and PCM"
Q_dot_cool_PCM=C_dot_a*(T_a_ex-T_a_su) "to determine T_a_ex"
"4.3.3. Transient energy balance"
dUdtau=Q_dot_heat_PCM- Q_dot_cool_PCM
DELTAu=integral(dUdtau;tau;tau_i;tau_f)
DELTAu=m_PCM*(h_PCM-h_PCM_0) "dh=du+Pdv+vdP: P constant and Pdv
    negligible = >dh=du"
h_PCM=interpolate3('PCMcurve';'h';'temp';temp=T_pcm) "Specific
    enthalpy of PCM, [kg/m^3]"
h_PCM_0=interpolate3('PCMcurve';'h';'temp';temp=T_pcm_i)
m_PCM=V_PCM*rho_PCM "mass of PCM, [kg]"
V_PCM=(W_mod+2*e_PCM)*(L_mod+2*e_PCM)*(h_mod+e_pcm)-W_mod*L_mod*h_mod
    "Volume of PCM, [m^3]"
$integralTable tau:60; T_mod; T_mod_surf; T_PCM; T_a_ex;
    Q_dot_cool_PCM;epsilon
```

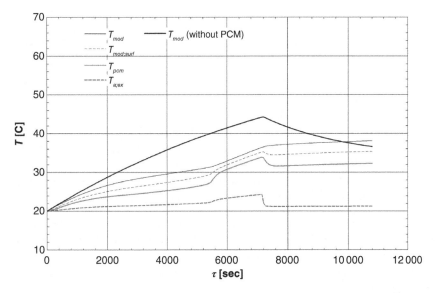

Figure 4.44 Time evolution of temperatures of the module, of the module surface, of the PCM, and of the air at the outlet of module.

Results are represented in Figure 4.44. The range of temperature during which PCM melts is visible. For indication, the evolution of the module temperature when no PCM is used is indicated. This illustrates that the melting slows down the increase of the module temperature. Without PCM, the module temperature would increase over 40 °C. After 2 hours, the activation of air cooling first decreases the PCM temperature, but not enough to solidify it again. The PCM temperature then keeps on slightly increasing.

Among other possible technologies, one can mention thermoelectric conditioning. The system is similar to an active air-based system, except that the air is cooled down or heated up by means of a thermoelectric module. The Peltier effect, one of the thermoelectric effects, is the conversion of an electric current into a difference of temperature. A direct current flows through a specific material made up of semiconductors. One of the faces of the material cools down, the other one heats up. The function is reversible. It allows through a closed loop ventilation circuit to cool (heat) the modules of the battery but requires a second air circuit and a fan to evacuate the heat (cold) produced by the system. The cooling power can be of several hundreds of W. Diffusers or fins are added to increase the thermal exchange with the outdoor air.

4.5 E-Motor and Power Electronics Cooling

To ensure a good operation of the vehicle but also the lifespan of the whole set of electronic and electrotechnical components highly sensitive to the temperature, the thermal management of the powertrain must be ensured. The electric powertrain (e-powertrain) made up of the electric motor and the inverter, the battery charger, and the battery must be thermally managed and requires a cooling demand to limit its level of operating temperature in such a way to optimize its reliability and its efficiency. The different components of the e-powertrain show different thermal needs according to their natures or their operating modes. Future development will probably allow to increase the acceptable operating temperatures.

The following sections describe the heat generation sources in power electronics and electric motors as well as thermal management systems.

4.5.1 Power Electronics

As explained hereunder, power electronics contains all components that convert the electricity from one form to another. The main power electronics components are as follows:

- Rectifiers convert the alternating current (AC) to direct current (DC). They are for instance used in battery chargers, which allow the battery to be charged from the grid. They are also used to charge the battery during regenerative braking.
- DC-DC converters are used to decrease or increase the voltage. For instance, they allow the low-voltage (12–42 V) on-board circuit (HVAC, radio, etc.) to be fed from the battery. They are also used between the fuel cell and the battery pack (Emadi et al., 2008).
- Inverters convert direct current to alternating current. They are, for instance, used between the battery and the electric motor to supply the latter with alternating current.

Conversion efficiency of power electronics is around 90%. Conversion losses result in heat production and require the power electronics to be cooled down by a dedicated cooling system. For a charging power of 40 kW, the rate of heat dissipated is 4 kW. Such cooling systems represent a considerable part of the power electronics cost and volume, which justifies the need to improve its performance and packaging.

Power electronics contains semi-conductor components (IGBT and diode dies). These components show temperature limitations. One of these limitations is the chip-junction temperature, which is the highest temperature of the semi-conductor. It depends on the ambient temperature and on the thermal resistance between the junction and the ambience and to the heat rate to dissipate. The maximum junction temperature is specified by the manufacturer and will influence the design of the dissipation system and selection of the heat sink. For instance, the maximum junction temperature of an IGBT module is around 125°C (Remsburg and Hager, 2007). The purpose of the cooling system is not to overpass this temperature limit and not to overpass a limit for the variation of the junction temperature.

The technical challenges associated with the cooling of power electronics silicon-base dies is the large heat flux, typically around 300 W m^{-2} (Remsburg and Hager, 2007), which showed an important increase during the last years.

A typical configuration of power electronics cooling system, inspired from Valenzuela et al. (2005) and Remsburg and Hager (2007), is represented in Figure 4.45. The power electronics dies are soldered to a direct bounded copper (DBC) substrate. The DBC is soldered to a copper base plate that acts as a heat spreader. Its surface is 5 to 8 times the surface of the die (Valenzuela et al., 2005).

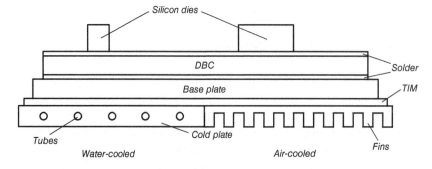

Figure 4.45 Schematic representation of power electronics cooling system. Source: Inspired from Valenzuela et al. (2005) and Remsburg and Hager (2007) (configuration (c) is inspired from Calyos technical documentation).

The base plate is attached to a cold plate. A TIM, such as thermal grease is used to ensure a good thermal contact between the base plate and the cold plate. It allows for compensating surface irregularities. The TIM thermal resistance represents a considerable fraction of the total resistance (Remsburg and Hager (2007) mention 29%). In other configurations, the DBC can be soldered directly to the cold plate. The cold plate is cooled down either by liquid coolant or by air. In the former case, the base plate is in contact with a plate with tubes fed by liquid coolant (glycol-water) that transfers heat toward the front-end of the vehicle in a radiator that is cooled down by ambient air. In the latter case, the base plate is covered by fins. Air-cooled base plates are rather used for lower powers (Valenzuela et al., 2005).

For water-cooled systems, the coolant inlet temperature is around 50–70°C. If the overall thermal resistance is reduced, the coolant temperature can be increased.

Heat emitted by electronics mainly depends on the electric power that pass through it. Calculators and DC/DC converters release a rather small quantity of heat, while inverter electronics that drives the electric motor and the charger that charges the battery have very high currents and generate heat at a rate of 1 to 3 kW on a small surface with a high power density transferred to the cooler.

The electronics can also be cooled down by a heat pipe system. A heat pipe is a tube filled with a fluid in liquid–vapor equilibrium. The choice of the fluid depends on the operating temperature of the system (it could be water, ammonia or other fluids). Water can be used in power electronics applications with the same concept as plates used for water cooling. Heat from power electronics vaporizes the liquid inside the tube, the local pressure inside the heated tube increases, vapor reaches the opposite side of the tube where it is condensed by a cold heat sink as air. Liquid circulates along the tube wall and goes back to the evaporating section, usually by gravity (thermosyphon heat pipe, as shown in Figure 4.46a) or by a capillary mechanism on the wall created by grooves or by a wick (Figure 4.46b). Vapor and liquid phases flow in counter-current. The advantage of capillary action is that the heat pipe can work in any direction. Grooved heat pipes create low capillary pressure but show low pressure losses associated with fluid flow. They are adapted to long horizontal heat pipes. Among wick heat pipes, one can distinguish sintered heat pipes (porous fritted material) and screen mesh heat pipes. Sintered heat pipes build higher capillary pressures but show large pressure losses. They are adapted to short heat pipes that can work in any orientation (they are less sensitive to gravity).

Using heat pipes for cooling is a potential solution with the following advantages in comparison with a water-cooling solution:

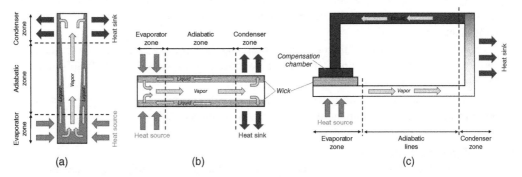

Figure 4.46 Different configurations of heat pipes (configuration (c) is inspired from Calyos technical documentation).

- The circulation of the heat transfer fluid does not require any mechanical device (the system is said to be passive). A coolant loop requires a pump.
- The system is much more compact since liquid/vapor phases flow in a small tube in counter-current.
- The performance of the heat pipe is much larger since the convective heat transfer coefficients in evaporation/condensation are much more important ($1000 - 50\,000$ W m^{-2}K^{-1}) than those in forced single-phase convection ($1000 - 2000$ W m^{-2}K^{-1}). It allows to extract heat from a confined zone.
- Reduced cost and low consumption.

A third configuration of passive two-phase heat transfer system, depicted in Figure 4.46c, is the loop heat pipe. Vapor and liquid flow through a piping describing a loop. Under capillary action, a wick ensures the capillary pressure to overcome pressure losses. Such configuration of heat pipe allows for increasing the distance between the evaporator and condenser zones. The purpose of the compensation chamber is twofold: (i) account for the variation of the volume of the liquid when operating temperatures varies (and therefore prevent liquid to accumulate at the condenser exhaust; (ii) ensure that the wick is wetted with liquid. Working fluids could be for instance ethanol, methanol or even R1233zd(e) (Nicolle et al., 2021). The distance between the evaporator and condenser could reach a few meters, making this solution interesting for vehicle applications. The condenser can be air-cooler or water-cooled. In the latter case, heat can be recovered for cabin heating.

4.5.2   Electric Motor (e-Motor)

The role of the electric machine is to convert the electric energy into mechanical energy (motor mode) or inversely the mechanical energy into electrical energy (generator mode). Since these conversions are not perfect, part of the energy provided to the machine is dissipated as heat and contributes to its heating.

An electric machine is made up of the following four fixed or moving parts:

- *The frame*: peripheral part of the machine supporting all the other fixed parts: stator and flanges. It can be adapted for the cooling of the machine: circulation of a heat transfer fluid, fins on the external surface, opening for ventilation, etc.
- *Flanges and bearings*: components located at the 2 extremities of the machine, ensuring the connection between the rotor and the stator. They can be equipped with openings for the ventilation of the machine.
- *The stator*: fixed active part of the machine comprising the laminated stator core with slots that contain copper wire coil windings. The coil temperature cannot generally exceed 180°C, making the coil the most sensitive part.
- *The rotor*: rotating part of the machine fixed to a shaft allowing to transmit or recover the mechanical energy. This component differs with the motor technology.

4.5.2.1   Types of Electric Motors

There exist several types of electrical machines that, from a thermal standpoint, differ essentially by their cooling techniques and by the geometrical and thermophysical characteristics of the stator and the rotor.

Two main categories of electric machines can be distinguished: the asynchronous machines (also called induction machines) and the synchronous machines.

4.5.2.1.1 Induction Machines In *induction motors,* the stator is made of a stack of flat annular plate laminations of ferro-magnetic materials. The periphery of the stator comprises slots that create teeth. Winding coils are placed in the stator slots. Feeding the stator winding with AC creates a rotating magnetic field (RMF), which rotates at the synchronous speed.

The rotor is also made of laminations. There are two main types of induction motors: the squirrel cage motors and the wound rotor motors. In a squirrel cage arrangement, the rotor comprises conductive aluminum bars nestled inside the grooves of the laminations and connected at both extremities by short-circuiting end-rings of the same material. According to Faraday's law of electromagnetic induction, the RMF induces a current in the bars or windings of the rotor. In wound rotor motors, wire windings are used instead of bars. Subject to Lorentz forces, the rotor starts to rotate. Because of Lenz's law, it rotates in the direction of the stator RMF (the production of induced current will be opposed by reducing the relative velocity between the stator RMF and the rotor). The rotor reaches a rotational speed function of the mechanical torque applied to it. In motor mode, the rotor rotational speed is slightly lower than the synchronous speed. The relative difference between the synchronous speed and the rotor speed is the slip. A slip of a few percent is necessary to have a relative movement between the RMF and the rotor conductors and to induce current in the latter. The slip self-adjusts to have the necessary induced current and torque to overcome the load torque.

4.5.2.1.2 Synchronous Machines In *synchronous machines*, the stator also contains a winding fed by a multi-phase AC (multiphase AC electromagnets) which creates an RMF. This RMF is synchronized with the frequency of the input current.

The rotor is magnetized independently and not through current induction. The rotor contains permanent or electro-magnets. The opposite poles of the magnetized rotor and of the RMF will attract each other, so that the rotor will be magnetically locked with the RMF. As a consequence, the rotation of the rotor is in step with the RMF and creates a second RMF, synchronized with the stator RMF.

Among synchronous machines, one can distinguish:

– *Permanent magnet synchronous machines.* The rotor contains permanent magnets that create a constant magnetic field. The poles of the rotor lock to the RMF and turn at the synchronous speed. The orientation and the position of magnets that equip the rotor laminations can vary from one rotor to another. If the magnets are surface mounted, the use of a protection layer in glass fiber, called sleeve, may be necessary on the outer surface of the rotor to contain the magnets when they are subject to centrifugal forces associated with the rotor rotation. This is not necessary for machines with buried magnets, as most of synchronous machines used in vehicle applications.
– *Synchronous reluctance machines.* The stator is a stack of ferromagnetic plate laminations, shaped in such a way to obtain a given number of teeth on which coils are wound. The turns of the wires in coils are well organized and compact. All coils are physically independent from each other. One of the consequences is to get an important filling of the slots and to limit the size of windings ends at both extremities of the machine. The toothed rotor is only made up of ferromagnetic laminations and does not have neither magnets nor rotor cage. This machine shows the advantage of a large simplicity. Alternatively, and more popular in vehicle applications, the rotor can be smooth and presenting holes of more or less complex shapes working as flux barriers and directing the magnetic flux.
– *The wound-rotor synchronous machine.* The stator of this machine has also a three-phase winding that creates an RMF. The rotor has winding coils nestled in slots. The rotor is excited by a direct current power supply by means of slip rings and creates a constant magnetic field similarly to

permanent magnets. As explained previously, opposite poles of the RMF and rotor will attract and get locked to that the rotor rotates at the RMF field (i.e., synchronous speed). This is the technology usually used for alternators in vehicles. More recently, machines have been proposed with a contactless supply of the rotor (by induction).

- *The double excitation synchronous machine.* The excitation of the rotor for this type of machine is created by two different sources. The first one is from permanent magnets and the second one is from field coils as in the previous technology. There are different types of double excitation machines, differing from the configuration of the two excitation sources. This type of machine can look like a wound-rotor synchronous machine in which permanent magnets have been added to increase the performance.
- Machines hybridizing a reluctance motor and a permanent magnet motor are also used in vehicles.

The two main types of electric motors used in hybrid and electric vehicles are the permanent magnet motors and to a lesser extent the induction motor. Both motors are fed with 3-phase alternating current. Both motors can work in the generator mode during regenerative braking.

Highly integrated powertrains using high-speed electric motor seem to have a high potential to reduce cost, space and weight.

4.5.2.2 Losses in Electric Motors

An electric motor converts electrical power into mechanical power. The conversion efficiency is defined as the ratio of the mechanical shaft power output to the electric power input to the motor.

$$\eta_{em} = \frac{\dot{W}_{sh}}{\dot{W}_{el}} = \frac{\dot{W}_{el} - \dot{Q}_{loss}}{\dot{W}_{el}} \tag{4.11}$$

The average efficiency of an electric motor is around 80%, while the maximum efficiency that can be reached on specific points is 95%. At the maximum speed, the dissipated heat rate (not accounting for power electronics) is of the order of 5–6 kW.

In the motor or generator mode, the energy losses represent heat sources that contribute to the heating of the machine. Their individual contribution varies as a function of the torque and the rotational speed. The main sources of losses in electric motors are as follows:

- The Joule losses in windings of the stator and of the rotor in the case of induction motors. The Joule losses occur in the conductors of electric circuits through which circulates electric current and in magnets if any.

  For a low frequency current or a direct current, the current density is supposed to be homogeneous within the wire cross-section. Any conductor of electric resistance R carrying an RMS current I_{RMS} produces Joule losses equal to

$$P_{Joule} = R \cdot I_{RMS}^2 \tag{4.12}$$

  For a high-frequency current, an induction effect, called skin effect, yields a current concentration near the surface of the conductors causing an effective increase of the resistance of the conductor.
- The iron losses in the stator and rotor cores.

  Iron losses are located in the parts of the machine subject to variable in time magnetic fields: laminations of the stator and laminations of the rotor. Therefore, a fixed field at the stator would not yield iron losses in the stator. These losses are proportional to the volume of the ferromagnetic material inside which they arise. One distinguishes two main types of losses the amplitude of which is difficult to evaluate.

The Eddy current (Foucault's current) losses are due the time-variations of the magnetic field that induce in the ferromagnetic materials electromotive forces and create currents that will cause a heat dissipation by Joule effect. To limit those losses, the magnetic parts of the rotor and of the stator are made of a stack of laminations parallel with the direction of the magnetic flux.

Hysteresis losses are related to the evolution of the crystalline structure of the material of the laminations. The flux variations act on the orientation of the magnetic domains in the ferromagnetic materials and induce a heating.

– The friction losses in bearings and friction losses between rotating surfaces and air.

These mechanical losses are generated by friction losses at the interface between shaft and its bearing and by aeraulic losses that encompass dissipations due to shear forces or friction of air inside the air gap on the rotor surface and aeraulic losses generated by the cooling fans.

The evaluation of the iron losses is generally achieved by measuring total losses during tests on the electric machine and by subtracting mechanical losses (friction and aeraulic) evaluated as a function of the regime by running the machine without any load and by subtracting Joule losses calculated as a function of measured currents and phase resistances.

Permanent magnets motors are more efficient than induction motors because they do not comprise windings in the stator (Polikarpova, 2014).

4.5.2.3 Operating Temperature Range of e-Motors

The losses described in the previous section result in heat production $\dot{Q}_{loss}$ that tends to increase the temperature of the motor. This temperature has to be maintained underneath given limits imposed by the components. Among other things, the conductivity of copper used in windings decreases with the temperature and hence the whole Joule losses increase. At higher temperatures, the insulation of the wires of the windings can be deteriorated.

During the transient increase of electric motor power, the temperature of sensitive zones can quickly increase to reach maximum temperatures. In order to ensure the good operation of the machine and its reliability, it is necessary to limit the temperature of several components:

– Magnets that get demagnetized over a given temperature limit.
– The different electric insulators that deteriorate until losing their insulating performance.
– Greases used in bearings that also lose their lubricating performance over a given temperature.

Comparing the thermal features of electric powertrain to those of conventional thermal powertrain:

– The rate of heat to dissipate is of the order of 10 to 20 times lower, since the energy efficiency is much higher. Accounting for heat dissipated by the A/C condenser, the total rate of heat to reject is only of 1/5 to 1/10.
– The maximum temperature of the windings is typically around 180°C. The maximum temperature that magnets can withstand is around 150°C and that of electronic functions is 120°C, while in a thermal engine, the maximum temperature of combustion chamber walls is around 250°C (except the exhaust pipe which could be hotter).
– The temperatures of the intermediate cooling fluids in a thermal engine vehicle are of the order of 110 and 135°C for the cooling water and lubricating oil, respectively. It is also possible to have a low-temperature cooling loop to cool down the CAC and the water-cooled condenser. An electric powertrain uses a cooling water loop with temperature ranging from 55°C to 80°C to cool down electronics and machines. Using oil solutions for motors and heat pipes for electronics, heat can be extracted with improved performance where it is produced and hence at much higher temperature level of the heat transfer fluid (making ultimately easier the heat transfer to the vehicle outdoor air).

4.5.2.4 E-Motor Cooling System

To maintain the temperature of the motor components under these limits, generated heat has to be transferred to the ambient, mainly through two heat transfer modes:

- Heat transfer by conduction in the structure of the machine,
- Heat transfer by convection, external or internal, natural or forced toward a heat transfer fluid (air, glycol water, oil, etc.). Natural convection alone often leads to an excessive heating of some components. It is then necessary to prefer a cooling by forced convection.

Several cooling principles are possible:

1. Passive cooling by air: no active mechanism is installed for the cooling ensured by air around the machine and can be improved by the use of fins increasing the heat transfer surface area between walls and air.
2. Active cooling by ventilation: one or several fans generally directly driven by the rotor shaft allow for a forced circulation of air through the internal structure of the machine.
3. Active cooling by the coolant (water, oil) in the motor housing: the water circulation is ensured by the mechanical pump of the thermal engine (for a hybrid vehicle) or by an independent electric pump.
4. Oil cooling in the internal structure of the machine in the close vicinity of coils.
5. Active cooling by ventilation in closed circuit and by coolant flow: air is blown by a rotor-driven fan through ducts in the rotor and is then cooled by the water-cooled housing.
6. Active cooling with oil spraying on windings ends and cooling in an oil sump exposed to surrounding air.

4.5.2.4.1 Air Cooling of Open Machines Air cooling of electric motor can be considered if the electric powertrain rejects a low rate of heat and if the use of an intermediate fluid increases the temperature difference between components and ambient air. Air can be drawn from the air flowing through the front-end that represents 10 times the air flow displaced by the machine fan. Its implementation is rather easy and not expensive, as it does not require other fluids than air available in the vicinity of the machine. However, this cooling technique is often limited to motors with low specific powers, because of the limited capacity of air to extract and transport heat from the walls to cool down.

It, however, shows several drawbacks, including efficiency and acoustics issues. The design of the fan (diameter, geometry) and of the aeraulic circuit (grille, shape of the fins) are important parameters to limit the global noise. At high rotational regime, the mechanical losses of a fan driven by the machine can appear to be significant.

For «open» machines, cooling is ensured by air taken from the machine surroundings with a constant renewal rate and displaced by one or two fans. This cooling mechanism is used for automotive alternators.

In the case of a single-flow ventilation, air sucked by the fan fixed on the shaft enters on one side, flows through the electric machine and leaves it on the opposite side. Air is drawn directly from the surroundings of the machine or from another zone of the motor compartment by means of a duct, which creates additional pressure losses requiring the use of a more powerful fan.

In the case of a double-flow ventilation, two airflows sucked by the fans on both sides of the machine through holes drilled in the flanges enter axially and leaves the machine radially. The cooling of the machine is more homogeneous.

The main flows allowing for the evacuation of most of the losses dissipated by the machine with an important convective heat exchange are those generated by the rotation of the fans. The airflow in the motor bay generated by ram effect (due to the motion of the vehicle) or by the activation of the fan unit allows renewing the air around the machine and limiting the temperature of the air sucked by the fans.

Convective heat exchanges between the air of the air gap and the walls of the machine due to the rotation of the rotor are even more important that the air gap is small. They mainly enhance the evacuation of the losses dissipated in the rotor. In order them to be advantageous to the cooling of the machine, it is necessary to have a flow in the axial direction to improve the heat transfers from the air gap to the airflow on both extremities of the machine. Without an axial flow, the convective heat transfers between the machine walls and the air gap allow only for the heat transfers between the rotor and the stator and vice-versa.

The last zone of convective heat transfer with the air zone is located between the walls of the rotor and the flanges.

In the case of the presence of a thermal engine, the hot zone around the exhaust gas manifold must be avoided for the location of any component cooled by air.

Also, the most severe regimes from the air temperature standpoint are the vehicle standstills after a full load driving and the urban driving when underhood air flow rates are rather low.

4.5.2.4.2 Water Cooling As explained in previous section, small electric engines are typically cooled down by air with forced convection achieved by means of a fan. Larger engines are generally cooled down by a liquid coolant loop. Coolant is usually glycol water, but oil can also be employed. It should also be mentioned that two-phase coolant could also be used. Liquid cooling is better adapted to large heat fluxes associated with large torque density (Caricchi et al., 1996). As illustrated in Figure 4.47, such engines comprise a water jacket, which consists of a cooling circuit in the periphery of the stator.

Water cooling in the machine casing is generally used in closed machines. Thermal losses must transit through the casing to be evacuated by the coolant. This mode of cooling shows the advantage not to be dependent on the rotational speed of the rotor, to allow for a better acoustic insulation and to prevent any ventilation losses.

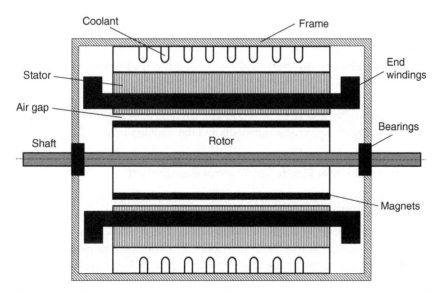

Figure 4.47 Schematic representation of a Permanent Magnet motor. Source: Adapted from Bertin et al. (2000).

Water-cooling shows better performance than air-cooling, but cooling does not involve direct contact as with oil.

The cooling of the machine involves two important convective exchange zones:

- The exchanges between the walls of the casing and the water flow.
- The exchanges in the motor air gap for the heat transfers from the rotor to the stator and vice-versa.

The motor casing, with a water jacket, allows for the distribution of the coolant flow around the whole external surface of the stator in such a way to optimize the repartition of the heat fluxes inside the machine and minimize the resulting thermal gradients.

The homogeneity of the cooling is also influenced by the heating of the coolant between the inlet and the outlet, which has to be as limited as possible and which is dependent on the flow rate. Thermal bridges can be used under the form of resins allowing to improve the heat transfer between end-windings and the water jacket. The efficiency of the cooling depends on the physical and geometrical characteristics of the water jacket.

Even though the convective heat transfer coefficient is large, the most important heat dissipations corresponding to Joule losses have to go through the thermal barriers resulting from insulating materials inside the structure and necessary to the electrical insulation of the machine. These thermal barriers yield a heating-up of the components and an important temperature difference between water and the motor hot-spot (end-winding). Iron losses in stator laminations are quite easily evacuated because of the proximity of the water jacket. The rotor iron losses transit more or less easily through the air gap as a function of the rotor regime that impacts convective heat exchange between air, stator, and rotor.

4.5.2.4.3 *Oil Cooling* Alternatives to the cooling by dry air with forced convection are the cooling by air with liquid mist (spray cooling) and the cooling by flowing liquid or by oil impinging jets. These solutions allow for largely increasing the heat transfer coefficients. The liquid can be oil. The coil windings of large power voltage transformers can also be cooled down by an oil bath. Oil, with its dielectric properties, can be in direct contact with all heating elements inside an electric machine (windings, end-windings that constitute the inactive part from an electromagnetic standpoint, laminations of the stator and of the rotor, magnets, flanges). Oil therefore allows for evacuating the heat from an electric machine with good heat transfer coefficients and with the different elements without needing given components such as insulators used when water is considered as the coolant of the housing.

Oil cooling mode is much better than air cooling and sometimes more efficient than water cooling when oil is used in the very close vicinity of slots and winding ends allowing for limiting the temperature difference between the latter components and the cooling fluid.

4.5.3 Combined e-Motor and Power Electronics Thermal Management

Trends are to combine electric motor, gearbox, and power electronics in a common housing with a common cooling and lubrication system to reduce the overall cost compared with the components which would have their own independent solutions for cooling and lubrication. This integration offers other potential advantages, such as using heat lost from the motor and power electronics to preheat the gearbox lubricant in cold ambient conditions to reduce gearbox losses and then a more efficient system.

4.6 Overall Thermal Energy Management of Electrified Vehicles

Full-electric or electrified vehicles represent a necessary solution to reduce CO_2 emissions, pollutants, and the fuel consumption. However, the development of those vehicles showing additional components to the thermal powertrain (such as the battery, the electric motor and the power electronics that emit heat) and additional working cycles (by extending the A/C loop to the heat pump mode and by introducing battery charging when the vehicle is at a standstill) yields additional challenges on the thermal functions (specific low temperature ranges, lifespan of more solicited components, development of new technologies) and requires more and more complex thermal systems architectures. Adding new cooling loops at different temperature levels and new heat exchangers is necessary.

4.6.1 Fluids Loops and their Connections

As explained throughout this book, the different components and systems to be heated or cooled are integrated inside fluid loops: air, glycol-water coolant, refrigerant, oil, etc. For instance, in the example of the PHEV thermal system illustrated in Figure 4.48, four thermal loops are necessary according to the different temperatures and operating points of the components: the high temperature loop, the low temperature loop, the A/C loop for the cabin and the battery loop.

The *high temperature coolant loop* answers the need to ensure the cooling during the operation of the internal combustion engine at full load when the temperature generated inside the combustion chamber reaches extreme values. The operation of this loop with the thermostat is described in Chapter 2. The high temperature loop is connected to the high temperature (HT) radiator. This loop also feeds the heater core HC for cabin heating during thermal engine mode.

The *low temperature coolant loop* is used to cool the power electronics and the electric powertrain. It is connected to the low temperature (LT) radiator. As explained in Chapter 2, this loop can also host a WCAC (the advantages of which have been described in Chapter 2).

The purpose of the *refrigerant loop* is twofold. First, it is used for cooling or heating the cabin. Secondly, it can be used for cooling the battery pack. Cabin cooling is ensured through the inner evaporator (IEV), while heating is ensured by the water cooled condenser (WCD). As explained in

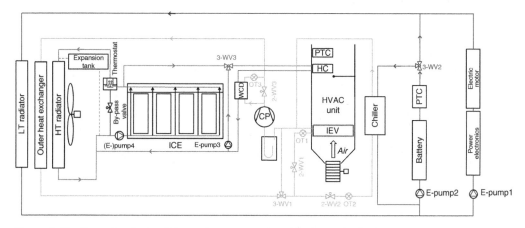

Figure 4.48 Example of overall thermal management architecture of a PHEV with three heat exchangers in the front-end module.

the heat pump section, the three-way valve 3-WV1 is used for switching from cooling to heating mode. Battery cooling is achieved by the chiller. In the example of Figure 4.48, the two-way valve 2-WV2 must be opened to feed the chiller with refrigerant. The refrigerant loop is more complex compared to the LT and HT cooling loops, because of physical phenomena of evaporation and condensation and refrigerant charge distribution among components.

The *battery loop* can be isolated by means of the three-way valve 3-WV2. In this case, the loop is cooled down by the chiller. If outdoor temperature is low enough, the loop can be cooled down by the low-temperature radiator. In this latter case, the chiller is not activated and coolant flows through the LT radiator and then through two branches in parallel: the battery branch and the PE/EM branch.

If cooled by water coolant, the specification regarding the cooling of a component comprises information on the minimal water flow rate, the maximal water flow rate, pressure losses associated to the internal water circuit of the component, the maximal temperature not to exceed, the maximal temperature difference between outlet and inlet, the maximal pressure of water, the water content in the component and the rate of heat generated by the component. In the driving mode, the rate of heat generated by each component significantly vary with the vehicle velocity profile.

Each loop could have its own control strategy and be individually controlled. However, the current cooling systems consume a large quantity of energy to control the temperature with a significant impact on the vehicle overall driving range. Therefore, the overall thermal management has gained significant importance and is necessary to optimize the energy performance and the autonomy of the electrified vehicles in electric mode, the available battery power, the durability, and the emissions. It brings together in a same reflection, the vehicle heat producers and consumers as well as their operating cycles. This reflection then allows designing the different necessary fluid loops (refrigerant, air, low-temperature glycol water, high-temperature glycol water, etc.) and connecting them together in such a way to meet the needs and to minimize the consumption and maximize the vehicle driving range. The thermal loops are becoming more and more connected and coupled and require a smart supervisory overall thermal management system with an optimal control strategy in such a way to efficiently control the operating temperatures of the components and systems (batteries, electric motors, power electronics, thermal engine, cabin, etc.). That allows to optimize their operation and guarantee their durability while limiting the cooling system overall energy consumption and achieving an identical thermal comfort.

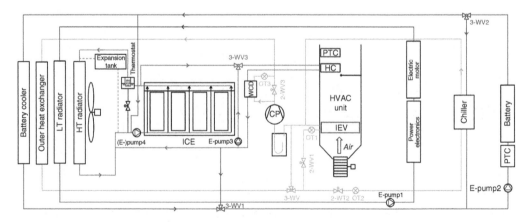

Figure 4.49 Example of overall thermal management architecture of a PHEV with four heat exchangers in the front-end module.

As an example, the refrigerant loop can deliver additional heating and cooling capacities to other end-users than the cabin or valorize the thermal losses of components in heat pump mode.

Another example of connection is illustrated in Figure 4.49. This architecture, inspired from Pischinger et al. (2014), is related to a PHEV. It uses four different heat exchangers in the front-end module. Adjusting the position of the valve 3-WV2, the battery can be cooled either by the chiller or by the battery cooler. As suggested by Pischinger et al. (2014), in combustion engine drive, the battery can be heated by the high temperature loop (using 3 way-valve 3-WV1). In purely electric drive, the electrical heater PTC is used.

4.6.2 Front-End Module Configuration

The arrangement of the different heat exchangers in the vehicle front end module differs as a function of the different heat rates to dissipate.

For *hybrid* vehicles, this zone can comprise a refrigerant/air exchanger (Outer Heat Exchanger in Figure 4.48 and Figure 4.49), a low-temperature radiator and a high-temperature radiator stacked downwards the active grille air shutters (described in Chapter 2). An air-cooled charge air cooler (ACAC) can also be integrated in the front-end module. Because of the fan in the front-end module and to the motion of the vehicle, air circulates and flows through the shutters and through the heat exchangers up to the high-temperature radiator. These hot heat exchangers induce a heating-up along the path of the outdoor air as a function of the dissipated heat rates. To allow the heat exchangers working at lower temperatures benefiting for a large enough temperature difference with air, the high-temperature radiator is placed further down on the air circuit. The refrigerant/air exchanger is preferentially located upstream the low-temperature radiator on a hybrid vehicle (as shown in Figure 4.49). One can also find an architecture with these two exchangers in parallel, each of them having a smaller cross-sectional area and an increased thickness and a similar overall volume.

On a *full-electric* vehicle, the surface of the low temperature radiator is as important as that of a high-temperature radiator in a thermal engine vehicle, since the temperature difference between air and water is much smaller, even though the rate of heat to evacuate is much lower. The low-temperature radiator is generally placed upstream the refrigerant/air exchanger in order to prevent the "hot mask" in condenser mode. However, it yields itself a hot mask for the refrigerant/air exchanger and filters water splashing. These latter two effects are favorable to the refrigerant/air exchanger working in evaporator mode (in heat pump mode), by increasing the air temperature upstream the evaporator and therefore delaying the frost formation.

4.6.3 Pumps and Fan-Motor Assembly

The electrified vehicles use one (or several) electric water pump(s) that impose recommendations on the pressure drops of the components of the electric powertrain to decrease the consumed power. EV also use a fan-motor assembly of brushless technology with increased durability. Potential new functionalities appear for the water pumps and the fan-motor assembly:

o Stationary charging (slow and fast): water pump and fan-motor assembly
o Cabin pre-conditioning: fan-motor assembly
o Battery cooling by air conditioning or free-cooling: fan-motor assembly
o Heating by heat pump: fan-motor assembly

References

Bandhauer, T.M., Garimella, S., and Fuller, T.F. (2011). A critical review of thermal issues in Lithium-ion batteries. *Journal of The Electrochemical Society* 158 (3): R1–R25.

Bernardi, D., Pawlikowski, E., and Newman, J. (1985). A general energy balance for battery systems. *Journal of The Electrochemical Society* 132 (1): 5–12.

Bertin, Y., Videcoq, E., Thieblin, S., and Petit, D. (2000). Thermal behavior of an electrical motor through a reduced model. *IEEE Transactions on Energy Conversion* 15 (2).

Buller, S., Thele, M., De Doncker, R.W.A.A., and Karden, E. (2005). Impedance-based simulation models of supercapacitors and Li-ion batteries for power electronic applications. *IEEE Transactions on Industry Applications* 41 (3): 742–747. https://doi.org/10.1109/TIA.2005.847280.

Caricchi, F., Crescimbini, F., and Di Napoli, A. (1996). Prototype of innovative wheel direct drive with water-cooled axial-flux PM motor for electric vehicle applications. In: *Applied Power Electronics Conference and Exposition*, 764–770. APEC.

Chan, C.C. (2007). The state of the art of electric, hybrid, and fuel cell vehicles. *Proceedings of the IEEE* 95 (4): 704–718.

Daccord, R., and Wang, J. (2021). The end of range anxiety thanks to fast charging and immersion cooling in BEVs. Presentation at the 4th Global NEV Thermal Management Summit 2021 & China Fuel Cell Vehicle Forum 2021. December 1–3, 2021, Shanghai, China.

Davila, C., Paulus, N., and Lemort, V. (2022). Experimental Investigation of a Micro-CHP Unit Driven by Natural Gas for Residential Buildings. In: *Proceedings of the 7th International High Performance Buildings Conference at Purdue*. Purdue, United States – Indiana: Purdue University, paper 2390.

Dechesne, B., Gendebien, S., Martens, J. et al. (2014). Designing and testing an air-PCM heat exchanger for building ventilation application coupled to energy storage. In: *2014 Purdue Conferences Proceedings*.

Emadi, A.Y., Lee, J., and Rajashekara, K. (2008). Power electronics and motor drives in electric, hybrid electric, and plug-in hybrid electric vehicles. *IEEE Transactions on Industrial Electronics* 55 (6): 2237–2245. https://doi.org/10.1109/TIE.2008.922768.

Feng, L. and Hrnjak, P. (2016). Experimental and numerical study of a Mobile reversible air conditioning heat pump system. *Proceedings of the 16th Internal Refrigeration and Air Conditioning Conference at Purdue*. July 11–14, 2016.

van Gils, R., Danilov, D., Notten, P. et al. (2014). Battery thermal management by boiling heat-transfer. *Energy Conversion and Management*. 79: 9–17. https://doi.org/10.1016/j.enconman.2013.12.006.

Girmscheid, F. (2014). Battery cooling concepts for PHEV and BEV. Halla Visteon Climate Control Corp.

ICCT, International Council on Clean Transportation Europe (2020), *European Vehicle Market Statistics*, Pocketbook 2020/21.

Karimi, G. and Li, X. (2012). Thermal management of lithium-ion batteries for electric vehicles. *International Journal of Energy Research* 37: 13–24. https://doi.org/10.1002/er.1956.

Kolp, E., Huber, C., and Jossen, A. (2016). Designing Thermally Safe Battery Packs for Electric Vehicle Applications. *Presentation in Int. Forum Automotive Thermal Management 2016*. Chester, U.K.

Leighton, D. (2015). Combined fluid loop thermal management for electric drive vehicle range improvement. *SAE International Journal of Passenger Cars—Mechanical Systems* 8 (2): doi: 10.4271/2015-01-1709.

Liu, G., Ouyang, M., Lu, L. et al. (2014). Analysis of the heat generation of lithium-ion battery during charging and discharging considering different influencing factors. *Journal of Thermal Analysis and Calorimetry* 2014 (116): 1001–1010.

Luo, J., Zou, D., Wang, Y. et al. (2022). Battery thermal management systems (BTMs) based on phase change material (PCM): a comprehensive review. *Chemical Engineering Journal* 430 (Part 1): 132741.

Ma, S., Jiang, M., Tao, P. et al. (2018). Temperature effect and thermal impact in lithium-ion batteries: a review. *Progress in Natural Science: Materials International* 28 (6): 653–666.

Nicolle, T., Kapaun, F., Lasserre, P., Piaud, B., Dupont, V., and Ybanez, L. (2021). Using loop heat pipe solutions, and a dielectric fluid, to cool SiC MOSFET power modules for aircraft systems. In *Proceedings of Joint 20th IHPC and 14th IHPS*, Gelendzhik, Russia.

Noura, N., Boulon, L., and Jemeï, S. (2020). A review of battery state of health estimation methods: hybrid electric vehicle challenges. *World Electric Vehicle Journal* 11: 66.

Paulus, N. and Lemort, V. (2022). Field-test performance of solid oxide fuel cells (SOFC) for residential cogeneration applications. In: *Proceedings of the 7th International High Performance Buildings Conference at Purdue*. Purdue, United States – Indiana: Purdue University.

Pesaran, A., Vlahinos, A., and Burch, S. (1997). Thermal performance of EV and HEV battery modules and packs. In: *Proceedings of the 14th international electric vehicle symposium*, Orlando, Florida; December 15–17.

Pesaran, A., Santhanagopalan, S., and Kim, G.H. (2013). Addressing the impact of temperature extremes on large format Li-ion batteries for vehicle applications, *Presented at: Proceedings of the 30th International Battery Seminar*, Ft. Lauderdale, Florida.

Petzl, M., Kasper, M., and Danzer, M.A. (2015). Lithium plating in a commercial lithium-ion battery – a low-temperature aging study. *Journal of Power Sources* 275: 799–807.

Pischinger, S., Genender, P., Klopstein, S., and Hemkemeyer, D. (2014). Challenges in thermal management of hybrid and electric vehicles. *ATZ* 116: 36–40.

Polikarpova, M. (2014). Liquid cooling solutions for rotating permanent magnet synchronous machines. Acta Universitatis Lappeenrantaensis 597.

Remsburg, R. and Hager, J. (2007). Direct integration of IGBT power modules to liquid cooling arrays. *23rd international electric vehicle symposium and exposition*, December 2–5, 2007; pp. 366–391.

Smith, K., Markel, T., and Pesaran, A. (2008). *System Tradeoffs in Temperature, Lifetime, and Cost for Plug-In Hybrid Electric Vehicle Batteries*, U.S. Dept. of Energy Milestone Report (2008). Golden, Colorado, USA: National Renewable Energy Laboratory.

Solai, E., Guadagnini, M., Beaugendre, H. et al. (2022). Validation of a data-driven fast numerical model to simulate the immersion cooling of a lithium-ion battery pack. *Energy* 249: 123633.

Valenzuela, J., Jasinski, T., and Sheikh, Z. (2005). Liquid cooling for high-power electronics. *Power Electronics Technology* .

Voelker, P. (2014). Trace degradation analysis of lithium-ion battery components. *R&DMagazine* .

Wang, Y., Sauer, D.-U., Koehne, S., and Ersoez, A. (2014). Dynamic modeling of high temperature PEM fuel cell start-up process. *International Journal of Hydrogen Energy* 39 (33): 19067–11907.

Index

Thermal Energy Management in Vehicles, First Edition. Vincent Lemort, Gérard Olivier, and Georges de Pelsemaeker.
© 2023 John Wiley & Sons Ltd. Published 2023 by John Wiley & Sons Ltd.
Companion website: www.wiley.com/go/lemort/thermal